Nonlinear Partial Differential Equations

Contributors

W. F. AMES
RICHARD BELLMAN
SAMUEL Z. BURSTEIN
W. J. CUNNINGHAM
R. C. DI PRIMA
JOHN R. FERRON
I. FLÜGGE-LOTZ
KAZUTOSHI FUJIMURA
ARTHUR G. HANSEN
DONALD H. HYERS
ROBERT KALABA
MILTON LEES
WILLIAM A. NASH
D. W. PEACEMAN
J. R. A. PEARSON
M. H. PROTTER
GUIDO SANDRI
NORMAN J. ZABUSKY

Nonlinear Partial Differential Equations

A SYMPOSIUM ON METHODS OF SOLUTION

Edited by

W. F. AMES

DEPARTMENT OF STATISTICS AND COMPUTER SCIENCE
AND
DEPARTMENT OF MECHANICAL ENGINEERING
UNIVERSITY OF DELAWARE
NEWARK, DELAWARE

1967

ACADEMIC PRESS New York and London

ACADEMIC PRESS, INC.
111 Fifth Avenue, New York, New York 10003

United Kingdom Edition published by
ACADEMIC PRESS, INC. (LONDON) LTD.
Berkeley Square House, London W. 1

LIBRARY OF CONGRESS CATALOG CARD NUMBER: 66-30116

Second Printing, 1969

PRINTED IN THE UNITED STATES OF AMERICA

List of Contributors

Numbers in parentheses refer to the pages on which the authors' contributions begin.

W. F. AMES,† Department of Statistics and Computer Science and Department of Mechanical Engineering, University of Delaware, Newark, Delaware (55)

RICHARD BELLMAN, Departments of Mathematics, Engineering, and Medicine, University of Southern California, Los Angeles, California (43)

SAMUEL Z. BURSTEIN, Courant Institute of Mathematical Sciences, New York University, New York, New York (279)

W. J. CUNNINGHAM, Department of Engineering and Applied Science, Yale University, New Haven, Connecticut (141)

R. C. DI PRIMA, Department of Mathematics, Rensselaer Polytechnic Institute, Troy, New York (19)

JOHN R. FERRON, Department of Chemical Engineering, University of Delaware, Newark, Delaware (203)

I. FLÜGGE-LOTZ, Division of Engineering Mechanics, Stanford University, Stanford, California (109)

KAZUTOSHI FUJIMURA, Department of Chemical Engineering, University of Delaware, Newark, Delaware (203)

ARTHUR G. HANSEN, Engineering College, Georgia Institute of Technology, Atlanta, Georgia (1)

DONALD H. HYERS, Department of Mathematics, University of Southern California, Los Angeles, California (125)

† *Present address:* Department of Mechanics and Hydraulics, University of Iowa, Iowa City, Iowa.

ROBERT KALABA, The Rand Corporation, Santa Monica, California (43)

MILTON LEES, Department of Mathematics, Case Institute of Technology, Cleveland, Ohio (193)

WILLIAM A. NASH, Department of Engineering Science and Mechanics, University of Florida, Gainesville, Florida (291)

D. W. PEACEMAN, Esso Production Research Company, Houston, Texas (171)

J. R. A. PEARSON, Department of Chemical Engineering, University of Cambridge, Cambridge, Great Britain (73)

M. H. PROTTER, Department of Mathematics, University of California, Berkeley, California (161)

GUIDO SANDRI, Aeronautical Research Associates of Princeton, Inc., Princeton, New Jersey (259)

NORMAN J. ZABUSKY, Re-entry Physics Department, Bell Telephone Laboratories, Whippany, New Jersey (223)

Foreword

The purpose of the seminar which engendered this volume was the consideration in depth of current knowledge and research in methods of solution for nonlinear partial differential equations. A spectrum of applied mathematicians, engineers, and physicists assembled at the University of Delaware, Newark, Delaware, December 27–29, 1965, for this review of the present state of the subject.

The seminar was designed with the dual goals of exposition and the presentation of research results. The sessions were divided into four Symposia: Analytic Methods, Approximate Methods, Numerical Methods, and Applications. The attendants were welcomed on behalf of the University by President J. A. Perkins. Dean E. W. Comings opened the technical program. The seminar was arranged by Professor W. F. Ames of the Department of Mechanical and Aerospace Engineering and the Department of Statistics and Computer Science. Technical sessions were chaired by Professors R. Remage, A. Rogers, D. Lamb, and B. Seidel.

The nineteen lecturers were chosen because of their research contributions and for their ability as expositors. The eighty active participants, invited on a geographical basis, were especially chosen to ensure a stimulating, profitable discussion of research in nonlinear theory. Future lines of research and teaching were subjects of many informal discussions.

One technical lecture and two dinner lectures were not available for inclusion in this volume; they were:

Transients in Counter-Current Separations Equipment, R. L. Pigford, University of Delaware

History of Computers and Future Applications, J. Mauchly, Mauchly Associates

The Modern Direction of Engineering, G. Fisher, Cornell University.

Thanks are due to all the lecturers, participants, and guests. I am especially grateful to those who responded by traveling from a great distance. The support of the National Science Foundation through grant GE-9750 and the University of Delaware is gratefully acknowledged.

W. F. Ames

Preface

The substantial progress of the eighteenth and nineteenth centuries in the construction of effective theories for physical phenomena were, in a large measure, due to a linear principle, that of (linear) superposition. Generally speaking, these theories are only a first approximation, for nature casts her phenomena in nonlinear form. Formulations which omit or suppress nonlinear terms often lead to inadequate or faulty results.

Considerable mathematical effort is being expended upon the qualitative theory of nonlinear partial differential equations. Existence and uniqueness theorems abound for *some* classes of equations. The development of methods of solution has lagged and even much of that effort has been in numerical analyses.

The foundations of the subject need greater attention. One of the basic questions—that of *nonlinear superposition*—has recently been explored by Jones and Ames in a paper of that title appearing in the *Journal of Mathematical Analysis and Applications* **17**, 44–47 (1967). Both linear and nonlinear equations of specific forms are examined. Perhaps the most striking result of that work is that the linear equation $U_x + U_y = U$ has a *noncountable infinity of nonlinear superposition principles.* If U_i, $i = 1, 2, \ldots, N$, are solutions then $[\sum_{i=1}^{P} U_i^n]^{1/n}$, $n \neq 0$, n real, are solutions. As $n \to 0$ one finds that $[\prod_{i=1}^{N} U_i]^{1/N}$ are solutions.

Before substantial progress can occur in this difficult area, guiding general principles must be developed. The linear theory has its Fourier analysis, Sturm–Liouville theory, and the like. The nonlinear theory is badly fragmented, with most techniques of an *ad hoc* nature.

This volume, being only a small chapter in the largely unwritten history, can serve both as a potential inspiration and as a plea for help. The challenge is here. Those who accept it do so with the forewarning that the road will not be an easy one. Yet we must recall the words of Sir James Jeans "to travel hopefully is better than to arrive."

June 1967 W. F. AMES

Contents

Integral Equations for Nonlinear Problems in Partial Differential Equations

DONALD H. HYERS

Electrical Problems Modeled by Nonlinear Partial Differential Equations

W. J. CUNNINGHAM

Difference Methods and Soft Solutions

M. H. PROTTER

Numerical Solution of the Nonlinear Equations for Two-Phase Flow through Porous Media

D. W. PEACEMAN

An Extrapolated Crank–Nicolson Difference Scheme for Quasilinear Parabolic Equations

MILTON LEES

Heat Transfer to the Endwall of a Shocktube. A Variational Analysis

JOHN R. FERRON AND KAZUTOSHI FUJIMURA

A Synergetic Approach to Problems of Nonlinear Dispersive Wave Propagation and Interaction

NORMAN J. ZABUSKY

Uniformization of Asymptotic Expansions

GUIDO SANDRI

High Order Accurate Difference Methods in Hydrodynamics

SAMUEL Z. BURSTEIN

Nonlinear Problems in the Dynamics of Thin Shells

WILLIAM A. NASH

Nonlinear Partial Differential Equations

Generalized Similarity Analysis of Partial Differential Equations

ARTHUR G. HANSEN

Engineering College
Georgia Institute of Technology
Atlanta, Georgia

Introduction

The term "similarity transformation" of a partial differential equation shall be defined to be a transformation of independent and dependent variables occuring in the equation such that the number of independent variables appearing in the transformed equation is at least one less than in the original equation. If a partial differential equation has two independent variables, a similarity transformation would transform the equation into an ordinary differential equation. In fact, the major application of similarity transformations has been the reduction of certain classes of nonlinear partial differential equations to ordinary differential equations.

The term "similarity analysis" will be defined to mean an analysis of a partial differential equation which has as its goal, the determination of possible similarity transformations.

From an application standpoint, one is usually interested in more than simply reducing the number of independent variables of an equation. A more demanding problem is the analysis of a partial differential equation with boundary and initial values. One then seeks to find similarity transformations that transform the equation and the boundary and initial values such that the transformed problem may be solved more readily.† It is possible (as we shall soon see) to find a similarity transformation of a partial differential equation such that the equation is reduced to an ordinary differential equation but such that the boundary values are not compatible with the transformed equation.

† From here on the term "boundary values" will be understood to mean both boundary values and initial values.

One of the classic examples of the application of a similarity analysis was presented by Blasius [1] who succeeded in reducing the system of nonlinear partial differential equations for flow over a flat plate to a single ordinary differential equation. Since the publication of the Blasius analysis, a wealth of literature has appeared in the area of fluid mechanics and convective heat transfer concerned with similarity analyses of specific problems.

More recently there has been increased interest in finding general methods of obtaining similarity solutions of classes of partial differential equations or in finding necessary and sufficient conditions for the existence of similarity solutions. A summary of typical techniques for finding similarity solutions of boundary value problems is presented in two recent books [2, 3]. These books also present bibliographies of the literature in the field of similarity analyses. The interested reader is referred to these bibliographies for a general perspective of past work. No attempt will be made to present a general review of literature in this article.

The purpose of the presentation to follow will be to describe the methods of similarity analysis that have been predominantly used to date. An assessment will then be made of the relative merits of these methods and, in particular, advantages and disadvantages described. On the basis of the perspective thus gained a particular method will be advocated for finding similarity solutions of boundary value problems. It should be recognized at the outset that the type of analysis which will ultimately be advocated is a subjective choice of the writer. It may well be that another writer would choose a method different than the one which will be described. The reason may simply be that familiarity with a given analysis makes its application in boundary value problems a simple matter for a particular individual. Thus, one should recognize that there is no "best method" of analysis. Nevertheless, there are specific problems that arise in the solution of boundary value problems which are independent of the type of analysis. These problems will be outlined and examined in the light of the particular classes of similarity analyses to be described.

Types of Similarity Analyses

There are four widely used types of similarity analyses which one encounters in the literature. These are named as follows:

(a) Free parameter
(b) Separation of variables
(c) Group theoretic
(d) Dimensional analysis

Although descriptions of these methods may be found in the literature (see, for example, Hansen [2] and Ames [3], a brief review of each of the methods will be given.

Free Parameter Analysis

The so-called free parameter analysis is a method of finding similarity solutions of partial differential equations by assuming that the dependent variables occuring in the partial differential equation can be expressed as functions of "similarity parameters," the number of which is one less than the total number of independent variables that occur in the original partial differential equation. In particular, if the original equation or system of equations contains only two independent variables, one proceeds to express each of the dependent variables as a function of a single parameter. The idea behind this technique is that the very nature of a similarity analysis is to reduce the number of independent variables by at least one. Therefore, at the outset it is assumed that this is possible. However, the heart and core of the method is the selection of the particular transformation of the dependent variable. The choice of transformation is determined by the boundary values of the original problem. One attempts to choose the transformation of the dependent variable in such a way that the boundary values transform into meaningful boundary values in terms of the new set of dependent variables. To illustrate this last point, suppose again that the number of independent variables were two in the original equation. Typical boundary value problems would usually specify boundary values for the dependent variables that were functions of the independent variables and, as such, would be variable. However, if the original partial differential equation is reduced to an ordinary differential equation, the boundary values for the ordinary differential must be constant to be meaningful. Likewise, the new independent variable must have constant values in the specification of boundary conditions. That point will be discussed shortly.

Having selected a transformation which not only reduces the number of independent variables, but also seems to provide a proper transformation of the boundary values, one proceeds to carry out the transformation in the original partial differential equation and to determine what restrictions on the form of the similarity parameters and on functions which might occur in the original boundary values are necessitated by the requirement that the final transformed equation contain only functions of the newly defined similarity parameters. Satisfaction of these restrictions determines whether or not a similarity solution is possible. The particular procedure is perhaps best illustrated by example and we shall do this directly. The example selected will also point out one of the basic weaknesses in the use of the free parameter type of analysis.

The example that we shall choose is a classical flow over a flat plate originally solved by Blasius [1]. Physically the problem is the following: a viscous fluid with a velocity U_0 flows over a flat plate as shown in Fig. 1.

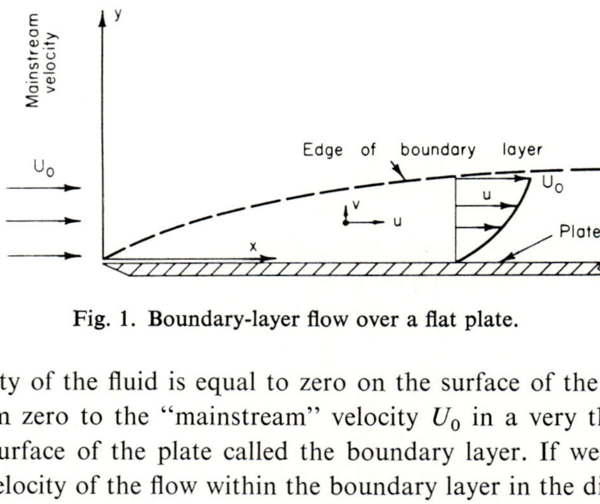

Fig. 1. Boundary-layer flow over a flat plate.

The velocity of the fluid is equal to zero on the surface of the plate and varies from zero to the "mainstream" velocity U_0 in a very thin region near the surface of the plate called the boundary layer. If we designate by u the velocity of the flow within the boundary layer in the direction of flow and by v the velocity in the boundary layer normal to the flow, the following equations can be written which describe the variation of u and v:

$$u \frac{\partial u}{\partial x} + v \frac{\partial u}{\partial y} = \nu \frac{\partial^2 u}{\partial y^2} \tag{1}$$

$$\frac{\partial u}{\partial x} + \frac{\partial v}{\partial y} = 0 \tag{2}$$

where ν = constant is the kinematic viscosity of the fluid. The boundary conditions on the problem are the following:

$$y = 0 \qquad u = v = 0 \qquad \lim_{y \to \infty} u = U_0$$

Before proceeding further, let us eliminate one of the dependent variables, v, from Eq. (2). We may solve Eq. (2) for v and obtain the following result:

$$v = -\int_0^y \left(\frac{\partial u}{\partial x}\right) dy \tag{3}$$

Taking the expression for v in Eq. (3) and substituting it into Eq. (1) yields the following single equation for the variable u:

$$\mu \frac{\partial u}{\partial x} - \frac{\partial u}{\partial y} \int_0^y \left(\frac{\partial u}{\partial x}\right) dy = \nu \frac{\partial^2 u}{\partial y^2} \tag{4}$$

Equation (4) is now a single equation for the dependent variable u which in turn is a function of the independent variables x and y. We shall now attempt to reduce this equation to an equation which involves a single independent variable, η, as follows: we wish to express u as a function of η in such a way that the boundary conditions which have been specified for the original equation carry over into meaningful boundary conditions in the transformed equation. One particular way of doing this is to let u/U_0 be a function of $\eta = \eta(x, y)$:

$$\frac{u}{U_0} = F'(\eta) \tag{5}$$

where $F'(\eta)$ is the derivative of an arbitrary function $F(\eta)$. (The derivative of a function $F(\eta)$ has been used as a matter of convenience.) The parameter η is simply a function of x and y and is yet unspecified in form. The boundary conditions which were given for the original problem become as follows under the transformation given by (5):

$$\begin{aligned} F' &= \frac{u}{U_0} = 0 \qquad \text{for} \quad \eta = \eta_0 \\ F' &= \frac{u}{U_0} = 1 \qquad \text{for} \quad \eta = \eta_\infty \end{aligned} \tag{6}$$

where $\eta_0 = \eta(0, y)$, $\eta = \eta(x, \infty)$. We also need to bring in the boundary condition on v but shall postpone that step for the present.

It is necessary that the quantities η_0 and η_∞ be constant if they are going to be values of independent variables in a final ordinary differential equation. As yet we have no assurance that this will be the case and it is precisely this point that describes a weakness in the free parameter type of analysis as we shall shortly see.

For the sake of completeness we also introduce a variable ζ such that $\zeta = x$.

Substitution of the transformation (5) into Eq. (4) yields the following result:

$$F'F'' \frac{\partial \eta}{\partial x} - \frac{\partial}{\partial x}\int_0^y F'\, dy\, F'' \frac{\partial \eta}{\partial y} = \frac{v}{U_0}\left[F'''\left(\frac{\partial \eta}{\partial y}\right)^2 + F''\frac{\partial^2 \eta}{\partial y^2}\right] \tag{7}$$

We shall now examine Eq. (7) to see what the restrictions are on the form of η which will enable Eq. (7) to be completely reduced to an ordinary differential equation. Before doing that, however, let us complete the transformation of Eq. (7) to an equation wholly in terms of η and ζ. We obtain

$$-F''\left(\frac{\partial \eta}{\partial y}\right)\frac{\partial}{\partial \zeta}\int_0^n F'\left(\frac{\partial \eta}{\partial y}\right)^{-1} d\eta = \frac{v}{U_0}\left[F'''\left(\frac{\partial \eta}{\partial y}\right)^2 + F''\left(\frac{\partial^2 \eta}{\partial y^2}\right)\right] \tag{8}$$

One cannot proceed further at this stage with making some assumption about $\partial\eta/\partial y$. A very simple assumption would be to assume that $\partial\eta/\partial y = g(\zeta)$. Under this assumption Eq. (8) reduces to

$$F''' - \left(\frac{U_0}{vg^3}\frac{dg}{d\zeta}\right)FF'' = 0 \tag{9}$$

The problem of reducing Eq. (9) to an ordinary differential equation in η reduces to merely choosing

$$\frac{U_0}{vg^3}\frac{dg}{d\zeta} = \text{const} = -\frac{1}{2} \tag{10}$$

where the $-\frac{1}{2}$ is chosen for convenience. Integrating Eq. (9) yields:

$$g = \sqrt{\frac{U_0}{v\zeta} + C_2}$$

where C_2 is a constant of integration and may be taken to be zero without loss of generality. We may now write (note $\zeta = x$):

$$\frac{\partial \eta}{\partial y} = \sqrt{\frac{U_0}{\nu x}}$$

and, thus,

$$\eta = y\sqrt{\frac{U_0}{\nu x}} + f(x) \tag{11}$$

where we set $f(x) = 0$ so that when $y = 0$ (a value associated with a boundary condition) η is not variable but rather equals the constant zero.

The form of v in terms of transformed variables can be found as follows:

$$v = -\frac{\partial}{\partial x}\int_0^y u\, dy = -\frac{\partial}{\partial x}\int_0^\eta F' U_0 \sqrt{\frac{\nu x}{U_0}}\, d\eta = \frac{1}{2}\sqrt{\frac{\nu U_0}{x}}\left[\eta F' - F + F(0)\right]$$

The value of F at $\eta = 0$, $F(0)$, is arbitrary. We shall choose $F(0) = 0$. We finally have the following differential equation with boundary conditions to evaluate (see Eq. 6),

$$F''' + \frac{FF''}{2} = 0 \qquad F(0) = F'(0) = 0 \qquad \lim_{\eta \to \infty} F' = 1 \tag{12}$$

where $\eta_0 = 0$ and $\eta_\infty = \infty$ have been employed in Eq. (6) from the form of η given by Eq. 11).

The goal of reducing the original partial differential equation to an ordinary differential equation has thus been achieved and the final step is to solve Eq. (12). Equation (12) is called the Blasius equation and the solution is given in a number of texts (cf., Schlichting, [1]).

While we have obtained appropriate forms for the transformed equation and associated boundary conditions, other possibilities may exist. For example, suppose that we return to Eq. (8) and simply choose

$$\frac{\partial \eta}{\partial y} = \text{const} = C_3 \tag{13}$$

The term on the left side of the equation becomes†

$$-F'' \frac{\partial}{\partial \zeta}(F(\eta) - F(0)) = 0$$

The final form of Eq. (8) is

$$F''' = 0 \tag{14}$$

and we have the solution

$$F = a\eta^2 + b\eta + c \tag{15}$$

The variable η is found from Eq. (13) to be

$$\eta = C_3 y + f(x) \tag{16}$$

Clearly, as $y \to \infty$, $\eta \to \infty$, and as a consequence, $F'(\eta) \to \infty$ which violates the boundary condition described in (6). This fact, however, is somewhat beside the point. The major difficulty encountered with the expression for η given by Eq. (16) is that for $y = 0$, $\eta = f(x)$ and is non-constant unless $f(x)$ is constant. (This latter case is normally ruled out for if it were true, $u = u(y)$ rather than $u = u(x, y)$.) This means that the values for the independent variables appearing in the original boundary conditions do not carry over as constant values in the boundary conditions for the ordinary differential equations. Let it suffice for now to point out that forms of η which are expressed as sums of functions of the original independent variable are usually not acceptable. Rather, product forms are required.

In summary, the chief weakness of the free-parameter method is that it only specifies an acceptable form for the new dependent variables but does not place any restrictions on the transformed independent variables. The proper form for the independent, however, must ultimately be chosen.

Separation of Variables Method

The weakness of the free-parameter method is recognized and corrected in "separation of variables method" first discussed by Abbott and Kline

† Note that ζ and η are treated as independent variables and that as a result, the derivative vanishes. This would not be the case if the derivative with respect to x were taken.

[4]. Quite simply, the approach employed is to set down at the outset transformations of the independent variables. For example, in the boundary-layer problem just discussed one would specify

$$\zeta = x \qquad \eta = y\,g(x)$$

It would follow that the values $y = 0$ and $y = \infty$ occuring in the specification of the boundary conditions carry over into constant values $\eta = 0$ and $\eta = \infty$ for the transformed variable. The next step in the analysis is to determine whether or not the original dependent variable can be expressed as a separable product of functions of the new variable. This follows the classical separation of variables method for solving partial differential equations. Thus in the boundary-layer problem, one might specify

$$u = H(\zeta)F'(\eta)$$

and employ this expression in Eq. (4) in an attempt to separate the variables and obtain a pair of ordinary differential equations for the functions H and F'. As it happens, the boundary conditions on u must carry over into constant boundary conditions for the function F' as was discussed in the previous section. This implies that $H(\zeta) = H(x)$ cannot be arbitrary but must have a form compatible with expressions occuring in the original boundary conditions. This was recognized in the free parameter method and, in fact, the function multiplying F' was chosen to be the mainstream velocity U_0.

We now see that both the free-parameter method and the separation of variables method have flaws in that they each take into account only part of the restrictions imposed by the boundary conditions. It is therefore recommended that in solving problems of the type discussed, that both restrictions on dependent and independent variables be recognized at the outset. Thus, one would specify for the boundary layer problem

$$u = U_0\,F'(\eta)$$

and

$$\eta = y\,g(x)$$

One would now be required to find acceptable forms of the unknown functions (in this case $g(x)$). This method will hereafter be referred to as the "restricted free-parameter method." While, formally, one might claim that such a transformation is not "general," analyses published to date are all based on transformations of this type.

As a matter of review, the following procedure is therefore advocated in reducing a partial differential equation in two independent variables to an ordinary differential equation:

(1) Anticipate that the dependent variable is ultimately to be expressed as a function of a single variable, η. Therefore, choose the dependent variable as a product of a function of η and a function which occurs in the boundary conditions so that, ultimately, constant boundary conditions on the function of η will be possible.

(2) Choose the parameter η in such a way that the values of independent variables occuring in the original boundary conditions become constant values for η.

(3) Solve for unknown functions occuring in the specification of dependent and independent variables such that an ordinary differential equation in η is obtained. One could also separate variables following the method of Abbott and Kline.

The Group Theory Approach

The most mathematically sophisticated approach to finding similarity transformations is based on concepts from the theory of transformation groups. One suspects that this approach will eventually lead to more general developments in similarity analyses. For the time being, we will consider only simple applications. Before proceeding, a few definitions are in order.

(1) A set of continuous functions, $Z^i = f^i(z^1, z^2, \ldots, z^m, a)$ which transform a set of variables $z^1, z^2, \ldots, z^m$ into a set of variables $Z^1, Z^2, \ldots, Z^m$ and which (a) form a group of transformations† and (b) depend upon a parameter, a, is called a continuous transformation group T_a. A function $g(z^1, \ldots, z^m)$ is called an *absolute invariant* of the transformation group T_a if

$$g(Z^1, \ldots, Z^m) = g(z^1, \ldots, z^m) \tag{17}$$

under the transformations.

† The reader is referred to Hansen [2] and Ames [3] for greater detail and a more complete discussion of the group theory principles.

(2) A function $\Phi(z^1, \ldots, z^m)$ is *conformally invariant* under T_a if

$$\Phi(Z^1, \ldots, Z^m) = F(z^1, \ldots, z^m; a)\, \Phi(z^1, \ldots, z^m)$$

If

$$\Phi(Z^1, \ldots, Z^m) = F(a)\, \Phi(z^1, \ldots, z^m)$$

the function Φ is said to be constant conformally invariant.

(3) The system of partial differential equations

$$\phi_j\left(x^1, \ldots, x^m, u^1, \ldots, u^n, \ldots, \frac{\partial^k u^1}{\partial(x^1)^k}, \ldots, \frac{\partial^k u^n}{\partial(x^m)^k}\right) = 0$$

where x^i and u^i are independent and dependent variables, respectively, is said to be invariant under the one-parameter transformation group

$$X^i = f^i(x^1, \ldots, x^m; a) \tag{18a}$$

$$U^j = h^j(u^1, \ldots, u^n; a) \tag{18b}$$

if

$$\phi_j\left(X^1, \ldots, X^m, U^1, \ldots, U^n, \ldots, \frac{\partial^k U^1}{\partial(X^1)^k}, \ldots, \frac{\partial^k U^n}{\partial(X^m)^k}\right) = 0 \tag{19}$$

Employing these definitions, we may now state the following by theorem:

Suppose that a system of partial differential equations

$$\phi_j\left(x^1, \ldots, x^m, u^1, \ldots, u^n, \ldots, \frac{\partial^k u^n}{\partial(x^m)^k}\right) = 0$$

is invariant under the transformation group T_a defined by Eq. (18). Then, solutions of the system that are invariant under T_a can be expressed as solutions of a system of equations having a number of independent variables one less than the original system. The solutions may be found as solutions of the system:

$$\Phi_j\left(\eta_1, \ldots, \eta_{m-1}, F_1, \ldots, F_n, \ldots, \frac{\partial^k F_1}{\partial \eta_1{}^k}, \ldots, \frac{\partial^k F_n}{\partial \eta_{m-1}^k}\right) \tag{20}$$

where, η_i are absolute invariants of $X^i = f^i(x^1, \ldots, x^m; a)$ and the functions $F_i(\eta_1, \ldots, \eta_{m-1})$ may be expressed as:

$$F_i(\eta_1, \ldots, \eta_{m-1}) = g_i(x^1, \ldots, x^m, u^1, \ldots, u^n) = g_i(X^1, \ldots, X^m, U^1, \ldots, U^n) \tag{21}$$

The functions $g_i(x^1, \ldots, u_n)$ are absolute invariants of both the original dependent and independent variables.

According to this theorem we may reduce a system of partial differential equations in two independent variables to a system of ordinary differential equations in one independent variable if we can but find a variable η which is absolutely invariant under a transformation of the independent variables and if we can find a function of both the dependent and independent variables that is an absolute invariant under the transformation group for all variables. The questions that naturally occur are these:

(1) What types of one-parameter transformation groups should one employ?

(2) How does one find absolute invariants η_i under the transformation group chosen?

(3) Given that expressions for the new dependent variables are of the form

$$F_i = g_i(x^1, \ldots, x^m, u^1, \ldots, y^n)$$

and that the g_i are absolute invariants, how might they be determined?

Answers to the above questions are by no means unique. In fact, current research is concerned with the range of possible answers for certain classes of partial differential equations. For our purpose, however, we will provide some very simple answers which have proven to encompass answers found by using other techniques. We shall begin by describing one of the simplest of all transformation groups, the so-called linear group.

Let us reconsider the pair of equations for laminar boundary-layer flow over a flat plate, Eqs. (1) and (2). In terms of the designation for variables given earlier, let $x^1 = x$, $x^2 = y$, $u^1 = u$, and $u^2 = v$. The linear one-parameter transformation group which transforms these variables into a corresponding set of variables $\bar{x}$, $\bar{y}$, $\bar{u}$ and $\bar{v}$ is

$$\bar{x} = a^n x \qquad \bar{y} = a^m y \qquad \bar{u} = a^p u \qquad \bar{v} = a^q v \tag{22}$$

where a is the parameter for the group. In order to apply the key theorem stated earlier we need to have a transformation group for which the differential equations under consideration are invariant. In essence this means that we need to determine values for n, m, p, and q which insure this.† Proceeding to substitute expressions (22) with Eq. (1) and (2) gives:

$$a^{-2p+n}\,\bar{u}\,\frac{\partial \bar{u}}{\partial \bar{x}} + a^{-q-p+m}\,\bar{v}\,\frac{\partial \bar{u}}{\partial \bar{y}} = \nu a^{-p+2m}\,\frac{\partial^2 \bar{u}}{\partial \bar{y}^2} \tag{23a}$$

$$a^{-p+n}\,\frac{\partial \bar{u}}{\partial \bar{x}} + a^{-q+m}\,\frac{\partial \bar{v}}{\partial \bar{y}} = 0 \tag{23b}$$

To establish invariance it is only necessary to choose the exponents of "a" in such a way that the transformed equations look exactly the same in terms of the new variables as they did with the old. We need only to equate these exponents and cancel out the common powers of a. Equating exponents leads to the following system of equations:

$$-2p + n = -q - p + m = -p + 2m$$

$$-p + n = -q + m$$

Solving this set of equations gives

$$q = -m \qquad p = n - 2m$$

Next, we turn our attention to finding an absolute invariant η that is invariant under the transformations of the independent variable. We choose a simple power expression as a possible functional form for η:

$$\eta = yx^s$$

It would be possible to have a power of y other than unity but nothing is gained by doing so.

The invariance of η is satisfied by

$$\eta = yx^s = \bar{y}\bar{x}^s = a\,y^m(a^n x)^s$$

Therefore, we require that

$$s = -\frac{m}{n}$$

The new independent variables to replace u and v are now sought and they are to be such that they are invariant under the complete set of

† It may be that the only permissible values are $n = m = p = q = 0$ in which case the method is abandoned for the group chosen.

transformations. We choose as possible forms:

$$g_1 = ux^r$$
$$g_2 = vx^t$$

Powers of y might be inserted into these expressions but once again it can be simply reasoned that nothing is gained.

The invariances of g_1 and g_2 requires that

$$ux^r = \bar{u}\bar{x}^r = (a^{n-2m}\,u)\,(a^n x)^r$$
$$vx^t = \bar{v}\bar{x}^t = (a^{-m}\,v)\,(a^n x)^t$$

Equating exponents to zero gives

$$n - 2m + nr = 0 \qquad \text{or} \qquad r = -1 + 2\mathrm{m/n}$$
$$-m + nt = 0 \qquad \text{or} \qquad t = \mathrm{m/n}$$

Lastly, according to the theory, the functions g_1 and g_2 can be considered to be functions of η and the partial differential equation should reduce to a set of ordinary differential equations for these functions. We therefore let

$$H(\eta) = g_1 = ux^r = ux^{-1+2m/n}$$
$$G(\eta) = g_2 = vx^t = vx^{m/n}$$

or

$$u = x^{1-2m/n}\,H(\eta)$$
$$v = x^{-m/n}\,G(\eta)$$

It should now be possible to substitute these expressions into the original set of equations to obtain ordinary differential equations for H and G. Let us proceed to do this.

The continuity equation, Eq. (2), becomes

$$\frac{\partial u}{\partial x} + \frac{\partial v}{\partial y} = \left(1 - \frac{2m}{n}\right)x^{-2m/n}\,H + x^{1-(2m/n)}\,H'\,\frac{\partial \eta}{\partial x} + x^{-m/n}\,G'\,\frac{\partial \eta}{\partial y} = 0 \qquad (24)$$

Now,

$$\frac{\partial \eta}{\partial x} = \frac{\partial}{\partial x}\,yx^{-m/n} = -\frac{m}{n}yx^{-(m/n)-1} = -\frac{m}{n}\frac{\eta}{x}$$

$$\frac{\partial \eta}{\partial y} = x^{-m/n}$$

Therefore, Eq. (24) becomes

$$\left(1 - \frac{2m}{n}\right) x^{-2m/n} H + x^{-2m/n} H'\left(-\frac{m}{n}\right)\eta + x^{-2m/n} G' = 0 \tag{25}$$

Assuming $x \neq 0$, we have

$$\left(1 - \frac{2m}{n}\right)H - \left(\frac{m}{n}\right)H'\eta + G' = 0 \tag{26}$$

At this point, let us choose $m/n = \frac{1}{2}$ to simplify our expression.† We then obtain

$$G' = \frac{H'\eta}{2}$$

Next we substitute our expressions for u and v into Eq. (1) with the following result (choosing $m/n = \frac{1}{2}$):

$$-\frac{HH'}{2}\frac{\eta}{x} + x^{-1/2} GH'x^{-1/2} = \nu H''x^{-1} \tag{27}$$

Again, for $x \neq 0$, we have

$$GH' - \frac{HH'}{2}\eta = \nu H'' \tag{28}$$

Thus, we have two ordinary differential equations for the functions G and H as was anticipated.

The results can be put in more conventional form by letting $H = F'$. Then, from Eq. (28),

$$\begin{aligned} G &= \int \frac{F'\eta}{2}\, d\eta + \text{const} \\ &= \tfrac{1}{2}(\eta F' - F) + \text{const} \end{aligned} \tag{29}$$

Equating the integration constant to zero and substituting the expression for G into Eq. (27) gives

$$\tfrac{1}{2}(\eta F' - F)F'' - \frac{F'F''}{2}\eta = \nu F'''$$

† This is admittedly a special choice but not too restrictive. A second interesting choice is to pick $m = 0$.

or

$$\nu F''' + \frac{FF''}{2} = 0 \tag{30}$$

By a proper change of variables the constant ν can be eliminated and the classical Blasius equation, Eq. (12), can be obtained.

We have not as yet determined what the boundary conditions are for the transformed equations. It is therefore necessary to determine whether or not the boundary conditions on the original equations transform into meaningful boundary conditions on the final differential equation. Fortunately in the present case, this is possible and we may obtain a set of boundary conditions exactly comparable to those specified for Eq. (12) by a redefinition of η as

$$\eta = yx^{-1/2} \sqrt{\frac{U_0}{\nu}}$$

and u as

$$u = U_0 F'(\eta)$$

It would be possible to check out the compatibility of boundary conditions under an assumed transformation in advance of actually determining the precise forms of the various invariants.

The linear group is not the only commonly used group of transformations. Other groups employing exponential rather than power expressions for the parameter "a" may be employed. (Additional information on this topic may be found in Hansen [2].)

Two major approaches to finding similarity transformations have been outlined up to this point. It has also been pointed out that the first two types of analyses, the "free-parameter method" and the "separation of variables" method may as well be combined into a single method. The so-called "dimensional analysis method" (see Hansen [2]) can be established as being a special group theoretical method and hence may be set aside.

Of these methods, then, which one might be recommended? To some extent, this is a matter of choice. The restricted free-parameter method focuses attention on the boundary values of a problem from the outset and insures that the transformations will, in all probability, have a proper form. The chief drawback is the determination of unknown functions which must be obtained as solutions to differential equations (although these equations are usually quite simple). The group theory approach is by far the easiest to employ and only a few elementary algebraic steps are required. The drawback to this method is the initial selection of the one-parameter transformation group. In practice, all similarity analyses published in the literature could have been obtained by using either the linear group already discussed or the group involving exponentials (the so-called spiral group) which has been alluded to earlier. The question of the suitability of these transformations to properly carry over the boundary conditions of the partial differential equations to a set valid for ordinary differential equations is largely academic. By the choice of the form of the linear group, for example, one notices that considerations which apply to the restricted free-parameter method are valid for the group theory method.

In summary, it would appear that the method which is at once the easiest to use and furthermore has the desirable features of other methods (namely, the ability to check out boundary conditions) is the group theory method. This method also has the potential of simple extension and generalization. Future research should indicate if generalizations for practical problems in applied mathematics are indeed possible.

References

1. Blasius, H., Grenzschichten in Flüssigkeiten mit kleiner Reibung, *Z. Math. Physik*, **56**, 1 (1908); also, NACA TM 1256.
2. Hansen, A. G., "Similarity Analyses of Boundary Value Problems in Engineering." Prentice-Hall, Englewood Cliffs, New Jersey, 1964.
3. Ames, W. F., "Nonlinear Partial Differential Equations in Engineering." Academic Press, New York, 1965.
4. Abbott, D. E., and Kline, S. J., Simple Methods for Classification and Construction of Similarity Solutions of Partial Differential Equations, AFOSR Rept. TN-60-1163 (1960).

Vector Eigenfunction Expansions for the Growth of Taylor Vortices in the Flow between Rotating Cylinders

R. C. DI PRIMA

Department of Mathematics
Rensselaer Polytechnic Institute
Troy, New York

I. Introduction

The stability of a viscous flow between concentric rotating cylinders is one of the classical problems in the theory of hydrodynamic stability. Taylor (1923) in his original investigation observed experimentally that at sufficiently high speeds of the inner cylinder the laminar circumferential flow (Couette flow) becomes unstable, the instability yielding a steady secondary motion in the form of toroidal vortices (Taylor vortices) spaced regularly along the axis of the cylinder.

An analysis of the stability of Couette flow based on the linearized equations for small disturbances leads to an eigenvalue problem for the critical speed of the inner cylinder in the form of a dimensionless combination T, the Taylor number. It is a function of the parameters $\mu = \Omega_2/\Omega_1$ and $\eta = R_1/R_2$ which describe the basic velocity and the geometry, and the dimensionless wave numbers describing the disturbance. Here Ω_1, Ω_2 and R_1, R_2 are the angular velocities and radii of the inner and outer cylinders, respectively. For the most part, attention has been restricted to the case in which the disturbance is axisymmetric; however, Krueger *et al.* (1966) have recently shown that for $1 - \mu$ small and μ sufficiently negative ($\mu \cong -1$) the critical conditions correspond to a nonaxisymmetric disturbance.

In the present paper we will only consider values of η and μ for which we can expect the critical speed to correspond to an axisymmetric disturbance and the resulting motion for $T > T_c$, where T_c the critical value of T, to be the axisymmetric Taylor vortex motion mentioned earlier.

There is a substantial amount of experimental evidence that indicates this is the case for a wide range of values of μ and η, certainly for $0 \leq \mu \leq \eta^2$ and for $T - T_c$ not too large.† The reader is referred to Taylor's own paper, and also to a recent paper by Coles (1965) for a comprehensive account of the growth of the Taylor vortex motion and its subsequent instability with increasing T. In addition, we shall primarily be concerned with the development of the finite-amplitude Taylor vortex motion for $T > T_c$; although as a necessary prerequisite for the present theoretical analysis the linear stability problem will be discussed in Section 3. For a complete discussion of the formulation of the linear stability problem for Couette flow, and its many ramifications, see the treatise by Chandrasekhar (1961) or a brief review paper by Di Prima (1963) and the references given in these papers.

According to linear theory, the disturbance grows exponentially with time for $T > T_c$. However, as has been pointed out, a definitie equilibrium vortex motion is observed for T slightly greater than T_c, the circulation in the vortex increasing with $T - T_c$. To explain this equilibration process and to compute the secondary motion it is necessary to turn to the full nonlinear equations of motion. Using an energy-balance method, and making several approximations, Stuart (1958) was able to obtain a surprisingly good estimate of the amplitude of the vortex motion. The first detailed and conclusive calculations, however, were carried out by Davey (1962), who based his analysis on an expansion procedure developed by Stuart (1960) and Watson (1960). This will be discussed in Section 4. Stuart's (1958) and Davey's (1962) computed values of the torque acting on the inner cylinder for several geometries were in very good agreement with the experimental measurements of Donnelly (1958). In addition, the calculations of Davey are in good agreement with the more detailed observations and measurements of the secondary motion by Donnelly and Schwarz (1965) and Snyder and Lambert (1966).

More recently Eckhaus (1965) has studied nonlinear problems in hydrodynamic stability by expanding the velocity field in the eigenfunctions of the corresponding linear stability problems. Here we extend the Eckhaus procedure by developing a formal *vector* eigenfunction expansion which

† For $\mu > \eta^2$, it has been conclusively shown that the flow is stable to axisymmetric disturbances (see Chandrasekhar (1961), Section 70), and probably to nonaxisymmetric disturbances, though the latter has not been proved.

can be used to study the growth of Taylor vortices in the flow between rotating cylinders (see Sections 3 and 5). While the procedure is considerably different from that developed by Stuart and Watson and used by Davey, the final results are the same, and as part of the present analysis we compare and contrast the two procedures. Though the present procedure is mathematically only a formal one, it does show the formulation of the stability problem as an initial value problem. In addition, it provides some information about the class of initial disturbances that can be considered.

2. The Governing Equations

Let r, θ, and z denote the usual cylindrical coordinates, and let u_r, u_θ, and u_z denote the components of velocity in the increasing r, θ, and z directions, respectively. Consider two infinitely long concentric circular cylinders with the z-axis as their common axis, with radii R_1 and R_2 ($> R_1$), and rotating with angular velocities Ω_1 and Ω_2, respectively. The equations of motion for a viscous, incompressible fluid admit the exact steady solution (Couette flow)

$$u_r = u_z = 0, \qquad u_\theta = V(r) = Ar + \frac{B}{r}, \qquad \partial p/\partial r = \rho\, V^2/r, \tag{1}$$

where p is the pressure and ρ is the density. The constants A and B are determined by the boundary conditions $u_\theta = R_1\Omega_1$ at $r = R_1$ and $u_\theta = R_2\Omega_2$ at $r = R_2$.

To study the stability of this flow we superimpose a general disturbance on the basic solution given by Eq. (1). Then substituting in the Navier–Stokes equations of motion and the equation of continuity, yields a set of nonlinear partial differential equations for the perturbations in the velocities and the pressure. For a fixed geometry ($\eta = R_1/R_2$) and a fixed velocity distribution ($\mu = \Omega_2/\Omega_1$), we try, first, to determine the critical conditions for which there exist a nondecaying solution of these equations satisfying the boundary conditions at $r = R_1$ and $r = R_2$; and then to determine what happens when the critical conditions are exceeded. As mentioned in the introduction we will limit our discussion to axisymmetric motions.

Let $u'(r, z, t)$, $v'(r, z, t)$, and $w'(r, z, t)$ denote the disturbances in the

r, θ, and z components of the velocity, respectively and $p'(r, z, t)$ the disturbance in the pressure, where t is the time. It is convenient to introduce the following dimensionless variables and parameters:

$$r = R_0 + xd, \quad z = \zeta d, \quad t = (d^2/\nu)\tau, \quad \xi(x) = \delta/(1 + \delta x),$$
$$\delta = d/R_0, \quad \alpha = \frac{-2Ad}{R_0\Omega_0}, \quad T = \frac{-4A\Omega_0 d^4}{\nu^2}, \tag{2}$$
$$\Omega = \Omega_0\Omega_l(x), \quad u' = \frac{\nu}{d}u, \quad w' = \frac{\nu}{d}w, \quad v' = \frac{2Ad}{\sqrt{T}}v, \quad p' = \rho\frac{\nu^2}{d^2}\varpi.$$

Here $d = R_2 - R_1$; R_0 is a reference length which is taken as $(R_1 + R_2)/2$ if the cylinders rotate in the same direction and as R_1 if the cylinders rotate in opposite directions; $\Omega(r)$ is the basic angular velocity; Ω_0 is a reference angular velocity which is taken as $(\Omega_1 + \Omega_2)/2$ if the cylinders rotate in the same direction and as Ω_1 if the cylinders rotate in opposite directions; ν is the kinematic viscosity; and T is the Taylor number. In terms of these dimensionless variables and parameters the three momentum equations and the continuity equation are:

$$(DD^* + \partial^2/\partial\zeta^2 - \partial/\partial\tau)u - \sqrt{T}\,\Omega_l(x)v - D\varpi = uDu + w\frac{\partial u}{\partial\zeta} - \frac{\alpha}{2}\frac{v^2}{1 + \delta x}, \tag{3}$$

$$(DD^* + \partial^2/\partial\zeta^2 - \partial/\partial\tau)v - \sqrt{T}u = uDv + w\frac{\partial v}{\partial\zeta} + \xi(x)uv, \tag{4}$$

$$(D^*D + \partial^2/\partial\zeta^2 - \partial/\partial\tau)w - \partial\varpi/\partial\zeta = uDw + w\frac{\partial w}{\partial\zeta}, \tag{5}$$

$$D^*u + \partial w/\partial\zeta = 0, \tag{6}$$

where

$$D = \frac{\partial}{\partial x}, \quad D^* = \frac{\partial}{\partial x} + \xi(x) = \frac{1}{1 + \delta x}D(1 + \delta x). \tag{7}$$

The choice of dimensionless variables (2) is particularly convenient if one wants to consider the so-called small-gap problem, $\delta \to 0$, i.e., $\eta \to 1$. In this case we hold the dependent and independent variables and the parameter T fixed and let $\delta \to 0$; then $\xi = 0$, $D^* = D$, and $\Omega_l(x)$ becomes a linear function of x. If in addition we consider only the case in which the cylinders rotate in the same direction, $\Omega_l(x)$ can be replaced by its avarage value 1 with only a very small error for a range of μ as large as $-\frac{1}{4} \leq \mu < 1$. The consequences of these two approximations are (i) Eqs. (3)–(6) reduce to constant coefficient partial differential equations, and (ii) the linear stability problem, when written as a system of two equations, is a selfadjoint eigenvalue problem. While the first consequence does lead to a considerable simplification in any analytic calculations, it is the second which leads to a simplification in the mathematical theory. This will be discussed further in Section 3.

In the present analysis we will consider the general problem, i.e., we will not make the small-gap approximation. However, we will restrict our attention to the case $\mu = \Omega_2/\Omega_1 \geq 0$ (or at least not too negative) so we choose $R_0 = (R_1 + R_2)/2$ and $\Omega_0 = (\Omega_1 + \Omega_2)/2$. Then $-\frac{1}{2} \leq x \leq \frac{1}{2}$, and the boundary conditions are

$$u = v = w = 0 \qquad \text{at} \quad x = \pm\tfrac{1}{2}. \tag{8}$$

In addition we have initial conditions of the form

$$u(x,\zeta,0) = u^{(0)}(x,\zeta), \quad v(x,\zeta,0) = v^{(0)}(x,\zeta), \quad w(x,\zeta,0) = w^{(0)}(x,\zeta), \tag{9}$$

where $u^{(0)}$, $v^{(0)}$, and $w^{(0)}$ must satisfy certain differentiability conditions. In particular, since we are dealing with a continuous fluid $u^{(0)}$ and $w^{(0)}$ must satisfy the continuity equation (6). The class of initial conditions that we are willing to consider (or can consider) determines the class of disturbances against which we can investigate the stability of the basic flow.

If the Taylor number is sufficiently small, we anticipate that all initially infinitesimal disturbances will decay with increasing time; but with increasing T a point will be reached at which at least one disturbance will grow. Experimental evidence indicates that when this critical value of T is exceeded, a steady secondary motion in the form of toroidal vortices spaced regularly along the axis of the cylinders occurs. Thus we restrict our attention to a *single* disturbance which is periodic in the ζ-direction with period $2\pi/a$. While this is restrictive (certainly there is no reason to believe that only periodic disturbances are present) it is general enough to allow us to obtain results in accord with the experimental observations. Clearly it would be desirable to consider more general disturbances, and show that aperiodic disturbances die out or lead to periodic motions. Also it would be desirable to consider the interaction of several disturbances of different wavelengths in the ζ-direction. In this regard, note that within the framework of linear theory there is no interaction between disturbances so we need only consider one, and fortunately experimental evidence indicates that the secondary motion is periodic with a wavelength which varies only very slightly with increasing T beyond the critical value. In order to simplify the ensuing discussion as much as possible, we consider only motions for which u, v, and $\bar{\omega}$ are even functions of ζ, and w (according to the continuity equation) is odd. While there is no difficulty in considering a general periodic motion, the secondary motion does appear to be of this form (see Snyder and Lambert (1966)).

Thus we assume†

$$\begin{aligned} v(x, \zeta, \tau) &= v_0(x, \tau) + \sum_{n=1}^{\infty} v_n(x, \tau)\cos na\zeta, \\ \bar{\omega}(x, \zeta, \tau) &= \bar{\omega}_0(x, \tau) + \sum_{n=1}^{\infty} \bar{\omega}_n(x, \tau)\cos na\zeta, \\ u(x, \zeta, \tau) &= u_0(x, \tau) + \sum_{n=1}^{\infty} nau_n(x, \tau)\cos na\zeta, \\ w(x, \zeta, \tau) &= \sum_{n=1}^{\infty} w_n(x, \tau)\sin na\zeta. \end{aligned} \tag{10}$$

Substituting for u, v, w, and $\bar{\omega}$ in Eqs. (3)–(6), equating coefficients of $\cos na\zeta$ and $\sin na\zeta$, eliminating w_n by using the continuity equations and $\bar{\omega}_n$ by using the third momentum equation we obtain an infinite system of coupled partial differential equations for the *vectors* $\mathbf{Q}_n(x, \tau) = (u_n(x, \tau), v_n(x, \tau))$, and the velocities‡ $u_0(x, \tau)$ and $v_0(x, \tau)$.

First it follows from the continuity equation and the boundary conditions (8) that $u_0 \equiv 0$. This means that for a disturbance of the form (10), u_0 must be initially zero and that it remains zero for all τ. From the second momentum equation we find

$$(-DD^* + \partial/\partial\tau)v_0 = S_0(u_1, u_2, \ldots, v_1, v_2, \ldots) = -\tfrac{1}{2}(D^* + \xi) \sum_{n=1}^{\infty} nau_n v_n. \tag{11}$$

† To obtain a certain symmetry in the final equations, it is convenient to write $nau_n(x, \tau)$ for the coefficient of $\cos na\zeta$ in the series expansion for u.

‡ The equations for $n = 0$ are different from those for $n \geq 1$, and we treat u_0 and v_0 separately.

The equations for the $\mathbf{Q}_n$ are

$$\left(L_n + \frac{\partial}{\partial \tau} M_n\right)\mathbf{Q}_n = \mathbf{S}_n(v_0, v_1, v_2, \ldots, u_1, u_2, \ldots), \qquad n \geq 1 \qquad (12)$$

where the L_n and M_n are the real, linear, matrix differential operators

$$L_n = \begin{pmatrix} (DD^* - n^2a^2)^2 & na\sqrt{T}\Omega_l(x) \\ na\sqrt{T} & -(DD^* - n^2a^2) \end{pmatrix},$$

$$M_n = \begin{pmatrix} -(DD^* - n^2a^2) & 0 \\ 0 & 1 \end{pmatrix}, \qquad n \geq 1, \qquad (13)$$

and the vector $\mathbf{S}_n$ is a complicated quadratic function of the u_n and v_n. It is given in the appendix.

The boundary conditions are

$$v_0 = 0; \qquad u_n = v_n = Du_n = 0 \qquad \text{at} \quad x = \pm\tfrac{1}{2}, \qquad n \geq 1. \quad (14)$$

The initial conditions consistent with the form of the disturbance (10) are

$$v_0(x, 0) = v_0^{(0)}(x); \quad u_n(x, 0) = u_n^{(0)}(x), \quad v_n(x, 0) = v_n^{(0)}(x), \qquad n \geq 1. \quad (15)$$

3. The Linear Problem

To determine the critical value of T above which an infinitesimal disturbance of the form (10) will grow and the corresponding critical value of a, it is permissible to neglect the quadratic terms in Eqs. (11) and (12). In this case the equations for v_0 and each of the $\mathbf{Q}_n$ uncouple, and we can consider each separately. We look for solutions of the form

$$v_0(x, \tau) = f_0(x)e^{-\sigma_0\tau}; \qquad \mathbf{Q}_n(x, \tau) = \mathbf{q}_n(x)e^{-\sigma_n\tau} = \begin{pmatrix} f_n(x) \\ g_n(x) \end{pmatrix} e^{-\sigma_n\tau}. \qquad (16)$$

Substituting for v_0 and $\mathbf{Q}_n$ in Eqs. (10) and (11), with the nonlinear terms neglected, yields the eigenvalue problems

$$(-DD^* - \sigma_0)f_0 = 0, \qquad f_0 = 0 \qquad \text{at} \quad x = \pm\tfrac{1}{2}, \qquad (17)$$

and

$$(L_n - \sigma_n M_n)\mathbf{q}_n = 0, \qquad f_n = Df_n = g_n = 0 \qquad \text{at} \quad x = \pm\tfrac{1}{2}, \qquad (18)$$

for σ_0 and the σ_n, respectively.

If there exists a solution of any of these equations for which at least one of the σ's has a negative real part, the flow is unstable. Since the eigenvalue problem (17) for σ_0 does not depend on a or T, it can be solved once and for all. As we shall soon see there are no solutions with $\mathrm{Re}(\sigma_0) < 0$. Next consider the eigenvalue problem (18) for $n = 1$. We anticipate that for each value of a, $0 < a < \infty$, there will be a certain value of T above which there will be solutions for eigenvalues σ_1 with negative real part. The minimum of these values of T over a determines a critical value of a, a_c, and of T, T_c, above which there exist growing solutions $\mathbf{Q}_1(x, \tau)$. From the form of the operators L_n and M_n, it is clear that the critical conditions for the eigenvalue problem (18) with $n \neq 1$ are $na = a_c$ and $T = T_c$. Thus for $T > T_c$ the flow is certainly unstable, and for $T < T_c$ the flow is stable to a class of infinitesimal initial disturbances of the form (10) whose Fourier components (i.e., $v^{(0)}(x)$, $\mathbf{Q}_n^{(0)}(x)$) can be represented in terms of the eigenfunctions of the corresponding eigenvalue problems (17) and (18). We turn to these problems now.

Consider first the eigenvalue problem (17) for σ_0. This is a standard Sturm–Liouville problem (see Ince 1926), Chapters 10 and 11, and Coddington and Levinson (1955, Chapters 7 and 8). Briefly, there exist infinitely many (necessarily) real positive eigenvalues σ_{0m}, $m = 0, 1, 2, \ldots$, which have no cluster point except $+\infty$, and which can be ordered $\sigma_{0m+1} \geq \sigma_{0m}$, and whose corresponding eigenfunctions $\{f_{0m}(x)\}$ have exactly m zeros in the interval $-\frac{1}{2} < x < \frac{1}{2}$. Further it easily follows from the differential equation (17) that the $f_{0m}(x)$ satisfy the orthogonality relation

$$(f_{0m} \,|\, f_{0p}) = \int_{-1/2}^{1/2} (1 + \delta x) f_{0m}(x) \tilde{f}_{0p}(x)\, dx = 0, \qquad m \neq p, \tag{19}$$

where a tilde denotes a complex conjugate (note that $(1 + \delta x) > 0$). In order to have a notation consistent with that needed for the eigenvalue problem (18), we have introduced a complex-valued inner product; however the $f_{0m}(x)$ are real, and can be normalized so that $(f_{0m} \,|\, f_{0m}) = 1$. We assume such a normalization in the rest of the paper. Finally any function $v_0^{(0)}(x)$, which is twice continuously differentiable on $-\frac{1}{2} \leq x \leq \frac{1}{2}$ and satisfies the boundary conditions $v_0^{(0)}(\pm\frac{1}{2}) = 0$, can be expressed as a uniformly convergent series:

$$v_0^{(0)}(x) = \sum_{m=0}^{\infty} (v_0^{(0)} \,|\, f_{0m}) f_{0m}(x), \qquad -\tfrac{1}{2} \leq x \leq \tfrac{1}{2}. \tag{20}$$

More generally, if $v_0^{(0)}(x)$ is square-integrable, it can be represented by the series (20), but now equality is defined by convergence in a norm $\| f \|_0^2 = (f|f)$. Thus the class of initial conditions $v_0^{(0)}(x)$ that we can consider for $v_0(x, \tau)$ is the class of square-integrable functions on $-\frac{1}{2} \leq x \leq \frac{1}{2}$. The solution of Eq. (11) with the nonlinear terms neglected which satisfies the boundary conditions (14) and the initial conditions (15) is given by

$$v_0(x, \tau) = \sum_{m=0}^{\infty} (v_0^{(0)} \,|\, f_{0m}) f_{0m}(x) e^{-\sigma_{0m}\tau}. \tag{21}$$

Since all of the σ_{0m} are positive, $v_0(x, \tau) \to 0$ as $\tau \to \infty$ irrespective of the initial conditions. When we consider the nonlinear problem for $T > T_c$, we will find this is no longer true.

Next consider the eigenvalue problem (18). First note that the space of two dimensional vectors $\mathbf{q}_n(x)$ with first and second components four times and two times continuously differentiable on $-\frac{1}{2} \leq x \leq \frac{1}{2}$, respectively, and satisfying the boundary conditions (18) form a pre-Hilbert space with inner product

$$(\mathbf{q}_n \,|\, \mathbf{q}_n') = \int_{-1/2}^{1/2} (1 + \delta x)(f_n \tilde{f}_n' + g_n \tilde{g}_n')\, dx. \tag{22}$$

It is easy to show that the eigenvalue problem (18) is not self-adjoint because of the off-diagonal terms in L_n; thus, in general, the eigenvalues will be complex.

The adjoint problem is

$$(L_n^+ - \tilde{\sigma}_n M_n^+)\, \mathbf{q}_n^+ = 0, \quad f_n^+ = Df_n^+ = g_n^+ = 0 \quad \text{at} \quad x = \pm\tfrac{1}{2}, \tag{23}$$

where

$$L_n^+ = \begin{pmatrix} (DD^* - n^2a^2)^2 & na\sqrt{T} \\ na\sqrt{T}\Omega_i(x) & -(DD^* - n^2a^2) \end{pmatrix}, \qquad M_n^+ = M_n. \tag{24}$$

Note that M_n is self-adjoint, and further is a positive definite operator:

$$(M_n \mathbf{q}_n \,|\, \mathbf{q}_n) = \int_{-1/2}^{1/2} (1 + \delta x)(|D^* f_n|^2 + n^2a^2 |f_n|^2 + |g_n|^2)\, dx \geq 0. \tag{25}$$

It is easy to show that $(M_n \mathbf{q}_n \,|\, \mathbf{q}_n')$ has the properties of an inner product, and our pre-Hilbert space can be completed under a norm, $\| \mathbf{q}_n \|_n^2 = (M_n \mathbf{q}_n \mid \mathbf{q}_n)$, see Mikhlin (1965), Section 5.

Our discussion of the eigenvalue problem (18) and its adjoint (23) will be purely formal. (A detailed discussion of the eigenvalues, the corresponding eigenfunctions, and eigenfunction expansions will be given in a subsequent paper.) We assume that there exists a denumerable infinity of eigenvalues σ_{nm}, $m = 0, 1, 2, \ldots$, and that these eigenvalues have no cluster point in the finite plane, and further can be ordered so that

$$\mathrm{Re}(\sigma_{nm+1}) \geq \mathrm{Re}(\sigma_{nm}),$$

where Re denotes real part. Such an ordering is suggested by the fact that physically we expect for T small that all the σ_{nm} have real parts greater than zero. The corresponding eigenfunctions will be denoted by $\mathbf{q}_{nm}(x)$. Note that since L_n and M_n are real it follows that if σ_{nm} is complex then $\tilde{\sigma}_{nm}$ is also an eigenvalue.

The adjoint problem will have eigenvalues $\tilde{\sigma}_{nm}$ with eigenfunctions $\overset{+}{\mathbf{q}}_{nm}(x)$. The biorthogonality relation

$$(L_n \mathbf{q}_{nm} \mid \overset{+}{\mathbf{q}}_{np}) = (\mathbf{q}_{nm} \mid L_n{}^{+} \overset{+}{\mathbf{q}}_{np}) = (M_n \mathbf{q}_{nm} \mid \overset{+}{\mathbf{q}}_{np}) = (\mathbf{q}_{nm} \mid M_n{}^{+} \overset{+}{\mathbf{q}}_{np}) = 0, \qquad m \neq p \tag{26}$$

follows easily from Eqs. (18) and (23). Further, the eigenfunctions $\mathbf{q}_{nm}$ and $\overset{+}{\mathbf{q}}_{nm}$ can be normalized, and we assume this to be the case in the rest of the paper, so that

$$(M_n \mathbf{q}_{nm} \mid \overset{+}{\mathbf{q}}_{nm}) = (\mathbf{q}_{nm} \mid M_n{}^{+} \overset{+}{\mathbf{q}}_{nm}) = 1; \qquad (L_n \mathbf{q}_{nm} \mid \overset{+}{\mathbf{q}}_{nm}) = (\mathbf{q}_{nm} \mid L_n{}^{+} \overset{+}{\mathbf{q}}_{nm}) = \sigma_{nm}. \tag{27}$$

Finally we expect from the corresponding theory when L_n is self-adjoint but not positive-definitive,† and from the form of the norm $\| \mathbf{q}_n \|_n^2$ as defined by Eq. (25); that functions $\mathbf{Q}_n^{(0)}(x) = (u_n^{(0)}(x), v_n^{(0)}(x))$ with $Du_n^{(0)}$ and $v_n^{(0)}$ square integrable can be expressed as

$$\mathbf{Q}_n^{(0)}(x) = \sum_{m=0}^{\infty} A_{nm} \mathbf{q}_{nm}(x). \tag{28}$$

Equality is meant in the sense of the norm defined by Eq. (25). By using

---

† For the case of a single equation $Ly = \sigma\, My$, L and M self-adjoint, M positive definite see Kestens (1955).

the orthogonality relations (26) and (27) it follows that

$$A_{nm} = (\mathbf{Q}_n^{(0)} \mid M_n^{+} \overset{+}{\mathbf{q}}_{nm}). \tag{29}$$

Hence the solution of Eq. (12) with the nonlinear terms neglected which satisfies the boundary conditions (14) and the initial conditions (15) of the class just mentioned is

$$\mathbf{Q}_n(x, \tau) = \sum_{m=0}^{\infty} (\mathbf{Q}_n^{(0)} \mid M_n^{+} \overset{+}{\mathbf{q}}_{nm})\mathbf{q}_{nm}(x)e^{-\sigma_{nm}\tau}. \tag{30}$$

Choosing the critical value of a for $n = 1$ as discussed at the beginning of this section, we have for $T < T_c$ (stability) $\mathbf{Q}_n \to 0$ as $\tau \to \infty$ for $n = 1, 2, \ldots$; however, for T slightly greater than T_c (instability) $\sigma_{10} < 0$ and hence $\mathbf{Q}_n \to 0$ as $\tau \to \infty$ for $n \geq 2$ but $\mathbf{Q}_1 \sim (\mathbf{Q}_1^{(0)} \mid M_1^{+} \overset{+}{\mathbf{q}}_{10})\mathbf{q}_{10}(x)e^{-\sigma_{10}\tau} \to \infty$ as $\tau \to \infty$.† How the nonlinear terms in Eq. (12) change this conclusion will be discussed in Section 5.

It is instructive to consider a simple example to illustrate the vector eigenfunction expansion (28). Consider the limiting case $\delta \to 0$, $\mu \to 1$, all other variables fixed. Then $D^* = D$, $\Omega_l(x) = 1$, the eigenvalue problem (18) is self-adjoint, and the σ_{nm} are necessarily real. (We might note that this provides a simple proof of the so-called principle of exchange of stabilities, i.e., that the instability is nonoscillatory.) Further, if we replace the boundary conditions (18) by

$$f_n = D^2 f_n = g_n = 0 \qquad \text{at} \quad x = \pm\tfrac{1}{2}, \tag{31}$$

we obtain an eigenvalue problem which we can solve exactly. This simplified problem is the classic Bénard problem of the instability of a layer of fluid heated from below with free-free boundaries. The Taylor number plays the role of the Rayleigh number, and the Prandtl number γ has been taken equal to one. Otherwise, replace the one in the definition of M_n by γ. See Chandrasekhar ((1961), Section 10) for a derivation of the equations and boundary conditions for the Bénard problem. Finally we

† We have assumed, as all numerical calculations indicate, that σ_{10} is real and simple and that no other eigenvalue has real part less than or equal to σ_{10}. A more precise definition of T slightly greater than T_c is $\sigma_{10} < 0$, $\text{Re}(\sigma_{1m}) > 0$ for $m \geq 1$, and $\text{Re}(\sigma_{nm}) > 0$ for $n \geq 2$. Also it may be necessary to add generalized eigenfunctions (see Friedman (1956), page 110) in the above expansions, but this does not affect the conclusions.

make a simple change of variables $y = x + \frac{1}{2}$ so that the interval of interest is $0 \le y \le 1$.

The boundary conditions (31) are automatically satisfied by

$$f_{nm} = A_{nm} \sin m\pi y, \qquad g_{nm} = B_{nm} \sin m\pi y. \tag{32}$$

It is easily verified by direct substitution that the vector $\mathbf{q}_{nm} = (f_{nm}, g_{nm})$ is a solution of $(L_n - \sigma_n M_n)\mathbf{q}_n = 0$ provided that σ_n is chosen as

$$\sigma_{nm}^{(1)} = (m^2\pi^2 + n^2a^2) + \frac{na\sqrt{T}}{(m^2\pi^2 + n^2a^2)^{1/2}} \quad \text{or}$$

$$\sigma_{nm}^{(2)} = (m^2\pi^2 + n^2a^2) - \frac{na\sqrt{T}}{(m^2\pi^2 + n^2a^2)^{1/2}}. \tag{33}$$

The corresponding eigenvectors are

$$A_{nm}^{(1)} \begin{pmatrix} 1 \\ \beta_{nm}^{(1)} \end{pmatrix} \sin m\pi y \qquad \text{and} \qquad A_{nm}^{(2)} \begin{pmatrix} 1 \\ \beta_{nm}^{(2)} \end{pmatrix} \sin m\pi y, \tag{34}$$

where

$$\beta_{nm}^{(1)} = \frac{-na\sqrt{T}}{(m^2\pi^2 + n^2a^2) - \sigma_{nm}^{(1)}}, \qquad \beta_{nm}^{(2)} = \frac{-na\sqrt{T}}{(m^2\pi^2 + n^2a^2) - \sigma_{nm}^{(2)}}. \tag{35}$$

For fixed n, and for each $m = 1, 2, \ldots$, we have two unequal eigenvalues $\sigma_{nm}^{(1)}$ and $\sigma_{nm}^{(2)}$; it is precisely these *two* eigenvalues for each m, leading to a double infinity of eigenvalues and vector eigenfunctions, that allow us to have an expansion of the form (28). Moreover, notice that while the vector eigenfunctions (34) have components which are linearly dependent, they themselves are linearly independent. Turning to the eigenfunction expansion, we have for the limiting case $\delta \to 0$, $\mu \to 1$, that the general solution of $[L_n + (\partial/\partial\tau)M_n]\mathbf{Q}_n = 0$ with the boundary conditions (31) is

$$\mathbf{Q}_n(y, \tau) = \sum_{m=1}^{\infty} \left\{ A_{nm}^{(1)} \begin{pmatrix} 1 \\ \beta_{nm}^{(1)} \end{pmatrix} \sin m\pi y \exp(-\sigma_{nm}^{(1)}\tau) \right.$$
$$\left. + A_{nm}^{(2)} \begin{pmatrix} 1 \\ \beta_{nm}^{(2)} \end{pmatrix} \sin m\pi y \exp(-\sigma_{nm}^{(2)}\tau) \right\}. \tag{36}$$

The $A_{nm}^{(1)}$ and $A_{nm}^{(2)}$ are determined by the initial conditions $\mathbf{Q}_n^{(0)}(y)$. Setting $\tau = 0$, multiplying each component of Eq. (36) by $\sin m\pi y$, and integrating from 0 to 1 gives two linear algebraic equations for the $A_{nm}^{(1)}$ and $A_{nm}^{(2)}$. The *identical result* is obtained by using the orthogonality relation (29) in the form

$$A_{nm}^{(1)} = (\mathbf{Q}_n^{(0)} \mid M_n{}^+\mathbf{q}_{nm}^{(1)})/(\mathbf{q}_{nm}^{(1)} \mid M_n{}^+\mathbf{q}_{nm}^{(1)}), \tag{37}$$

with a similar equation for $A_{nm}^{(2)}$. The definition of the $\mathbf{q}_{nm}^{(1)}$ is clear from Eq. (36), and the term in the numerator occurs because the $\mathbf{q}_{nm}^{(1)}$ have not been normalized. A similar comparison of a vector eigenfunction expansion and the corresponding orthogonality relation with a direct calculation using the component equations has been given by Dolph (1954) for a problem in beam vibrations.

4. The Growth of Taylor Vortices

We turn now to the problem of what happens for T slightly greater than T_c. According to linear theory there exist a continuum of wave numbers a including a_c for which periodic disturbances will grow exponentially with time. In contrast, as mentioned in the introduction, a steady secondary motion in the form of toroidal vortices spaced regularly along the axis of the cylinders is oberserved. We consider here not the selection process of a definite wave number a for $T > T_c$, but rather the process of equilibration (rather than exponential growth) of a single disturbance. Thus we fix† $a = a_c$, and consider what happens for T slightly greater than T_c. It is no longer permissible to neglect quadratic terms in the velocities, and we consider the full nonlinear equations (11) and (12) with the boundary conditions (14) and initial conditions (15).

For completeness and for comparison with the procedure to be discussed in the next section, we will first describe a procedure due to Stuart (1960) and Watson (1960) and applied to the present problem by Davey (1962). Our discussion is necessarily brief; the reader is referred to Davey's paper for a detailed exposition. In place of the general initial value problem they actually consider what happens to a disturbance which has the spatial form of the critical disturbance, but an unknown amplitude $A(\tau)$, i.e., the disturbance $A(\tau)\mathbf{q}_{10}(x)\cos a\zeta$. This fundamental disturbance will grow according to linear theory for $T > T_c$ (indeed $A \sim e^{-\sigma_{10}\tau}$ for $A \to 0$), and because of the nonlinear terms it will force its harmonics ($\cos na\zeta$) and a mean motion term. Using $\cos^2 a\zeta = (1 + \cos 2a\zeta)/2$, one finds

† The variation of the observed wavenumber of the vortex motion with T is slight.

that the first harmonic ($\cos 2a\zeta$) and the mean motion will be $O(A^2)$. Then the interaction of the fundamental with its first harmonic ($\cos a\zeta$ $\cos 2a\zeta$), will lead to a second harmonic ($\cos 3a\zeta$) and a correction to the fundamental ($\cos a\zeta$) which are $O(A^3)$. Continuing this process it can be shown that v_0 and $\mathbf{Q}_n$ have the formal series expansion

$$v_0(x, \tau) = A^2(\tau)\left[V_{00}(x) + \sum_{m=1}^{\infty} A^{2m}(\tau)V_{0m}(x)\right], \tag{38}$$

$$\mathbf{Q}_n(x, \tau) = A^n(\tau)\left[\mathbf{F}_{n0}(x) + \sum_{m=1}^{\infty} A^{2m}(\tau)\mathbf{F}_{nm}(x)\right], \tag{39}$$

where $\mathbf{F}_{nm}(x) = (U_{nm}(x), V_{nm}(x))$. Consistent with these series expansions, $dA/d\tau$ can be expanded as a series in powers of A:

$$dA/d\tau = a_0A + a_1A^3 + a_2A^5 + \cdots, \tag{40}$$

where $a_0, a_1, a_2, \ldots$ are functions of the parameters μ, η, a, T, and are to be determined. From the above discussion it is clear that a_0 is identified with $-\sigma_{10}$. We shall show how a_1 is computed and how the inclusion of the term a_1A^3 in Eq. (40) completely changes the prediction of linear theory.

If we substitute for v_0, $\mathbf{Q}_n$, and $dA/d\tau$ in Eq. (11) and (12), we obtain the following sequence of problems. Terms $O(A)$ give

$$(L_1 + a_0M_1)\mathbf{F}_{10} = 0, \quad U_{10} = DU_{10} = V_{10} = 0 \quad \text{at} \quad x = \pm\tfrac{1}{2}. \tag{41}$$

Terms $O(A^2)$ give

$$(-DD^* + 2a_0)V_{00} = -\frac{a}{2}(D^* + \xi)(U_{10}V_{10}), \tag{42}$$

$$\mathrm{V}_{00} = 0 \quad \text{at} \quad \mathrm{x} = \pm\tfrac{1}{2},$$

and

$$(L_2 + 2a_0M_2)\mathbf{F}_{20} = \mathbf{S}_{20},$$
$$U_{20} = DU_{20} = V_{20} = 0 \qquad \text{at} \quad x = \pm\tfrac{1}{2}, \tag{43}$$

where $\mathbf{S}_{20}$ is a quadratic function of U_{10} and V_{10}. It is given in the appendix. And finally, for our purposes, corresponding to terms $O(A)^3$ and $n = 1$:

$$(L_1 + 3a_0M_1)\mathbf{F}_{11} = -a_1M_1\mathbf{F}_{10} + \mathbf{S}_{13},$$
$$U_{11} = DU_{11} = V_{11} = 0 \qquad \text{at} \quad x = \pm\tfrac{1}{2}, \tag{44}$$

where $\mathbf{S}_{13}$ is a quadratic function of V_{00}, V_{10}, V_{20}, U_{10}, and U_{20}. It is given in the appendix.

The differential equation and boundary conditions (41) form an eigenvalue problem for a_0. As mentioned earlier in this section we choose

$$a_0 = -\sigma_{10}, \qquad \mathbf{F}_{10}(x) = \mathbf{q}_{10}(x). \tag{45}$$

(We continue to assume that σ_{10} is real.) Knowing $\mathbf{F}_{10}$ we can evaluate $\mathbf{S}_{20}$ and then solve Eq. (42) and (43) for $V_{00}(x)$ and $\mathrm{F}_{20}(x)$, which are used to evaluate $\mathbf{S}_{13}$ in Eq. (44). Next we determine the parameter a_1. The procedure that we use here is different from that used by Davey (1962), but, of course, the final result is the same. We assume that $\mathbf{F}_{11}(x)$ can be expanded in a series of the eigenfunctions $\mathbf{q}_{1m}(x)$, (see Eq. (28)).

$$\mathbf{F}_{11}(x) = \sum_{m=0}^{\infty} \alpha_{1m}\mathbf{q}_{1m}(x). \tag{46}$$

The α_{1m} are determined by substituting for $\mathbf{F}_{11}$ in Eq. (44), and then taking the inner product with the adjoint eigenfunction $\overset{+}{\mathbf{q}}_{1m}(x)$.† We obtain, using (26) and (27),

$$(\sigma_{1m} - 3\sigma_{10})\alpha_{1m} = -a_1(M_1\mathbf{q}_{10} \mid \overset{+}{\mathbf{q}}_{1m}) + (\mathbf{S}_{13} \mid \overset{+}{\mathbf{q}}_{1m}). \tag{47}$$

In particular, note that for $m = 0$, α_{10} is proportional to $(\sigma_{10})^{-1}$. To prevent $\mathbf{F}_{11}$ from becoming singular as $T \to T_c$ (recall $\sigma_{10} \to 0$ as $T \to T_c$), we choose a_1 so that the right hand side of Eq. (47) vanishes for $m = 0$. Thus

$$a_1 = (\mathbf{S}_{13} \mid \overset{+}{\mathbf{q}}_{10}) \tag{48}$$

since $(M_1\mathbf{q}_{10} \mid \overset{+}{\mathbf{q}}_{10}) = 1$. This is equivalent to the result given by Davey; however, the notation is considerably different. With a_1 determined, $\mathbf{F}_{11}$ can be computed either by using the series expansion (46) or by solving the nonhomogeneous problem (44) directly.

Finally turning to Eq. (40) for $A(\tau)$, and neglecting terms $O(A^5)$ it is easy to show that

$$A(\tau) \to (-a_0/a_1)^{1/2} = (\sigma_{10}/a_1)^{1/2} \qquad \text{as} \quad \tau \to \infty. \tag{49}$$

† $V_{00}(x)$ and $\mathbf{F}_{20}(x)$ could have been computed in a similar manner by writing $V_{00}(x) = \sum_{m=0}^{\infty} \alpha_{0m} f_{0m}(x)$, and $\mathbf{F}_{20}(x) = \sum_{m=0}^{\infty} \alpha_{2m}\mathbf{q}_{2m}(x)$.

The computation of a_1 by Davey shows that it is negative and $O(1)$. Thus for $T > T_c$ in which case $\sigma_{10} < 0$, we obtain an equilibrium amplitude $A_e = (\sigma_{10}/a_1)^{1/2}$, in contrast to the exponential growth $A(\tau) \sim e^{-\sigma_{10}\tau}$ as predicted by linear theory. Since $A_e = 0(|\sigma_{10}|^{1/2})$ it is clear that the terms $a_0 A$ and $a_1 A^3$ in Eq. (40) are $0(|\sigma_{10}|^{3/2})$ while terms neglected are $O(|\sigma_{10}|^{5/2})$ or higher. Thus

$$\mathbf{Q}_1(x, \tau) = \begin{pmatrix} u_1(x, \tau) \\ v_1(x, \tau) \end{pmatrix} \to A_e \mathbf{q}_{10}(x) + O(|\sigma_{10}|^{3/2}) \qquad \text{as} \quad \tau \to \infty. \tag{50}$$

Similarly v_0 and $\mathbf{Q}_2$ are determined correct to terms $O(A_e{}^2)$. Davey (1962) has computed A_e for several geometries and rotation rates. Agreement with experiment, either through torque measurements or measurements of A_e and the amplitudes of the harmonics is excellent, see Davey (1962), Donnelly and Schwarz (1965), and Snyder and Lambert (1966).

5. Eigenfunction Expansions

In the expansions (38) and (39) the form of the expansion in time was prescribed, i.e., we assumed a series expansion in powers of $A(\tau)$. Then the coefficients, which were functions of the spatial variable, were determined. We now consider a different procedure in which we expand the velocity field in known spatial functions, the eigenfunctions of linear stability problems, with unknown coefficients which depend upon the time τ. Thus

$$v_0(x, \tau) = \sum_{m=0}^{\infty} a_{0m}(\tau) f_{0m}(x), \qquad \mathbf{Q}_n(x, \tau) = \sum_{m=0}^{\infty} a_{nm}(\tau) \mathbf{q}_{nm}(x). \tag{51}$$

Substituting in Eq. (11) and (12), and taking the inner product with f_{0m} and $\overset{+}{\mathbf{q}}_{nm}$, respectively, we find

$$\frac{da_{0m}}{d\tau} + \sigma_{0m}\, a_{0m} = (S_0 \,|\, f_{0m}), \tag{52}$$

$$\frac{da_{nm}}{d\tau} + \sigma_{nm}\, a_{nm} = (\mathbf{S}_n \,|\, \overset{+}{\mathbf{q}}_{nm}). \tag{53}$$

This infinite system of nonlinear ordinary differential equations is to be solved subject to the initial conditions $a_{0m}(0) = c_{0m} = (v^{(0)} \,|\, f_{0m})$ and $a_{nm}(0) = c_{nm} = (\mathbf{Q}_n^{(0)} \,|\, M_n{}^+ \overset{+}{\mathbf{q}}_{nm})$. Notice that if the quadratic terms are neglected, then $a_{nm}(\tau) = c_{nm} \exp[-\sigma_{nm}\tau]$. Hence for $a = a_c$, and T slightly greater than T_c, all the $a_{nm} \to 0$ as $\tau \to \infty$, except a_{10} which becomes unbounded.

In order to construct asymptotic solutions of Eqs. (52) and (53), and hence of Eqs. (11) and (12), we limit ourselves to disturbances which are $O(\varepsilon)$ where ε is a small parameter not yet defined. Also we associate with each of the harmonics a scale δ_n; thus we make the formal transformations

$$v_0 = \varepsilon\delta_0\psi_0; \qquad \mathbf{Q}_n = \begin{pmatrix} u_n \\ v_n \end{pmatrix} = \varepsilon\delta_n \begin{pmatrix} \varphi_n \\ \psi_n \end{pmatrix} = \varepsilon\delta_n \mathbf{\Phi}_n, \tag{54}$$

where ψ_0 and the $\mathbf{\Phi}_n$ are to be $O(1)$. If ψ_0 and the $\mathbf{\Phi}_n$ are now expanded in series of the eigenfunctions f_{0m} and $\mathbf{q}_{nm}$,

$$\psi_0(x, \tau) = \sum_{m=0}^{\infty} A_{0m}(\tau) f_{0m}(x), \qquad \mathbf{\Phi}_n(x, \tau) = \sum_{m=0}^{\infty} A_{nm}(\tau)\mathbf{q}_{nm}(x), \tag{55}$$

we obtain the same set of Eqs. (52), (53) with the a_{nm} replaced by the A_{nm} and S_0 and the $\mathbf{S}_n$ defined in terms of ψ_0 and the $\mathbf{\Phi}_n$. A systematic investigation of these equations shows that $\delta_1 = 1$, $\delta_0 = \delta_2 = \varepsilon$, and $\delta_n = \varepsilon^{n-1}$ for $n \geq 3$. (See, for example, Eckhaus (1965), Chapter 7 for a similar expansion.) Alternatively we can proceed in the heuristic manner described in the previous section with ε identified as $A(\tau)$. Thus, assuming that the fundamental $\mathbf{Q}_1$ is the dominant term in the Fourier expansion, we associate with it a scale ε. Since mean motion terms and the first harmonic are generated by quadratic terms in the fundamental we can expect that $v_0 = O(\varepsilon^2)$ and $\mathbf{Q}_2 = O(\varepsilon^2)$. The interactions cascade to the higher harmonics, as described in the previous section.

Consider now the fundamental. Expanding in ε we find that the equations for $\mathbf{\Phi}_1(x, \tau)$ and the $A_{1m}(\tau)$ are

$$\left(L_1 + \frac{\partial}{\partial\tau} M_1\right)\mathbf{\Phi}_1 = \varepsilon^2\mathbf{s}_{13} + O(\varepsilon^4), \tag{56}$$

$$\frac{dA_{1m}}{d\tau} + \sigma_{1m}A_{1m} = \varepsilon^2(\mathbf{s}_{13} \mid \overset{+}{\mathbf{q}}_{1m}) + O(\varepsilon^4), \tag{57}$$

where $\mathbf{s}_{13}$ depends only upon ψ_0, $\mathbf{\Phi}_1$, and $\mathbf{\Phi}_2$. It is the $\mathbf{S}_{13}$ defined in the appendix with V_{00}, V_{10}, V_{20}, U_{10}, U_{20} replaced by ψ_0, ψ_1, ψ_2, φ_1, φ_2, respectively. Note that $\mathbf{s}_{13}$ depends upon τ. If $A_{10}(\tau)$ is to be $O(1)$ for all

τ, then it is clear from Eq. (56) that ε^2 must be $O(\sigma_{10})$, otherwise terms on the right hand side can be neglected and A_{10} would become unbounded as $\tau \to \infty$. Thus we *must identify* ε, the small expansion parameter, with $|\sigma_{10}|^{1/2}$. Then it further follows from Eq. (57) that $A_{1m} = C_{1m}\exp[-\sigma_{1m}\tau] + O(\varepsilon^2)$ for $m \geq 2$, where $A_{1m}(0) = C_{1m}$ and the $C_{1m} = O(1)$. Hence, $A_{1m} = O(\varepsilon^2)$ as $\tau \to \infty$ for $m \geq 2$. Thus

$$\mathbf{Q}_1(x, \tau) = \varepsilon\mathbf{\Phi}_1(x, \tau) = \varepsilon[A_{10}(\tau)\mathbf{q}_{10}(x) + O(\varepsilon^2)] \qquad \text{as} \quad \tau \to \infty. \tag{58}$$

For the purpose of comparison, note that the procedure of Stuart, Watson, and Davey gives $\mathbf{Q}_1 = A(\tau)\mathbf{q}_{10}(x) + O(A^3)$, (see Eqs. (39) and (45)), which shows that ε plays the role of $A(\tau)$ as an expansion parameter. Note further, that $A(\tau)$ is eventually identified as being proportional to $|\sigma_{10}|^{1/2}$ as $\tau \to \infty$, (see Eqs. (49) and (50)).

To determine $A_{10}(\tau)$, we must compute $(\mathbf{s}_{13} \mid \overset{+}{\mathbf{q}}_{10})$ for which we need ψ_0, $\mathbf{\Phi}_1$, and $\mathbf{\Phi}_2$. The vector $\mathbf{\Phi}_1$ is given by (58). The equations for ψ_0 and the A_{0m}, after expanding in ε, are

$$(-DD^* + \partial/\partial\tau)\psi_0 = A_{10}^2(\tau)S_{00}(x) + O(\varepsilon^2), \tag{59}$$

$$\frac{dA_{0m}}{d\tau} + \sigma_{0m}A_{0m} = A_{10}^2(\tau)\,(S_{00}(x) \mid f_{0m}(x)) + O(\varepsilon^2), \tag{60}$$

where $S_{00} = -(a/2)\,(D^* + \xi)\,(f_{10}g_{10})$. In evaluating the right hand side of the equation for ψ_0 we have neglected terms in $\mathbf{\Phi}_1$ which arise from the initial conditions, but die out exponentially in time. That is, we have taken the A_{1m} to be $O(\varepsilon^2)$ for all time instead of $A_{1m} = C_{1m}\exp[-\sigma_{1m}\tau] + O(\varepsilon^2)$ for $m \geq 2$. The inclusion of the terms arising from the initial conditions does not affect the validity of our asymptotic expansion in ε, nor does it change our results for $\tau \to \infty$. To keep what is already a very messy calculation as simple as possible we will neglect such terms throughout. Thus, our calculations are really only correct in the limit $\tau \to \infty$.

Solving the *linear* first order equation (60) for $A_{0m}(\tau)$ subject to the initial conditions† $A_{0m}(0) = C_{0m}$, and making use of the fact that

† In our construction of the asymptotic expansion in ε, we have assumed that $v_0 = O(\varepsilon^2)$ for all τ. Thus we must require that the initial conditions on v_0 are $O(\varepsilon^2)$, (the initial conditions on ψ_0 are $O(1)$). More generally the initial conditions on Q_n are $O(\varepsilon^n)$, $n \geq 1$.

$dA_{10}/d\tau = O(\varepsilon^2 A_{10})$, we find that

$$A_{0m}(\tau) = \frac{1}{\sigma_{0m}} A_{10}^2(\tau)\,(S_{00}(x)\,|\,f_{0m}(x)) + O(\varepsilon^2) \qquad \text{as} \quad \tau \to \infty. \tag{61}$$

Hence

$$v_0(x, \tau) = \varepsilon^2 \psi_0(x, \tau) = \varepsilon^2 A_{10}^2(\tau) \sum_{m=0}^{\infty} \frac{(S_{00}(x)\,|\,f_{0m}(x))}{\sigma_{0m}} f_{0m}(x) + O(\varepsilon^4),$$

$$= \varepsilon^2 A_{10}^2(\tau) f(x) + O(\varepsilon^4) \qquad \text{as} \quad \tau \to \infty. \tag{62}$$

Alternatively we can compute ψ_0 by simply looking for a solution of Eq. (59) of the form $\psi_0(x, \tau) = A_{10}^2(\tau)f(x)$. Since $dA_{10}/d\tau = O(\varepsilon^2 A_{10})$, $f(x)$ must satisfy

$$-DD^*f = S_{00}(x) + O(\varepsilon^2). \tag{63}$$

It is instructive to compare Eq. (63) with Eq. (42). Since $U_{10} = f_{10}$, $V_{10} = g_{10}$, and $a_0 = -\sigma_{10} = O(\varepsilon^2)$, the equations for $f(x)$ and $V_{00}(x)$ are identical except for terms $O(\varepsilon^2)$. Indeed, Davey neglects the term $2a_0 V_{00}$ when solving Eq. (42).

Next we turn to the equation for $\mathbf{\Phi}_2(x, \tau)$. Expanding in ε, we find that the equations for $\mathbf{\Phi}_2$ and the A_{2m} are

$$\left(L_2 + \frac{\partial}{\partial \tau} M_2\right)\mathbf{\Phi}_2 = A_{10}^2(\tau)\mathbf{S}_{20}(x) + O(\varepsilon^2), \tag{64}$$

$$\frac{dA_{2m}}{d\tau} + \sigma_{2m}A_{2m} = A_{10}^2(\tau)\,(\mathbf{S}_{20}(x)\,|\,\overset{+}{\mathbf{q}}_{2m}(x)) + O(\varepsilon^2), \tag{65}$$

where $\mathbf{S}_{20}(x)$ is the same as the $\mathbf{S}_{20}$ in Eq. (44), and defined in the appendix with U_{10} and V_{10} replaced by f_{10} and g_{10}, respectively. Proceeding as above we find that

$$A_{2m}(\tau) = \frac{1}{\sigma_{2m}} A_{10}^2(\tau)\,(\mathbf{S}_{20}(x)\,|\,\overset{+}{\mathbf{q}}_{2m}(x)) + O(\varepsilon^2) \qquad \text{as} \quad \tau \to \infty. \tag{66}$$

Hence

$$\mathbf{Q}_2(x, \tau) = \varepsilon^2 \mathbf{\Phi}_2(x, \tau) = \varepsilon^2 A_{10}^2(\tau) \sum_{m=0}^{\infty} \frac{(\mathbf{S}_{20}(x)\,|\,\overset{+}{\mathbf{q}}_{2m}(x))}{\sigma_{2m}} \mathbf{q}_{2m}(x) + O(\varepsilon^4)$$

$$= \varepsilon^2 A_{10}^2(\tau)\mathbf{F}(x) + O(\varepsilon^4) \qquad \text{as} \quad \tau \to \infty. \tag{67}$$

Again, alternatively, we can compute $\mathbf{\Phi}_2(x, \tau)$ by assuming a solution of Eq. (64) of the form $\mathbf{\Phi}_2(x, \tau) = A_{10}^2(\tau)\mathbf{F}(x)$, and then making use of the fact that $dA_{10}/d\tau = O(\varepsilon^2 A_{10})$. The vector $\mathbf{F}(x)$ would then be identical with $\mathbf{F}_{20}(x)$ in Eq. (43) if terms $O(\varepsilon^2)$ are consistently neglected there.

To determine $A_{10}(\tau)$ we turn to Eq. (57) for $m = 0$. Substituting for ψ_0, $\mathbf{\Phi}_1$ and $\mathbf{\Phi}_2$ in $\mathbf{s}_{13}$, we find $\mathbf{s}_{13} = A_{10}^3(\tau)\mathbf{S}_{13}(x) + O(\varepsilon^2)$, where $\mathbf{S}_{13}(x)$ is the expression in Eq. (44) and defined in the appendix with $V_{00}(x)$ and the vectors $\mathbf{F}_{10}(x)$ and $\mathbf{F}_{20}(x)$ replaced by $f(x)$, $\mathbf{q}_{10}(x)$ and $\mathbf{F}(x)$, respectively. We obtain

$$\frac{dA_{10}}{d\tau} + \sigma_{10}A_{10} = |\sigma_{10}|\, A_{10}^3\, (\mathbf{S}_{13} \,|\, \overset{+}{\mathbf{q}}_{10}) + O(\varepsilon^4). \tag{68}$$

But as we have just seen $V_{00}(x) = f(x)$, $\mathbf{F}_{10}(x) = \mathbf{q}_{10}(x)$, and $\mathbf{F}_{20}(x) = \mathbf{F}(x)$ except for terms $O(\varepsilon^2)$ which may be neglected. Thus $\mathbf{S}_{13}(x)$ in Eq. (68) is identical with the $\mathbf{S}_{13}$ of Eq. (44) and (48). Letting $a_1 = (\mathbf{S}_{13} \,|\, \overset{+}{\mathbf{q}}_{10})$, we find

$$A_{10}^2(\tau) \to \frac{\sigma_{10}}{|\sigma_{10}|\, a_1} + O(\varepsilon^2), \tag{69}$$

independent of the initial condition $A_{10}(O) = C_{10}$. Thus

$$\begin{aligned} \mathbf{Q}_1(x, \tau) &= \varepsilon A_{10}(\tau)\mathbf{q}_{10}(x) + O(\varepsilon^3) \\ &= A_e\mathbf{q}_{10}(x) + O(|\sigma_{10}|^{3/2}) \qquad \text{as} \quad \tau \to \infty, \end{aligned} \tag{70}$$

where $A_e = \sigma_{10}/|\sigma_{10}|^{1/2}a_1$, which is identical with the result obtained by the method of Stuart, Watson, and Davey (see Eq. (50)).

6. Discussion

In the preceding sections we have discussed the stability of flow between rotating cylinders to periodic disturbances of the form (10), and the subsequent growth and equilibration of such disturbances for speeds above critical. We might mention that there is no difficulty if the sin $na\zeta$ terms are also included in the series (10), the final results would be the same—the vortex motion simply being shifted in the z direction.

The Stuart, Watson, and Davey approach is based on expanding the coefficient† $\mathbf{Q}_n(x, \tau) = (u_n, v_n)$ of $\cos na\zeta$, $n = 1, 2, \ldots$, in a suitable power series of a small time dependent amplitude $A(\tau)$ of the fundamental disturbance ($n = 1$) which is growing exponentially according to linear theory for $T > T_c$. The amplitude $A(\tau)$ satisfies a nonlinear first order equation, the solution of which leads to the identification of $A(\tau)$ in the limit $\tau \to \infty$ as being proportional to $|\sigma_{10}|^{1/2} = \varepsilon$, where $-\sigma_{10}$ is the growth rate of the fundamental mode.

The second approach, which is a generalization of the work of Eckhaus, is based on expanding the $\mathbf{Q}_n(x, \tau)$ in series of the vector eigenfunctions $\{\mathbf{q}_{nm}(x)\}$ of the corresponding linear stability problems; see Eqs. (20), (30), (51), and (55). Such an expansion shows clearly that the class of initial periodic disturbances of the form (10) which can be considered are those for which each $\mathbf{Q}_n(x, 0)$ can be represented in terms of the eigenfunctions $\{\mathbf{q}_{nm}(x)\}$; see Section 3. The time-dependent coefficients $A_{nm}(\tau)$ satisfy a system of nonlinear first order ordinary equations of the form (52) and (53). It should be emphasized that given arbitrary initial conditions these equations may be integrated numerically to determine the $A_{nm}(\tau)$, and hence the velocity field. However, if one is willing to restrict himself to "small" initial disturbances and speeds slightly above critical it is possible to construct asymptotic solutions of the nonlinear equations for the $A_{nm}(\tau)$. Thus if ε is a measure of the amplitude of $\mathbf{Q}_1(x, \tau)$, the fundamental mode, then there exist asymptotic solutions with the $\mathbf{Q}_n = O(\varepsilon^n)$, and ε identified as $|\sigma_{10}|^{1/2}$. Further if the initial conditions are taken account of in the calculations (we did not do this, but see the remarks following Eq. (60)), the solutions are asymptotic in ε for all τ. Implicit in the Stuart, Watson, Davey approach is a restriction that $\mathbf{Q}_n = O(\varepsilon^n)$, (see Eqs. (38) and (39)). And through terms $O(\varepsilon^2)$ both procedures give the same result in the limit $\tau \to \infty$. Clearly it would be desirable to construct asymptotic solutions under less restrictive conditions, for example, that initially all of the harmonics are $O(\varepsilon)$.

† Note that $\mathbf{Q}_n$ is not precisely the coefficient of $\cos na\zeta$ in the Fourier expansion of the velocity $(u(x, \tau, \zeta), v(x, \tau, \zeta))$ because of the term na which was introduced in the expansion for u, (see equation (10)). Also for simplicity we will not bother with the mean motion terms $v_0(x, \tau)$ since we are only interested in summarizing the basic ideas.

Acknowledgments

The author would like to express his appreciation to Professors G. Habetler, L. Segel, and W. Eckhaus for their helpful comments, and to Mr. R. Grannick for his assistance with some of the analytical computations. This work was initiated while the author was a Fulbright Lecturer at the Weizmann Institute of Science, Rehovoth, Israel. It was partially supported by the Mechanics Branch of the Office of Naval Research, Contract Nonr 591-08, and by the Army Research Office, Durham, Contract DA 31-124 ARO-D-269.

APPENDIX

The first component of $\mathbf{S}_n$ is $\alpha nav_0v_n/(1 + \delta x)$ plus

$$\sum_{q=1}^{n-1}\left\{\frac{na}{2}\left[\frac{\alpha/2}{1+\delta x}v_qv_{n-q} - D^*[qau_q(n-q)\,au_{n-q}] + (na)(qa)u_qD^*u_{n-q}\right]\right.$$

$$\left. + \tfrac{1}{2}D[D^*(qau_qD^*u_{n-q}) - naD^*u_qD^*u_{n-q}]\right\}$$

$$+\sum_{q=n+1}^{\infty}\left\{\frac{na}{2}\left[\frac{\alpha/2}{1+\delta x}v_qv_{q-n} - D^*[qau_q(q-n)\,au_{q-n}] - (na)(qa)u_qD^*u_{q-n}\right]\right.$$

$$\left. - \tfrac{1}{2}D[D^*(qau_qD^*u_{q-n}) - naD^*u_qD^*u_{q-n}]\right\}$$

$$+\sum_{q=1}^{\infty}\left\{\frac{na}{2}\left[\frac{\alpha/2}{1+\delta x}v_qv_{q+n} - D^*[qau_q(q+n)\,au_{q+n}] + (na)(qa)u_qD^*u_{q+n}\right]\right.$$

$$\left. + \tfrac{1}{2}D[D^*(qau_qD^*u_{q+n}) + naD^*u_qD^*u_{q+n}]\right\}.$$

The second component of $\mathbf{S}_n$ is

$$-nau_nD^*v_0 - \sum_{q=1}^{n-1}\left\{\frac{qa}{2}(D^* + \xi)(u_qv_{n-q}) - \frac{na}{2}v_qD^*u_{n-q}\right\}$$

$$-\sum_{q=n+1}^{\infty}\left\{\frac{qa}{2}(D^* + \xi)(u_qv_{q-n}) + \frac{na}{2}v_qD^*u_{q-n}\right\}$$

$$-\sum_{q=1}^{\infty}\left\{\frac{qa}{2}(D^* + \xi)(u_qv_{q+n}) - \frac{na}{2}v_qD^*u_{q+n}\right\}.$$

The components of $2\mathbf{S}_{20}/a$, Eq. (42), are

$$\frac{\alpha}{1+\delta x} V_{10}^2 + U_{10}D^2D^*U_{10} - (D^*U_{10})(DD^*U_{10})$$
$$+ \xi U_{10}(2a^2U_{10} - DD^*U_{10})$$

and

$$V_{10}D^*U_{10} - U_{10}D^*V_{10}.$$

The components of $\mathbf{S}_{13}$, Eq. (43), are

$$\frac{a\alpha}{1+\delta x} V_{00}V_{10} + \frac{a\alpha/2}{1+\delta x} V_{20}V_{10} + \frac{a}{2} D[U_{10}D^{*2}U_{20} - 2U_{20}D^{*2}U_{10}$$
$$+ D^*U_{10}D^*U_{20} + \xi(2U_{20}D^*U_{10} - U_{10}D^*U_{20})]$$
$$+ \frac{a^3}{2}[-4D^*(U_{20}U_{10}) + U_{10}D^*U_{20} - 2U_{20}D^*U_{10}],$$

and

$$-aU_{10}D^*V_{00} - \frac{a}{2}[2U_{20}D^*V_{10} + U_{10}D^*V_{20} + V_{10}D^*U_{20} + 2V_{20}D^*U_{10}].$$

References

Chandrasekhar, S., 1961, "Hydrodynamic and Hydromagnetic Stability." Oxford Univ. Press, London and New York.

Coddington, E. A. and Levinson, N., 1955, "Theory of Ordinary Differential Equations." McGraw-Hill, New York.

Coles, D., 1965, Transition in circular Couette flow, *J. Fluid Mech.* **21**, 385–425.

Davey, A, 1962, The growth of Taylor vortices in flow between rotating cylinders, *J. Fluid Mech.* **14**, 336–368.

Di Prima, R. C., 1963, Stability of curved flows, *J. Appl. Mech.* **30**, 486–492.

Dolph, C. L., 1954, On the Timoshenko theory of transverse beam vibrations, *Quart. Appl. Math.* **12**, 175–187.

Donnelly, R. J., 1958, Experiments on the stability of viscous flow between rotating cylinders. I. Torque measurements, *Proc. Roy. Soc. London, Ser. A* **246**, 312–325.

Donnelly, R. J., and Schwarz, K. W., 1965, Experiments on the stability of viscous flow between rotating cylinders. VI. Finite amplitude experiments, *Proc. Roy. Soc. London, Ser. A* **283**, 531–556.

Eckhaus, W., 1965, "Studies in Nonlinear Stability Theory." Springer-Verlag, Berlin.

Friedman, B., 1956, "Principles and Techniques of Applied Mathematics." Wiley, New York.

Ince, E. L., 1926, "Ordinary Differential Equations." Longmans, Green, New York, and Dover, New York, 1956.

Kestens, J., 1955, The normal self-adjoint eigenvalue problem and upper and lower bounds for the eigenvalues with the iteration method, Ph. D. thesis, Brown Univ., Providence, Rhode Island.

Krueger, E. R., Gross, A., and Di Prima, R. C., 1966, On the relative importance of Taylor-vortex and nonaxisymmetric modes in flow between rotating cylinders' *J. Fluid Mech.* **24**, 521–538.

Mikhlin, S. G., 1965, "The Problem of the Minimum of a Quadratic Functional." (Translation of Problema Minimuma Kvadratichnogo Funktsionala, 1952.) Holden-Day, San Francisco.

Snyder, H. A., and Lambert, R. B., 1966, Harmonic generation in Taylor vortices between rotating cylinders, *J. Fluid Mech.* **26**, 545–562.

Stuart, J. T., 1958, On the nonlinear mechanics of hydrodynamic stability, *J. Fluid Mech.* **4**, 1–21.

Stuart, J. T., 1960, On the nonlinear mechanics of wave disturbances in stable and unstable parallel flows. Part 1. The basic behaviour in plane Poiseuille flow, *J. Fluid Mech.* **9**, 353–370.

Taylor, G. I., 1923, Stability of a viscous liquid contained between two rotating cylinders, *Phil. Trans. Roy. Soc. London, Ser. A* **223**, 289–343.

Watson, J., 1960, On the nonlinear mechanics of wave disturbances in stable and unstable parallel flows. Part 2. The development of a solution for plane Poiseuille flow and for plane Couette flow, *J. Fluid Mech.* **9**, 371–389.

New Methods for the Solution of Partial Differential Equations

RICHARD BELLMAN

Departments of Mathematics,
Engineering and Medicine
University of Southern California
Los Angeles, California

ROBERT KALABA

The Rand Corporation
Santa Monica, California

I. Introduction

Over the last ten years, we encountered a variety of conventional and unconventional functional equations in connection with various applications of the theories of dynamic programming, invariant imbedding, and quasilinearization. In this paper we wish to describe the fundamental ideas behind the methods we have developed and used to obtain numerical solutions. All of these methods have been constructed with a digital computer in mind.

The first set of methods is derived from the theory of dynamic programming [1–3]; the next is based upon a classical use of an infinite system of ordinary differential equations to treat partial differential equations, along the lines of Lichtenstein, Siddiqi, and others. This brings us into contact with the "closure techniques" of mathematical physics [4–6], and with differential approximation [7]. Following this, we consider the use of Laplace transform techniques, followed by numerical inversion of the Laplace transform [8]. Finally, we discuss the use of quadrature techniques, a method forced upon us by the exigencies of invariant imbedding, and the combination of this approach with perturbation expansions.

All that we have done has been consciously or subconsciously motivated by our desire to utilize the structural properties of both the underlying process and the superimposed analytic formulation. It is desirable to construct computational algorithms which automatically exhibit boundedness, nonnegativity, stability, and so forth. Little has been done in these directions; cf. however, [9, 10]. Furthermore, it will be clear from what follows that much remains to be done in connection with the rigorous justification of some of the techniques that are presented below, in connection with error estimates and with the acceleration of convergence, the reduction and minimization of computing time, and so on. Applications of these approaches will be found in the references cited.

2. Partial Differential Equations in Dynamic Programming

The theory of dynamic programming is a prolific source of partial differential equations. Conversely, many important classes of partial differential equations can be considered to arise in this fashion, and thus can be treated in novel ways.

Consider, for example, the problem of minimizing the functional

$$J(x, y) = \int_0^T g(x, y)\, dt, \tag{2.1}$$

where the scalar variables x and y are connected by the differential equation

$$\frac{dx}{dt} = h(x, y), \qquad x(0) = c. \tag{2.2}$$

Write

$$\min_y J(x, y) = f(c, T). \tag{2.3}$$

Then, proceeding formally, we obtain from the principle of optimality the nonlinear partial differential equation

$$f_T = \min_v \left[g(c, v) + h(c, v) f_c\right], \qquad f(c, 0) = 0, \tag{2.4}$$

cf. [1–3]. Under various assumptions concerning g and h, this equation can be established rigorously, cf. [11]. Carrying out the minimization over v, and thereby eliminating v, we obtain an equation of classical appearance,

$$f_T = \varphi(c, f_c). \tag{2.5}$$

This suggests, consistent with what we have stated above, that it might be profitable to study equations such as (2.5) by means of equations such as (2.4), and even directly in terms of the original variational problem.

In particular, this approach affords a new type of approximation not encountered in classical analysis, *approximation in policy space*. The method of successive approximations applied to (2.5) in any of a number of ways yields a sequence $\{f_n\}$ which converges to $f(c, T)$. Returning to (2.4), we see that we can also approximate to $v(c, T)$, the optimal policy. This latter procedure has many desirable features, as we shall indicate below.

In dealing simultaneously with both $f(c, T)$ and $v(c, T)$, we are exploiting the duality inherent in the original process. For a more detailed discussion, together with references to the theory of characteristics and techniques of Friedrichs and others, see [3].

Methods of the type sketched above have been employed by Lax and Bellman to treat the equation

$$f_T = \varphi(f) f_c, \tag{2.6}$$

and to obtain an alternate and simpler derivation of results of Lax [12]; see Kalaba [13] for the only published version of this approach.

Recently, the technique has been extensively applied by Conway–Hopf [14], to treat more general equations. Emphasis upon the policy function $v(c, T)$ yields a clearer understanding of the generalized solutions of (2.5).

All of these results can be extended and generalized by considering first the corresponding multidimensional variational problems, with and without constraints, then corresponding stochastic variational processes, cf. Fleming [15], then adaptive variational processes, and finally more general variational processes where the equations corresponding to (2.2) are themselves partial differential equations, and equations of more general evolutionary type, and the equations corresponding to (2.4) involve Gateaux derivatives, cf. [16]. In this way, among other things an extensive generalization of the theory of characteristics may be obtained.

3. Quasilinearization

If we are interested in an approximate solution of (2.5), we can proceed in the following fashion. Let $v_1(c, T)$ be a first approximation to the minimizing function in (2.4). Let $f_1(c, T)$ be determined as the solution of the *linear* equation

$$\frac{\partial f_1}{\partial T} = \left[g(c, v_1) + h(c, v_1)\frac{\partial f_1}{\partial c}\right], \qquad f_1(c, 0) = 0. \tag{3.1}$$

The next approximation $v_2(c, T)$ is determined by the condition that it minimize the function $[g(c, v) + h(c, v)\,(\partial f_1/\partial c)]$. Then $f_2(c, T)$ is determined as the solution of

$$\frac{\partial f_2}{\partial T} = \left[g(c, v_2) + h(c, v_2)\frac{\partial f_2}{\partial c}\right], \qquad f_2(c, 0) = 0. \tag{3.2}$$

It follows from this derivation that

$$\frac{\partial f_1}{\partial T} = \left[g(c, v_1) + h(c, v_1)\frac{\partial f_1}{\partial c}\right] \geq \left[g(c, v_2) + h(c, v_2)\frac{\partial f_1}{\partial c}\right]. \tag{3.3}$$

From this, we readily conclude that $f_1 \geq f_2$ for $T \geq 0$.

This general procedure can be applied to many other classes of nonlinear functional equations [17–19]. The establishment of monotonicity in general requires the modern theory of positive operators and maximum principles for partial differential equations; Collatz [20], Walter [21], Beckenbach–Bellman [22].

For certain classes of functional equations, the foregoing procedure yields the Newton–Raphson–Kantorovich approximation scheme; see [23] for numerous applications.

4. Novel Difference Techniques

Consider next the question of obtaining a computational solution of (2.5) via difference techniques. The discrete version of the continuous variational problem posed in Sec. 2 yields, when treated by means of dynamic programming, the functional equation

$$f(c, T) = \min_{v} \left[g(c, v)\,\Delta + f(c + h(c, v)\,\Delta,\, T - \Delta)\right], \tag{4.1}$$

$f(c, 0) = 0$, with $T = \Delta, 2\Delta, 3\Delta, \ldots, -\infty < c < \infty$. The equation in (4.1) may be regarded as a new finite-difference approximation to (2.4) and (2.5). It possesses the advantage of exhibiting explicitly only one grid variable. The values of $f(c, T)$ for fixed T are obtained by means of an interpolation formula.

There are now a number of interesting problems concerning the convergence of the solution obtained from (4.1) to the solution of (2.4) as $\Delta \to 0$. See [24].

5. Novel Difference Techniques

Having observed the form of the difference approximation in (4.1), it is natural to apply the same ideas to other classes of partial differential equations. It turns out that just as dynamic programming applied to discrete decision processes suggests (4.1) as a way of treating (2.4), so invariant imbedding applied to discrete transport processes suggests alternate difference schemes.

Consider, for example, the equation

$$u_t = uu_x, \qquad u(x, 0) = g(x), \tag{5.1}$$

which is so useful for testing computational procedures in view of both its simple explicit solution and the possibility of shocklike phenomena. In place of any of the usual difference methods, we can write

$$u(x, t + \Delta) = u(x + u(x, t)\,\Delta, t), \tag{5.2}$$

for $t = 0, \Delta, 2\Delta, \ldots, -\infty < x < \infty$. The values of $u(x, t)$ for fixed t are obtained by means of an interpolation; see [25, 26]. We can further ask for approximations which are valid to order Δ^k; [27].

The same idea can be applied to study general systems of hyperbolic partial differential equations and equations of the form

$$u_t = uu_x + \varepsilon u_{xx}, \tag{5.3}$$

the Burgers' equation [4]. In all these cases, the idea is to use a difference scheme which automatically preserves the boundedness and nonnegativity of the solution. Furthermore, we want algorithms which are obviously stable and which do not require the solution of large systems of linear algebraic equations. See [9, 10, 28].

6. Infinite Systems of Ordinary Differential Equations

Consider an equation such as the Burgers equation

$$u_t = uu_x + \varepsilon u_{xx}, \tag{6.1}$$

where $0 < x < 1$, $t > 0$, $u(x, 0) = g(x)$. A simple straightforward approach to the numerical solution of (6.1) is to write

$$u(x, t) = \sum_{-\infty < n < \infty} e^{2\pi inx} u_n(t), \tag{6.2}$$

and to obtain in this way an infinite system of ordinary differential equations

$$u_n' = -4\pi^2 n^2 \varepsilon u_n + g_n(u_0, u_1, u_{-1}, \cdots), \qquad u_n(0) = g_n, \tag{6.3}$$

$n = 0, \pm 1, \pm 2, \ldots$. The simplest truncation scheme is that of setting $u_n = 0$ for $|n| > N$. A number of other truncation methods can be applied yielding more accurate results for a given value of N; see [4].

In any case, with current and contemplated computers which can treat thousands and tens of thousands of simultaneous ordinary differential equations, many direct methods are quite successful. If it is not clear what value of N should be used, $N = 10, 50, 100$, we try all of the values in turn until the solution stabilizes. This is an example of adaptive computation, and there is no difficulty in writing computer programs which embody this feature.

The general problem of obtaining a *finite* system of ordinary differential equations whose solution can be used to approximate that of an equation of far more complex type is part of the theories of closure and differential approximation; see [5–7].

7. Laplace Transform Techniques

Consider an equation such as

$$k(x)u_t = u_{xx}, \qquad t > 0, \qquad 0 < x < 1, \tag{7.1}$$

$u(x, 0) = g(x)$, $u(0, t) = u(1, t) = 0$. Use of the Laplace transform reduces this to an ordinary differential equation subject to two-point boundary conditions. Thus the function

$$U(x, s) = \int_0^\infty e^{-st} u(x, t)\, dt$$

can readily be determined for any $s \geq 0$. The computational feasibility of this approach to the numerical solution of the foregoing partial differential equation thus depends upon our ability to invert the Laplace transform numerically. Although no uniformly applicable technique exists at present (nor can ever exist!), a number of powerful methods for selected problems are now available.

In treating an equation such as (7.1), we have the luxury of many equally efficient methods. In treating many other classes of time-dependent processes which give rise to rather formidable functional equations, such as

$$u_t + u_x = 1 + \int_0^t u(t_1, x)u_t(t - t_1, x)\, dt_1, \tag{7.2}$$

or

$$\begin{aligned} g_x + (u^{-1} + v^{-1})\left[\frac{\partial}{\partial t} + 1\right]g = \lambda\Bigg[\frac{1}{4v} &+ \frac{1}{2}\int_0^1 g(v, u', x, t)\frac{du'}{u'} \\ &+ \frac{1}{2v}\int_0^1 g(v', u, x, t)\, dv' \\ &+ \int_0^t dt' \int_0^1 g(v', u, x, t')\, dv' \\ &\times \int_0^1 g_t(v, u', x, t - t')\frac{du'}{u'}\Bigg], \end{aligned} \tag{7.3}$$

see [8, 29], it would seem that the use of the Laplace transform would be essential.

8. Quadrature Techniques

Many processes in mathematical physics give rise to functional equations of more complex structure than ordinary or partial differential equations. Typically transport processes introduce equations involving partial derivatives and integrals, e.g., the Boltzmann equation. In order to treat these equations using a digital computer, it is convenient to approximate to them by means of systems of ordinary differential equations. A simple and direct way of accomplishing this is by use of quadrature techniques.

Consider, for example, an equation encountered in the theory of radiative transfer in a slab,

$$R_x + \left(\frac{1}{u} + \frac{1}{v}\right)R = \lambda\left[1 + \frac{1}{2}\int_0^1 R(v, u', x)\frac{du'}{u'}\right]\left[1 + \frac{1}{2}\int_0^1 R(v', u, x)\frac{dv'}{v'}\right],$$

$$R(v, u, 0) = 0. \tag{8.1}$$

Using a quadrature formula of Gaussian type [29, 30], we write

$$\int_0^1 f(x)\,dx \cong \sum_{i=1}^N w_i f(x_i) \tag{8.2}$$

and then

$$\int_0^1 R(v, u', x)\frac{du'}{u'} \cong \sum_{i=1}^N R(v, u_i, x)\frac{w_i}{u_i}. \tag{8.3}$$

In this way, the numerical solution of the original integro-differential equation is reduced to that of the solution of the system of ordinary differential equations

$$\frac{df_{ij}}{dx} = -\left(\frac{1}{x_i} + \frac{1}{x_j}\right)f_{ij} + \lambda\left[1 + \frac{1}{2}\sum_{l=1}^N f_{il}\frac{w_l}{x_l}\right]\left[1 + \frac{1}{2}\sum_{l=1}^N f_{lj}\frac{w_l}{x_l}\right], \qquad f_{ij}(0) = 0. \tag{8.4}$$

Very accurate results can be obtained in this way.

If, in place of considering plane-parallel regions, we consider radiative transfer processes in spherical regions, we obtain a more complicated partial differential integral equation

$$\frac{\partial S}{\partial z} + \frac{1 - v^2}{zv}\frac{\partial S}{\partial v} + \frac{1 - u^2}{zu}\frac{\partial S}{\partial u} + \left(\frac{1}{v} + \frac{1}{u}\right)S - \frac{v^2 + u^2}{v^2u^2}\frac{S}{z}$$

$$= \lambda\left[1 + \frac{1}{2}\int_0^1 S(z, v, u')\frac{du'}{u'}\right]\left[1 + \frac{1}{2}\int_0^1 S(z, v', u)\frac{dv'}{v'}\right],$$

$$z > a, \quad S(a, v, u) = 0. \tag{8.5}$$

The point here is that in addition to the partial derivative with respect to z, the partial derivatives with respect to u and v also occur.

One approach to this problem consists in approximating the value of S_u and S_v by appropriate linear combinations of the values of S. In view of the integrals which occur, we select $u = v_i$, $v = v_i$, where v_i = ith root of $P_N(1-2x)$. The polynomials $P_0(1-2x)$, $P_1(1-2x)$, ..., form a set of orthogonal polynomials on the interval [0, 1]. We write

$$\frac{\partial S(z, v_i, u_j)}{\partial v} \cong \sum_{m=1}^{N} a_{im} S(z, v_m, u_j) \tag{8.6}$$

and

$$\frac{\partial S(z, v_i, u_j)}{\partial u} \cong \sum_{m=1}^{N} a_{jm} S(z, v_i, u_m). \tag{8.7}$$

The coefficients are chosen so that the equations above are exact for all polynomials of degree $N-1$ or less. A simple calculation shows that the coefficients are given by the expressions

$$a_{im} = \frac{Q_N'(x_i)}{(x_i - x_m)Q_N'(x_m)}, \qquad i \neq m, \tag{8.8}$$

$$a_{mm} = \frac{1 - 2x_m}{2(x_m^2 - x_m)}, \tag{8.9}$$

for $m = 1, 2, \ldots, N$. These are tabulated in [31].

Writing

$$S_{ij}(z) = S(z, v_i, u_j), \tag{8.10}$$

we approximate Eq. (8.5) by the system of ordinary differential equations

$$\begin{aligned} \frac{dS_{ij}}{dz} &+ \frac{1 - v_i^2}{v_i z} \sum_{k=1}^{N} a_{ik} S_{kj} + \frac{1 - v_j^2}{v_j z} \sum_{k=1}^{N} a_{jk} S_{ik} \\ &+ \left(\frac{1}{v_i} + \frac{1}{v_j}\right) S_{ij} - \frac{v_i^2 + v_j^2}{v_i^2 v_j^2} \frac{S_{ij}}{z} \\ &= \lambda \left[1 + \frac{1}{2} \sum_{k=1}^{N} S_{ik} \frac{w_k}{v_k}\right] \left[1 + \frac{1}{2} \sum_{k=1}^{N} S_{kj} \frac{w_k}{v_k}\right], \qquad z > a, \end{aligned} \tag{8.11}$$

together with the initial conditions

$$S_{ij}(a) = 0, \qquad i, j = 1, 2, \ldots, N. \tag{8.12}$$

Results of a sample calculation can be found in [31].

9. Perturbation Techniques

We may also approach the problem of solving Eq. (8.5) by a perturbation technique. First we introduce $x = z - a$. Then we expand $S(x, v, u)$ in a formal power series in a^{-1}, substitute in (8.5), and equate the coefficients of like powers of a^{-1}. For

$$S(x, v, u) = S^0 + a^{-1}S^1 + a^{-2}S^2 + \dots, \tag{9.1}$$

this yields the equations

$$S_x{}^0 + \left(\frac{1}{u} + \frac{1}{v}\right)S^0 = \lambda\left[1 + \frac{1}{2}\int_0^1 S^0(x, v', u)\frac{dv'}{v'}\right]\left[1 + \frac{1}{2}\int_0^1 S^0(x, v, u')\frac{du'}{u'}\right] \tag{9.2}$$

and

$$\begin{aligned} S_x{}^1 + \frac{1 - v^2}{v} S_v{}^0 + \frac{1 - u^2}{u} S_u{}^0 + \left(\frac{1}{u} + \frac{1}{v}\right) S^1 - \frac{u^2 + v^2}{u^2 v^2} S^0 \\ = \lambda\left[1 + \frac{1}{2}\int_0^1 S^0(x, v', u)\frac{du'}{u'}\right]\left[\frac{1}{2}\int_0^1 S^1(x, v', u)\frac{dv'}{v'}\right] \\ + \lambda\left[1 + \frac{1}{2}\int_0^1 S^0(x, v', u)\frac{dv'}{v'}\right]\left[\frac{1}{2}\int_0^1 S^1(x, v, u')\frac{du'}{u'}\right]. \end{aligned} \tag{9.3}$$

Equation (9.2) is just the equation for the slab problem. Equation (9.3) is resolved by producing S^0, $S_u{}^0$, and $S_v{}^0$ simultaneously with S^1.

Higher order terms could be calculated but results of some trial calculations [32] showed that we obtained agreement with the results of the earlier calculation using only two terms.

References

1. Bellman, R., "Dynamic Programming." Princeton Univ. Press, Princeton, New Jersey, 1957.
2. Bellman, R., and Dreyfus, S., "Applied Dynamic Programming." Princeton Univ. Press, Princeton, New Jersey, 1962.
3. Bellman, R., "Adaptive Control Processes: A Guided Tour." Princeton Univ. Press, Princeton, New Jersey, 1961.

4. Azen, S., Bellman, R., and Richardson, J. M., On New and Direct Computational Approaches to Some Mathematical Models of Turbulence, *Q. Appl. Math.* **23**, 55–67 (1965).
5. Bellman, R., and Richardson, J. M., On Some Questions Arising in the Approximate Solution of Nonlinear Differential Equations, *Q. Appl. Math.* **20**, 333–339 (1963).
6. Bellman, R. and Richardson, J. M., Self-consistent Solutions of Deterministic and Stochastic Nonlinear Differential Equations, *Symp. Proc., Kiev, USSR* (1963).
7. Bellman, R., "Perturbation Techniques in Mathematics, Physics, and Engineering." Holt, New York, 1964.
8. Bellman, R., Kalaba, R., and Lockett, J., "Numerical Inversion of the Laplace Transform: Applications to Biology, Economics, Engineering, and Physics." American Elsevier, New York, 1967.
9. Bellman, R., and Lehman, R. S., Invariant Imbedding, Particle Interaction, and Conservation Relations, *J. Math. Anal. Appl.* **10**, 112–122 (1965).
10. Bellman, R., Cooke, K. L., Kalaba, R., and Wing, G. M., Existence and Uniqueness Theorems in Invariant Imbedding—I: Conservation Principles, *J. Math. Anal. Appl.*, **10**, 234–244. (1965).
11. Berkovitz, L. D., Variational Methods in Problems of Control and Programming, *J. Math. Anal. Appl.* **3**, 145–169 (1961).
12. Lax, P., Hyperbolic Systems of Conservation Laws, II, *Comm. Pure Appl. Math.* **10**, 537–566 (1957).
13. Kalaba, R., On Nonlinear Differential Equations, the Maximum Operation, and Monotone Convergence, *J. Math. Mech.* **8**, 519–574 (1959).
14. Conway, E. D., and Hopf, E., Hamilton's Theory and Generalized Solutions of the Hamilton–Jacobi Equation, *J. Math. Mech.* **13**, 939–986 (1964).
15. Fleming, W. H., Some Markovian Optimization Problems, *J. Math. Mech.* **12**, 131–140 (1963).
16. Bellman, R., and Kalaba, R., Dynamic Programming Applied to Control Processes Governed by General Functional Equations, *Proc. Natl. Acad. Sci. US* **48**, 1735–1737 (1962).
17. Bellman, R., Functional Equations in the Theory of Dynamic Programming—V: Positivity and Quasi-Linearity, *Proc. Natl. Acad. Sci. US* **41**, 743–746 (1955).
18. Bellman, R., On the Representation of the Solution of a Class of Stochastic Differential Equations, *Proc. Am. Math. Soc.* **9**, 326–327 (1958).
19. Bellman, R., Quasilinearization and Upper and Lower Bounds for Variational Problems, *Q. Appl. Math.* **19**, 349–350 (1962).
20. Collatz, L., Applications of the Theory of Monotonic Operators to Boundary Value Problems, *in* "Boundary Problems in Differential Equations." Univ. Wisconsin Press, Madison, Wisconsin, 1960.
21. Walter, W., "Differential- und Integral Ungleichungen." Springer, Berlin, 1964.
22. Beckenbach, E. F., and Bellman, R., "Inequalities." Springer, Berlin, 1961.
23. Bellman, R., and Kalaba, R., "Quasilinearization and Nonlinear Boundary-Value Problems, American Elsevier, New York, 1965.

24. Bellman, R., Functional Equations in the Theory of Dynamic Programming—VI: A Direct Convergence Proof, *Ann. Math*, **65**. 215–223 (1957).
25. Bellman, R., Kalaba, R., and Kotkin, B., On a New Approach to the Computational Solution of Partial Differential Equations, *Proc. Natl. Acad. Sci. US*, **48**, 1325–1327 (1962).
26. Bellman, R., Cherry, I., and Wing, G. M., A Note on the Numerical Integration of a Class of Nonlinear Hyperbolic Equations, *Q. Appl. Math.* **16**, 181–183 (1958).
27. Bellman, R., Some Questions Concerning Difference Approximations to Partial Differential Equations, *Boll. d'Unione Math.* **17**, 188–190 (1962).
28. Bellman, R., and Cooke, K. L., Existence and Uniqueness Theorems in Invariant Imbedding—II: Convergence of a New Difference Algorithm, *J. Math. Anal. Appl.*, to appear.
29. Bellman, R., Kagiwada, H., Kalaba, R., and Prestrud, M., "Invariant Imbedding and Time-Dependent Transport Processes." American Elsevier, New York, 1964.
30. Bellman, R., Kalaba, R., and Prestrud, M., "Invariant Imbedding and Radiative Transfer in Slabs of Finite Thickness." American Elsevier, New York, 1963.
31. Bellman, R., and Kalaba, R., On a New Approach to the Numerical Solution of a Class of Partial Differential-Integral Equations of Transport Theory, *Proc. Natl. Acad. Sci. US* **54**, No. 5, pp. 1293-1296 (1965).
32. Bellman, R., Kagiwada, H., and Kalaba, R., Invariant Imbedding and Perturbation Techniques Applied to Diffuse Reflection from Spherical Shells, *The RAND Corporation*, Santa Monica, California, RM-4730-NASA (August, 1965).

Ad hoc Exact Techniques for Nonlinear Partial Differential Equations

W. F. AMES

Department of Statistics and Computer Science
and Department of Mechanical Engineering
University of Delaware
Newark, Delaware

I. Introduction

The phrase *ad hoc*, from the Latin, means "*for this case alone*" or "special." Much of the progress in the linear theory of partial differential equations has resulted from the use of various assumed (ad hoc) forms of solution. Examples of ad hoc procedures include the analytic techniques of *separation of variables*, *integral transformations*, and the *finite difference numerical techniques*. On the other hand the development of *general solutions* and the *method of characteristics* are not ad hoc but arise *naturally* from the specific system under consideration.

In specific nonlinear examples the ad hoc techniques of the linear theory may prove useful. However, we must recognize that their great utility rests primarily upon the principle of superposition. In accordance with this principle, elementary solutions of the pertinent mathematical equations can be combined to yield more flexible ones, namely ones which could satisfy the auxiliary conditions that arise from the particular physical phenomena. In nonlinear problems this principle no longer holds. Its loss and the lack of an effective replacement constitutes the major hurdle in the present chaotic state.

It is pertinent here to remark that we are preconditioned by education and training to think

(a) that all solutions are explicit;

(b) that boundary and initial conditions should be imposed at the very beginning of the problem and carried along;

(c) that, with a very few exceptions, *particular solutions* are more to be sought after than *general solutions*;

(d) that a constant times a solution is also a solution;

(e) that the sum of solutions is a solution;

(f) that the proper course in mathematical study is to learn all that went before.

As one comes to grips with nonlinear problems he soon realizes that each of these points must be discarded in its turn. And sometimes I curse my background and predisposition to the methods of mathematics based primarily on the linear operators—the study of which we devote *excessive time*.

Nevertheless, some information has been painfully extracted from nonlinear partial differential equations by the utilization of specific methods. In this paper we survey some of these ad hoc procedures including physical applications, introduce the concept of equation splitting and develop some applications.

2. Separation of Variables

Various authors have utilized a direct separation of variables. Among these we find the work of Oplinger [1, 2] in the nonlinear vibration of threads and Smith [3] in one-dimensional anisentropic flow. Tomotika and Tamada [4] and Tamada [5] study certain idealized two-dimensional transonic flows by means of separation and other assumed forms of solution (see Section 3). Other authors, notably Sedov [6], Lidov [7], and Keller [8] utilize the separation principle to obtain some exact solutions of gas flows. In all of these examples the usual *modus operandi* is to examine the form of the solutions obtainable by this process and *then* investigate what auxiliary (initial and boundary) conditions can be satisfied. Since superposition is no longer available it is trite to remark that solutions obtainable by this process, or by any other procedure, are not amenable to all forms of auxiliary conditions.

Oplinger's problem [1] is that of solving the "weakly" nonlinear hyperbolic equation

$$y_{tt} - c^2\left[1 + \frac{Ea}{2LT_0}\int_0^L y_x^{\,2}\,dx\right]y_{xx} = 0 \tag{1}$$

governing a string fixed at one end ($x = 0$) and subject to an *unspecified* periodic oscillation at the other end ($x =$ L). Upon attempting solutions of the form $y = F(x)\,G(t)$ it is evident that

$$\int_0^L {y_x}^2\,dx = G^2(t)\int_0^L \left(\frac{dF}{dx}\right)^2 dx = IG^2(t) \tag{2}$$

so that the separation equations for Eq. (1) become

$$F'' + v^2F = 0 \tag{3}$$

and

$$G'' + v^2c^2\left[1 + \frac{Ea}{2LT_0}IG^2\right]G = 0.$$

Here $E, a, L, T_0, c^2 = T_0/m$, and v^2 are respectively elastic modulus, cross sectional area, length, tension when the string is along the x axis, density m, and separation parameter. The solutions of these equations are

$$\begin{aligned} F(x) &= A\sin vx + B\cos vx \\ G(t) &= D\,\mathrm{cn}(ft, K) \end{aligned} \tag{4}$$

with $\mathrm{cn}(ft, K)$ a periodic elliptic function. Possible boundary conditions are severely limited by this form. Since $y(0, t) = F(0)G(t) = 0$ we find $B = 0$ and therefore

$$y(x, t) = E\sin vx\,\mathrm{cn}(ft, K). \tag{5}$$

With a driven right end of period t' and maximum amplitude α we ask that

$$y(L, t) = F(L)\,G(t) = \alpha G(t)$$

and

$$G(t + t') = G(t).$$

These requirements generate the solution

$$y(x, t) = \alpha\frac{\sin vx}{\sin vL}\mathrm{cn}(ft, K), \tag{6}$$

where

$$f = (t')^{-1}, \qquad K = \int_0^{\pi/2} [1 - k^2 \sin^2 \psi]^{-1/2} \, d\psi, \qquad d = c^2 v^2,$$

$$b = \frac{c^2 v^2 E a \alpha^2}{2T_0} \left[\frac{v^2 L + (v \sin 2vL)/2}{2 \sin^2 vL} \right], \qquad k^2 = \frac{b}{2(d + b)}.$$

It is clear that the form of $G(t)$ is restricted to the periodic elliptic function $\mathrm{cn}(ft, K)$.

As a second example let us consider the simplified boundary layer equations

$$uu_x + vu_y = \nu u_{yy}, \qquad u_x + v_y = 0. \tag{7}$$

Upon introducing the stream function ψ defined by $u = \psi_y$, $v = -\psi_x$ we have the equation

$$\psi_y \psi_{xy} - \psi_x \psi_{yy} = \nu \psi_{yyy}. \tag{8}$$

After making the ad hoc separation assumption $\psi = F(x)\,G(y)$ we find that F and G are determined from the equations

$$F' = \nu/\lambda \,(\text{constant}) \qquad \lambda G''' + GG'' - (G')^2 = 0. \tag{9}$$

Clearly, separation is possible but the resulting velocity component $v = -\psi_x = -F'(x)G(y) = -(\nu/\lambda)G(y)$ is independent of x! Various other examples are detailed in the book by Ames [9] on nonlinear partial differential equations.

A last remark of this section concerns the development of similarity solutions by means of separation of variables. This method, first introduced by Birkhoff [10], essentially depends upon the introduction of some unknown transformation function(s), suggested by possible similarity. The condition that the transformed partial differential equation should be solvable by the method of separation of variables and also satisfy the boundary conditions determines the unknown function(s). Since this procedure is usually replaceable by the group method and since they are detailed by Hansen in this volume and in Hansen [11] and Ames [9] we do no more here than sketch the idea. Consider the "power law" diffusion equation

$$r^{1-q} [r^{q-1} T^n T_r]_r = T_t, \qquad n > 0 \tag{10}$$

with q-dimensional spherical symmetry. If the similarity variable is assumed to have the form $r/R(t)$, a trial solution

$$T(r, t) = U(t)\,Y(r/R(t)), \tag{11}$$

with U, Y, and R unspecified, is attempted. As is easily seen if $R(t) = [U(t)]^{-A}$ then the equation separates into

$$\frac{x^{1-q}(d/dx)\,[x^{q-1}Y^nY']}{Y + AxY'} = \frac{U^{-2A}U'}{U^{n+1}} = -\lambda, \tag{12}$$

where $x = r/R(t)$. Further details are available in Ames [9].

3. Further Specific Forms

A wide variety of physical problems (see Ames [9], pp. 4–8) are governed by the "power law" diffusion equation

$$w_t = (w^n)_{xx}. \tag{13}$$

In addition to this equation we shall use the (hypothetical gas) transonic equation of Tomotika and Tamada [4] and Tamada [5] to illustrate various other specific forms. That equation is

$$(kw)_{\psi\psi} = [(kw)^2]_{\phi\phi} \tag{14}$$

which approximates the nearly uniform transonic flow of a real gas.

(a) Traveling Wave Solutions

If λ is a constant what form must f have so that $f(x - \lambda t)$ is a solution of Eq. (13)? Upon substitution we find that f must satisfy the differential equation

$$(f^n)'' + \lambda f' = 0, \tag{15}$$

where the prime indicates differentiation with respect to the grouping $\eta = x - \lambda t$. The first integration is immediate so that $(f^n)' + \lambda f = A$. The final integration is easily accomplished *if n is a positive integer* yielding the implicit solution

$$\sum_{j=0}^{n-2} \frac{(A/\lambda)^j u^{n-1-j}}{n-1-j} + \left(\frac{A}{\lambda}\right)^{n-1} \ln(u - A/\lambda) = \frac{\lambda}{n}(\lambda t - x + B), \tag{16}$$

where $A \neq 0$ and B are constants of integration. If $A = 0$ the integration generates the explicit form

$$u(x, t) = \left[\frac{\lambda(n-1)}{n}(\lambda t - x + B)\right]^{1/(n-1)}. \tag{17}$$

Traveling wave solutions of the form

$$kw = f(\phi + \lambda\psi) \tag{18}$$

are used by Tamada [5] to explore possible solutions of Eq. (14). Classical traveling wave solutions

$$\exp[i(wt - kx)] \tag{19}$$

are used extensively in various disciplines. Plasma physics, see e.g., Dolph [12] and Chandrasekhar [13], utilize them in the development of dispersion relations. Stuart [14, 16, 17] and Watson [15] utilize this concept to investigate stability problems in fluid mechanics.

(b) "Elementary" Expansions

For purposes of exploration two or three term (exact) expansions are sometimes useful. To illustrate the development of one of these consider the equation

$$w_t = (ww_x)_x \,. \tag{20}$$

For this we construct functions $F(t)$ and $G(t)$ and determine p so that

$$w = F(t) + G(t)x^p \tag{21}$$

is a solution. Upon substitution of Eq. (21) into (20) we find the relations

$$F' + G'x^p = FG\,p(p-1)x^{p-2} + G^2 p(2p-1)x^{2p-2}. \tag{22}$$

With $p = 2$ the terms in G and its derivatives may be grouped to obtain

$$F' - 2FG = (6G^2 - G')x^2. \tag{23}$$

This is clearly impossible unless $G' - 6G^2 = 0$ whereupon F may be determined from $F' - 2FG = 0$. Performing these integrations we find that

$$w(x, t) = (C - 6t)^{-1}\, x^2 + D(C - 6t)^{-1/6}, \tag{24}$$

where C and D are constants. From this result it is immediately obvious that such solutions have limited utility but may provide important clues as to the actual form of the solution—which no numerical result can do. Tamada [5] found that Eq. (14) possessed solutions of the form

$$\begin{aligned} &(i)\ kw = f(\phi) + g(\psi) \\ &(ii)\ kw = f_0(\psi) + f_1(\psi)\phi^2 \\ &(iii)\ kw = F(\phi + \psi^2) + 2\psi^2 \end{aligned} \tag{25}$$

as well as a traveling wave $f(\phi + \lambda\psi)$ and a separation $F(\phi)G(\psi)$. Case (iii) was extensively studied because it provides important information concerning the transition from Taylor flow to Meyer flow in a Laval nozzle.

(c) Two Arbitrary Functions

Burgers [18], Hopf [19] and Cole [20] have considered the Burger's equation

$$u_t + uu_x = \nu u_{xx} \tag{26}$$

which was developed as a mathematical model of turbulence. Setting $u = \psi_x$, integrating with respect to x, and discarding an arbitrary function of time we have

$$\psi_t + \tfrac{1}{2}\psi_x{}^2 = \nu\psi_{xx}. \tag{27}$$

We now attempt a solution in the form

$$\psi = f(t + g(x)), \tag{28}$$

where f and g are to be determined. Upon substitution into Eq. (27) we obtain, with $w = t + g(x)$,

$$f'(w)\,[1 - \nu g''(x)] = [g'(x)]^2\,[\nu f'' - \tfrac{1}{2}(f')^2] \tag{29}$$

which may be rewritten as

$$\frac{1 - \nu g''(x)}{(g')^2} = \frac{\nu f'' - \frac{1}{2}(f')^2}{f'} = \text{const} = c \tag{30}$$

by the standard separation argument. Thus f and g must satisfy

$$\nu f'' - \tfrac{1}{2}(f')^2 - cf' = 0$$
$$\nu g'' + c(g')^2 = 1. \tag{31}$$

In particular if $c = 0$ we have

$$g = \frac{1}{2\nu}x^2 + ax + b, \qquad f = -2\nu \ln(Aw + B),$$

where a, b, A, and B are arbitrary constants. Consolidating these results it is seen that

$$\psi = -2\nu \ln[A(t + x^2/2\nu + ax) + B], \tag{32}$$

is a solution. Integrations for arbitrary c can be carried out.

4. Assumed Relations between Dependent Variables

On occasion insight into the form of the solution may be obtained by assuming an unspecified relation between two (say) of the dependent variables. The resulting simplification in the equations can be striking. To illustrate these concepts we consider two examples.

(a) Boundary Layer Flow

Consider the boundary layer equations

$$uu_x + vu_y = \nu u_{yy}, \qquad u_x + v_y = 0 \tag{33}$$

for viscous flow over a semi-infinite flat plate. We shall investigate forms of the solution for which

$$v = F(u) \tag{34}$$

for general F. From the continuity equation we find $u_x = -F'(u)u_y$ and therefore the momentum equation becomes

$$[uF'(u) - F]\frac{\partial u}{\partial y} + \nu\frac{\partial^2 u}{\partial y^2} = 0 \tag{35}$$

—that is an ordinary nonlinear differential equation in y with parameter x.

We assume the following boundary conditions:

$$\begin{array}{llll} y = 0 & u = 0, \quad v = 0 & \text{so that} & F(0) = 0 \\ y \to \infty & u \to U \quad (\text{const}) & u_y \to 0. & \end{array} \tag{36}$$

If we select $F = \alpha u^{n-1}$, $n > 1$, Eq. (35) becomes

$$\nu u_{yy} + \alpha(n - 2)u^{n-1}u_y = 0$$

which integrates to

$$u_y + \frac{\alpha(n - 2)}{\nu n} u^n = A(x) \tag{37}$$

for $n > 1$, $n \neq 2$. When $n = 2$ the equation integrates to $u = 1/\nu[B_1(x)y + B_2(x)]$.

We examine only the general case. Since $u_y \to 0$ and $u \to U$ as $y \to \infty$ we have $A(x) = (\alpha(n - 2)/\nu n)\, U^n$ (const). With this value of $A(x)$ the final integration may be difficult. Setting $R = u/U$, $\delta = \alpha(n - 2)/\nu n$ Eq. (37) becomes

$$dR/dy + \delta U^{n-1}[R^n - 1] = 0 \qquad R(0) = 0 \tag{37a}$$

which can be integrated for integral n. However, the resulting implicit form is cumbersome, unless n is small, and not very useful. It appears to be much easier to integrate Eq. (37a) approximately or by series.

(b) Magnetogasdynamics

The governing equations for an inviscid, nonheat conducting, infinite electrical conductivity, ideal plasma are (see Pai [21])

$$\rho_t + (\rho u)_x = 0 \tag{38a}$$

$$\rho u_t + \rho u u_x + p_x + \mu_e H H_x = 0 \tag{38b}$$

$$H_t + (uH)_x = 0 \tag{38c}$$

$$\rho h_t + \rho u h_x - p_t + \mu_e u H H_x = 0. \tag{38d}$$

Here ρ, u, p, H, and h are density, velocity, pressure, magnetic field, and stagnation enthalpy ($h = c_p T + u^2/2$), respectively. Since Eqs. (38a) and (38c) are identical their solutions have the same functional form—that is we can write

$$H = A\rho = \frac{H_0}{\rho_0}\rho. \tag{39}$$

When the flow is anisentropic we write

$$p\rho^{-\gamma} = b\exp[S/c_v] = b\theta, \tag{40}$$

where $b = (p_0/\rho_0{}^{\gamma})\exp[-S_0/c_v]$ with S the entropy and $(\ )_0$ refers to values at the stagnation point. From these results we have

$$\left[\frac{\partial}{\partial t} + u\frac{\partial}{\partial x}\right]\ln\theta \equiv 0 \tag{41}$$

—i.e., there is no change in entropy along any line of flow.

Some critical speeds are useful:

First, the local speed of sound is defined as

$$c^2 = \left(\frac{\partial p}{\partial \rho}\right)_\theta = \frac{\gamma p}{\rho} = b\gamma\theta\rho^{\gamma-1}. \tag{42a}$$

Second, the local speed of Alfven's wave

$$V_H{}^2 = \mu_e H^2/\rho = \mu_e A^2\rho = \mu_e(H_0{}^2/\rho_0{}^2)\,\rho = V_{H_0}{}^2\frac{\rho}{\rho_0}. \tag{42b}$$

Third, the effective speed of sound

$$c_e{}^2 = c^2 + V_H{}^2. \tag{42c}$$

Generally, c_e is a function of both θ and ρ, except that for isentropic flow (θ = constant throughout the flow) c_e depends only on ρ.

Considerable information can be extracted from the Eqs. (38) by a method used by Pai [21]. For isentropic flow we write

$$\rho = f(u), \tag{43}$$

where f is to be determined. Then Eq. (38a) becomes

$$f'(u)u_t + f(u)u_x + uf'(u)u_x = 0$$

or

$$u_t + uu_x = -\frac{f}{f'}u_x. \tag{44}$$

Utilizing Eqs. (39) and (43) we write Eq. (38b) as

$$fu_t + fuu_x + (\gamma b\theta\rho^{\gamma-1})f'u_x + \mu_e A^2\rho f'u_x = 0. \tag{45}$$

But since the effective speed is defined as

$$c_e^2 = \gamma b\theta\rho^{\gamma-1} + \mu_e A^2\rho$$

we write Eq. (45) in the form

$$u_t + uu_x = -\frac{c_e^2 f'}{f} u_x. \tag{46}$$

Of course Eqs. (44) and (46) must agree so that

$$c_e^2\frac{f'}{f} = \frac{f}{f'} \tag{47}$$

or

$$\frac{f'}{f} = \frac{1}{\rho}\frac{d\rho}{du} = \pm\frac{1}{c_e} \tag{48}$$

which when integrated yields

$$u = \pm\int_{\rho_0}^{\rho}\frac{c_e(\rho)}{\rho}\,d\rho. \tag{49}$$

Actually, Eq. (48) when substituted into Eq. (44) gives us the proper information to proceed without the use of Eq. (49). Thus Eq. (44) becomes

$$u_t + (u \pm c_e)u_x = 0 \tag{50}$$

and Eq. (38c) becomes

$$\rho_t + (u \pm c_e)\rho_x = 0. \tag{51}$$

Direct integration by means of the Lagrange subsidiary equations gives

$$u = F[x - (u \pm c_e)t] \qquad \rho = G[x - (u \pm c_e)t], \tag{52}$$

where

$$u(x, 0) = F(x) \qquad \rho(x, 0) = G(x).$$

For waves whose amplitude is finite, the velocity of wave propagation is different at various positions of the flow field and the shape of the wave is distorted as the wave propagates. In particular for waves of the type $u = F[x - (u + c_e)t]$ the velocity of propagation $u + c_e$ at the crest is greater than at the trough. Therefore the crest will overtake the trough, a shock will form and the isentropic assumption breaks down.

(c) Some Further Assumptions

Banta [22] considers the lossless propagation of one dimensional finite amplitude sound waves and by means of an ad hoc transformation $U = \int h(c)\, dc$ relating velocity U and speed of sound c obtains an equation for U alone. A series solution for the initial value problem is then obtained.

5. Equations Equivalent to Linear Forms

Classes of nonlinear partial differential equations which are equivalent, under some transformation, to a linear equation can be derived by the following device. Let us subject the linear parabolic equation

$$v_t = \lambda \nabla^2 v \tag{53}$$

to the transformation

$$v = F(u). \tag{54}$$

Under this transformation Eq. (53) becomes

$$u_t = \lambda(F''/F')(\nabla u)^2 + \lambda \nabla^2 u. \tag{55}$$

This equation has some import in physical problems. Its one dimensional form

$$u_t = \lambda(F''/F')u_x^{\ 2} + \lambda u_{xx} \tag{56}$$

occurs in several engineering studies. The first of these is the modified Burger's equation (27) wherein we set $\nu = \lambda$, $u = \psi$ and take

$$\nu(F''/F') = -\tfrac{1}{2} \tag{57}$$

so that

$$v = F(\psi) = A \exp[-\psi/2\nu) + B. \tag{58}$$

Cole [20] obtained this transformation, which I call the *Hopf transformation*, by an alternate ingenious method.

Second, the burning of a gas in a rocket is described by Forsythe and

Wasow [23] in terms of the equation

$$\bar{u}_t = -\tfrac{1}{2}\bar{u}_x^{\,2} + \lambda \bar{u}_{xx} + d(x, t) \tag{59}$$

for $-\infty < x < \infty$, $t > 0$, $\lambda > 0$ and $d(x + 2\pi, t) = d(x, t)$. The auxiliary conditions are $\bar{u}(x, 0) = f(x)$, $\bar{u}(x + 2\pi, t) = \bar{u}(x, t)$, and $f(x + 2\pi) = f(x)$. The general case of this problem is not reducible to Eq. (56) but the case $d(x, t) = g(t)$ is reducible. Setting

$$\bar{u} = u(x, t) + \int_0^t g(r)\, dr \tag{60}$$

the equation for u becomes

$$u_t = -\tfrac{1}{2}u_x^{\,2} + \lambda u_{xx} \tag{61}$$

with the auxiliary conditions $u(x, 0) = f(x)$, $u(x + 2\pi, t) = u(x, t)$. The remarks concerning the modified Burger's equation (27) apply to Eq. (61)—that is under the inverse transformation (take $B = 0$, $A = 1$ without loss of generality)

$$u = -2\lambda \ln v \tag{62}$$

Eq. (61) transforms into the linear diffusion equation

$$v_t = \lambda v_{xx}\,! \tag{63}$$

In the general case, Eq. (59) becomes the linear equation

$$v_t = \lambda v_{xx} - (2\lambda)^{-1} d(x, t)v$$

subject to the auxiliary conditions

$$v(x, 0) = \exp[-(2\lambda)^{-1}f(x)]$$
$$v(x + 2\pi, t) = v(x, t)$$

when Eq. (62) is applied. *The rocket problem is therefore solvable by well understood linear numerical methods.*

The point to remember here is that by the proper choice of F one can generate a *class* of nonlinearities of the form

$$(F''/F')\,(\nabla u)^2 \tag{64}$$

or in the one dimensional case

$$(F''/F')u_x^{\,2}. \tag{65}$$

Moreover, equations of the form (55) are equivalent under F^{-1} to the *linear* diffusion Eq. (53).

Having chosen F to generate a specific equation (55), the next form in the "hierarchy" of nonlinear equations whose antecedent is the linear diffusion equation, can easily be generated. In fact, since Burger's original equation (26) is easily transformed to Eq. (27) we illustrate the next step from that equation. By setting $u = G(U)$ into Eq. (26) we find that U satisfies

$$U_t + G(U)U_x = \nu[G''/G']\,U_x^{\,2} + \nu U_{xx}. \tag{66}$$

Note that Burger's equation is recovered by taking $G = U$.

Extension to higher dimension is given in Cole [20] and Ames [9]. Alternate basic linear equations can be utilized, but by this time the method is clear. Of course, we may encounter difficulties in transforming the boundary conditions to the basic linear problem.

6. Equation Splitting

Past successes in the development of *particular solutions* for linear partial differential equations have served to downgrade the position of the *general* solution. Research into techniques for the development of general solutions reached a zenith around 1890. The treatise of Forsyth [24], recently reprinted by Dover, bears witness to the extensive efforts put forth prior to that time.

A solution of an nth order partial differential equation which contains n arbitrary functions is called a general solution. Clearly, a general solution gives the form of the broadest class of solutions. Further it is not ad hoc in character and is not restricted in its utility by any linearity assumption or superposition principle. A resurgence of interest in these solutions appears to be necessary before any real depth of knowledge can develop on the difficult problems that concern us here. We invoke the general solution in the development of the *splitting concept* which is the subject of this section.

The *splitting concept* is fundamentally simple—although not always simple to execute. What we do is disregard the inviolate nature of the equation(s) to be solved. It is then decomposed into parts, which are equated to a common factor, in such a manner that a general solution (containing the appropriate number of arbitrary functions) can be constructed for at least one part. The form of the arbitrary function(s) is then obtained by means of the requirement that the other part be satisfied.

To fix the ideas we consider the stream function form of the boundary layer equations

$$\psi_y \psi_{xy} - \psi_x \psi_{yy} = \nu \psi_{yyy}. \tag{67}$$

We split off the right hand side and consider the two equations

$$\psi_y \psi_{xy} - \psi_x \psi_{yy} = 0 \tag{68a}$$

and

$$\psi_{yyy} = 0. \tag{68b}$$

Note here that alternate expressions such as

$$\psi_y \psi_{xy} - \psi_x \psi_{yy} = \psi_y^{\,3} \tag{69}$$

are possible and do generate alternate solutions. The general solution of Eq. (68a) can be constructed by the Monge method (see Ames [9] and Forsyth [24]) to be

$$\psi = F[y + G(x)] \tag{70}$$

with the functions F and G arbitrary. What forms must F and G have so that Eq. (68b) is also satisfied? Considerable generality may sometimes be maintained if only one arbitrary function needs to be particularized. This is in fact the case here. For Eq. (70) to satisfy (68b) it is obvious that $F'''(\eta) = 0$, $\eta = y + G(x)$ and upon integration we get

$$\psi = F[y + G(x)] = a[y + G(x)]^2 + b[y + G(x)] + c, \tag{71}$$

where a, b, c are constants. Equation (71) is a solution of the boundary layer equation for arbitrary $G(x)$.

7. Equation Splitting and the Navier–Stokes Equations

We consider the more general problem specified by the Navier–Stokes equations in two dimensions. Upon introducing the stream function ψ defined by means of $u = \psi_y$, $v = -\psi_x$ the dimensionless equations take the form

$$\psi_y\psi_{xy} - \psi_x\psi_{yy} = -p_x + \psi_{yyy} + \mathrm{Re}^{-1}\psi_{yxx} \tag{72}$$

$$p_y + \mathrm{Re}^{-1}[\psi_x\psi_{xy} - \psi_y\psi_{xx} + \psi_{xyy}] = -\mathrm{Re}^{-2}\psi_{xxx}. \tag{73}$$

Here

$$u = \frac{\bar{u}}{U} \qquad v = \mathrm{Re}^{1/2}\,\frac{\bar{v}}{U} \qquad p = \frac{\bar{p} - \bar{p}_\infty}{\rho U^2} \qquad x = \bar{x}/L$$

$$y = \mathrm{Re}^{1/2}\,\bar{y}/L \qquad \mathrm{Re} = UL/\nu$$

are the dimensionless relations of Eqs. (72) and (73).

We suppose $p = p(y)$ and split Eq. (72) into the two parts

$$\psi_y\psi_{xy} - \psi_x\psi_{yy} = F(x, y, \psi, \psi_x, \psi_y, \psi_{yy}, \psi_{xy}, \psi_{xx}) \tag{72a}$$

$$\psi_{yyy} + \mathrm{Re}^{-1}\psi_{yxx} = F(x, y, \psi, \psi_x, \psi_y, \psi_{yy}, \psi_{xy}, \psi_{xx}) \tag{72b}$$

in such a way that the *general* solution of at least one of these can be developed. The form of the arbitrary functions is then determined by requiring that the other equation is also satisfied. There will remain certain arbitrary constants which we select in such a way that the y-momentum equation (73) is satisfied—that is so that $p = p(y)$.

It is clear that the choice made for F strongly influences the general solution obtained, the labor involved and the final result. For simplicity it is herein again chosen as zero, other forms have been used but are not yet published. With this choice our system becomes

$$p_x = 0 \tag{74a}$$

$$\psi_y\psi_{xy} - \psi_x\psi_{yy} = 0 \tag{74b}$$

$$\psi_{yyy} + \mathrm{Re}^{-1}\psi_{yxx} = 0 \tag{74c}$$

and p_y is defined by

$$p_y = -\mathrm{Re}^{-2}\psi_{xxx} - \mathrm{Re}^{-1}[\psi_x\psi_{yx} - \psi_y\psi_{xx} + \psi_{xyy}]. \tag{74d}$$

The general solution of Eq. (74b) again has the explicit form, with arbitrary ϕ and η,

$$\psi = \phi[y + \eta(x)] \tag{75}$$

although it is the usual situation that the general solution is *implicit*. In such a case the details of the computation are more complicated. Upon substituting Eq. (75) into Eq. (74c) we find, with $w = y + \eta(x)$ that

$$\phi'''(w)\,[1 + \alpha^2(\eta')^2] + \alpha^2\phi''(w)\eta''(x) = 0, \tag{76}$$

where $\alpha^2 = \mathrm{Re}^{-1}$. To eliminate the dependence on x, and thus determine η, set

$$1 + \alpha^2(\eta')^2 = A\,\alpha^2\eta'', \tag{77}$$

where A is an arbitrary constant. Equation (77) has the solution

$$\eta(x) = -A \ln \cos[(\mathrm{Re}^{1/2}x/A) + C] + c_1. \tag{78}$$

Then ϕ satisfies

$$\phi''' + A^{-1}\phi'' = 0$$

or

$$\phi(w) = \Gamma + \gamma A w + \varepsilon \exp[-w/A]. \tag{79}$$

Finally we see that

$$\begin{aligned}\psi(x, y) = \Gamma + \gamma A y - \gamma A^2 \ln \cos[(\mathrm{Re}^{1/2}x/A) + C]\\ + \varepsilon\, e^{-y/A} \cos[(\mathrm{Re}^{1/2}x/A) + C],\end{aligned} \tag{80}$$

where Γ, γ, A, C, ε are arbitrary constants.

Lastly, are there any values of these constants for which $p = p(y)$? From Eq. (74d) we find that this will be the case when $\gamma \equiv 0$. Hence

$$\begin{aligned}\psi &= \Gamma + \varepsilon \exp[-y/A] \cos[(\mathrm{Re}^{1/2}x)/A + C]\\ u &= -\varepsilon A^{-1} \exp[-y/A] \cos[(\mathrm{Re}^{1/2}x)/A + C]\\ v &= \varepsilon A^{-1}\,\mathrm{Re}^{1/2} \exp[-y/A] \sin[(\mathrm{Re}^{1/2}x)/A + C]\\ p(y) &= -\tfrac{1}{2}(\varepsilon A)^{-2} \exp[-2y/A] + D.\end{aligned} \tag{81}$$

With $C = 0$ this might be interpreted as the flow through a porous flat plate on the y axis. Note especially the appearance of the $\mathrm{Re}^{1/2}$.

We note that in this case only ($F = 0$) that the solution can also be obtained by the ad hoc assumption of separation.

References

1. OPLINGER, D. W., *J. Acoust. Soc. Am.* **32**, 1529 (1960).
2. OPLINGER, D. W., *Proc. Intern. Congr. Rheol. 4th, Providence, Part 2*, 231 (1965).
3. SMITH, P., *Appl. Sci. Res.* **A12**, No. 1, 66 (1963).
4. TOMOTIKA, S., and TAMADA, K., *Quart. Appl. Math.* **7**, 381 (1949).
5. TAMADA, K., Studies on the two-dimensional flow of a gas, with special reference to the flow through various nozzles, Ph. D. thesis, *Univ. Kyoto, Japan*, (1950).
6. SEDOV, L. I., *Dokl. Akad. Nauk SSSR* **90**, 735 (1953).
7. LIDOV, L. I., *Dokl. Akad. Nauk SSSR* **97**, 409 (1954).
8. KELLER, J. B., *Quart. Appl. Math.* **14**, 171 (1956).
9. AMES, W. F., "Nonlinear Partial Differential Equations in Engineering." Academic Press, New York, 1965.
10. BIRKHOFF, G., "Hydrodynamics." Princeton Univ. Press, Princeton, New Jersey, 1960.
11. HANSEN, A. G., "Similarity Analyses of Boundary Value Problems in Engineering." Prentice-Hall, Englewood Cliffs, New Jersey, 1964.
12. DOLPH, C. L., *in* "Nonlinear Problems" (R. E. Langer, ed.) pp. 13–46. Univ. Wisconsin Press, Madison, Wisconsin 1963.
13. CHANDRASEKHAR, S., "Plasma Physics." Univ. Chicago Press, Chicago, Illinois, 1960.
14. STUART, J. T., *J. Fluid Mech.* **9**, 353 (1960).
15. WATSON, J., *J. Fluid Mech.* **9**, 371 (1960).
16. STUART, J. T., *J. Aeron. Sci.* **23**, 86 (1956).
17. STUART, J. T., *Z. Angew. Math. Mech., Sonderheft*, S32 (1956).
18. BURGERS, J. M., *Advan. Appl. Mech.* **1**, 171 (1948).
19. HOPF, E., *Commun. Pure Appl. Math.* **3**, 201 (1950).
20. COLE, J. D., *Quart. Appl. Math.* **9**, 225 (1951).
21. PAI, S. I., *Proc. Midwestern Conf. Fluid Mech. 5th Ann Arbor*, p. 251 (1951).
22. BANTA, E. D., *J. Math. Anal. Appl.* **10**, 166 (1965).
23. FORSYTHE, G. E., and WASOW, W. R., "Finite Difference Methods for Partial Differential Equation," p. 141. Wiley, New York, 1960.
24. FORSYTH, A. R., *Theory of Differential Equations*, Vols. 5 and 6: "Partial Differential Equations." Dover, New York, 1959.

The Lubrication Approximation Applied to Non-Newtonian Flow Problems: A Perturbation Approach

J. R. A. PEARSON

Department of Chemical Engineering
University of Cambridge
Cambridge, Great Britain

A scheme for asymptotic-expansion solution of the equations governing flow of general elasticoviscous fluids through narrow channels between rigid boundaries is given. A dimensionless channel depth is taken as the small expansion parameter. A solution very similar to the classical lubrication approximation of Osbourne Reynolds and which uses only the first viscometric (material) function for a simple fluid is shown to be a valid first approximation, while a method for obtaining higher-order terms is outlined. The difficulties associated with open boundaries are discussed.

I. Introduction

The central problems in classical fluid mechanics reside in the nonlinearity of the equations of motion, even when the fluid is taken to be Newtonian, i.e., linear inelastic viscous. When more complicated rheological equations of state are used to represent the fluid, e.g., to take account of elasticoviscosity, these problems are made far more difficult. In classical fluid mechanics, analytic progress has usually depended on making one or more approximations, of which the most important are:

(1) The low Reynolds number or Stokes approximation; here inertia forces in the equations of motion are neglected completely and the problem becomes linear.

(2) The high Reynolds number approximation; here the flow field is supposed split into various regions, in the largest of which the motion is

supposed irrotational (the Euler approximation) or of constant vorticity, while large velocity gradients are confined to much smaller regions, such as boundary-layers, in which simplified forms of the equations of motion are relevant, e.g., the Prandtl boundary-layer equations.

(3) Geometrical symmetry approximations; here the flow patterns and therefore the boundary conditions are assumed to be two-dimensional or axisymmetric,† thus reducing the number of variables, both dependent and independent, in the equation of motion.

(4) Similarity approximations; here the flows are supposed self-similar when expressed in terms of suitably chosen similarity variables; the classical example of this is provided by the Blasius profile for boundary-layer flow, where the nonlinear aspect of the problem is reduced to one of solving a nonlinear ordinary differential equation.

Recent developments in the subject have sought to replace these simple approximate solutions by perturbation expansions, which may be regular and uniformly valid over the entire flow field or singular and nonuniformly valid. The expanding parameter, necessarily small, can be either a Reynolds number or its inverse, both of which are fixed by the boundary conditions on the flow and the physical parameters defining the fluid, or a coordinate describing position in the flow field. In the latter case, the expansions are almost certainly not uniformly valid; they often arise as a result of similarity approximations. In many situations, particularly those involving solid bodies immersed in fluid flows of infinite extent, analytic solution for successive terms in expansions depends upon a matching technique involving two or more separate expansions each valid in a different part of the flow field and with regions of overlap. These recent studies, though rather too complex to describe in any detail here (interested readers are referred to Van Dyke (1964) for a detailed treatment), have shown both that significant improvements to a first-order approximation can be obtained by taking second and even third-order terms in relevant perturbation expansions, and also that rather unexpected results are obtained when expansions are made uniformly valid.

The object in this paper is to apply the same principles to the lubrication approximation (which is concerned with flow in confined shallow channels), making a Reynolds-type solution the first term of an expansion based on

† In the case of compressible flow even one-dimensional flows can be considered.

a small parameter defined by the geometry (configuration) of the solid boundaries to the flow. Higher-order terms in the expansion will take account of (a) effects caused by the slow changes in the depth of the flow channel, (b) effects caused by the curvature of the boundary surfaces, and (c) effects caused by the elasticoviscous nature of many (non-Newtonian) fluids, though the effect of shear-dependent viscosity (under steady shear conditions) is accounted for in the lowest-order solution.

There have been previous extensions of the simple lubrication approximation, to deal with one or other of the three effects (a)–(c), but none of them has covered them all simultaneously, nor have the formal methods of perturbation expansion always been used. The work presented here can be regarded primarily as an investigation of the scope of the proposed scheme. The main result of immediate interest is concerned with the range of non-Newtonian behavior that can be naturally accommodated within the first-order term of the expansion, and it is here that a significant advance is achieved, allowing a class of problems, significant in an analysis of polymer melt processing, to be treated by numerical solution of ordinary differential equations.

2. The Lubrication Approximation

The classical theory of lubrication was developed (Fuller 1961, Pinkus and Sternlicht 1961, Tipei 1962) to explain the fluid mechanical behavior of lubricants acting as thin load-bearing films between almost-mating smooth metal surfaces moving relative to one another, usually without altering their instantaneous geometrical configuration.

The assumptions of the original theory are usually taken to be:

(1) the fluid is Newtonian, and of constant viscosity;
(2) the flow is laminar, and is dominated by viscous and pressure forces;
(3) the fluid is incompressible;
(4) the flow is steady;
(5) the pressure in the film is a function only of the coordinates measuring position on either surface, but not of the coordinate measuring position between opposing surfaces.

It is the last assumption that is characteristic of the lubrication approximation. It implies that the velocity profiles at right angles to the surfaces, and hence the mass flow along the surfaces, at any position in the film are functions only of the local pressure gradient, the local film thickness, the local surface boundary conditions, and the coefficient of viscosity. These relations together with the continuity condition and the stated pressure or mass flow boundary conditions then allow of a solution for the pressure and mass flow fields.

The justification for these five assumptions is usually that the films are very thin and so the length scales in the "plane" of the surfaces are large compared with this film thickness. Most later authors have assumed that the assumptions are independent, and have tried variously by relaxing one or other condition to account for the effects of

(a) non-Newtonian behavior; this may be due to elasticoviscosity or to temperature and pressure variations;

(b) inertia forces, even to the extent of considering turbulent flow in the films;

(c) compressibility;

(d) unsteadiness – this can include the case of oscillating fluctuations, or of the surfaces moving slowly together (squeeze films);

(e) pressure variations across the film, i.e., changes in the velocity profiles caused by variations in film thickness.

In most cases, order of magnitude arguments can be applied to compare one dynamical term with another, and so an adequate physical understanding of their relative effects can usually be achieved. However, difficulties arise (Langlois 1963a, 1964a) when the fluid is taken to be elasticoviscous, both because its "apparent" viscosity is a function of its previous deformation history and because it exhibits "normal forces" which question the validity of assumption (5). It is not that these difficulties are of a totally new kind, rather that the interdependence of the assumptions (1)–(5) become more complicated. For this reason, it proves useful to replace the more intuitive physical arguments by a formal mathematical expansion procedure.

3. Equations of State for Non-Newtonian Fluids

Rheological equations of state, in this context, are those equations relating the basic dynamical variable (the stress tensor) to the basic kinematic variables (the deformation tensor and its various time-derivatives) and certain thermodynamic field variables, such as temperature. Particular fluids are characterised by certain intrinsic tensor parameters which in general are chosen to be functions of the thermodynamic, but not the kinematic, variables, and by certain general tensor functions, which are not so restricted. Determination of these equations of state is regarded as the province of rheology, and has been approached from two fronts: that of theoretical continuum mechanics, and that of empirical model-building based on experimental observation. Despite early mathematical weaknesses in proposed models, most of the latter can now be made to conform to the strict requirements of continuum mechanics, and so they may be regarded as special cases of more general relations.

To allow of easy comparison with experiment and of feasible mathematical analysis, these empirical models are in practice made as simple as possible: specific forms for parametric functions are assumed, while expansion procedures are often employed to describe the fluid by means of sets of constants, perhaps temperature dependent. These expansion procedures limit the range of rates of deformation (or of flow patterns) to which they can be applied, and hence limit the range of non-Newtonian effects that can be predicted. On the whole, theoreticians tend to neglect observed effects that they cannot predict, and real situations that they cannot treat, while technologists tend to reject all theories that do not predict all of their observations, whatever the circumstances. This has led unfortunately to an unnecessary lack of contact, and has meant that relatively little progress has been made in applying the results of theoretical rheology to practically important nontrivial flow patterns.

For our later purposes, it is worth introducing and describing at this stage certain classes of constitutive equations (equations of state) that have been proposed by various authors.

Oldroyd (1950) was the first to elaborate a theory of materials where stress and deformation *histories*, rather than instantaneous values, are related. For isotropic materials, whose physical behavior is independent of the coordinate frame of reference used to view them, certain invariance requirements are imposed on these relations. Oldroyd chose to work with co- and contravariant tensor functions, defined both in a fixed frame of reference (x) and in a time varying frame of reference (ξ) fixed in the deforming body. In the latter frame at some time t, to quote Oldroyd

(1961), the stress history $\pi_{jl}(\xi, t')$ and temperature history $T(\xi, t')$ are related to deformation history by equations of state

$$\pi_{jl}(\xi, t') = \pi'_{jl}(\xi, t') - p''(\xi, t')\,\gamma_{jl}(\xi, t'), \tag{3.1}$$

where $p''(\xi, t')$ is an arbitrary isotropic pressure, together with an invariant integro-differential equation (or set of equations) relating the following functions of t':

$$\gamma_{jl}(\xi, t'), \qquad \pi'_{jl}(\xi, t'), \qquad T(\xi, t'), \qquad t - t'\ (t' < t).$$

Here $\gamma_{jj}(\xi, t')$ is the covariant metric tensor at time t' and the coordinate $\xi(\equiv \xi^1, \xi^2, \xi^3)$ occurs as an independent parameter. The derivatives arising are the local first, second, third etc., rate-of-strain tensors

$$\eta_{jl}{}^{(N)}(\xi, t') = \tfrac{1}{2}\partial^N \gamma_{jl}(\xi, t')/\partial t'^N \qquad (N = 1, 2, 3, \ldots). \tag{3.2}$$

Although this representation is suitable for the purposes of physical rheology, it proves unsuitable if combined with Euler's equations of motion, and so much of Oldroyd's paper is concerned with the representation of equations of state in a frame of reference (x) fixed in space. This leads to his convected derivative

$$\frac{\mathfrak{d} b^{\cdot\,\cdot\,k\,\cdot\,\cdot}_{\cdot\,i\,\cdot\,\cdot\,\cdot}}{\mathfrak{d} t} = \frac{\partial b^{\cdot\,\cdot\,k\,\cdot\,\cdot}_{\cdot\,i\,\cdot\,\cdot\,\cdot}}{\partial t} + v^m b^{\cdot\,\cdot\,k\,\cdot\,\cdot}_{\cdot\,i\,\cdot\,\cdot\,\cdot\,,\,m} + \sum v^m_{,i}\, b^{\cdot\,\cdot\,k\,\cdot\,\cdot}_{\cdot\,m\,\cdot\,\cdot\,\cdot}$$
$$- \sum{}' v^k_{,m}\, b^{\cdot\,\cdot\,m\,\cdot\,\cdot}_{\cdot\,i\,\cdot\,\cdot\,\cdot} + W v^m_{,m}\, b^{\cdot\,\cdot\,k\,\cdot\,\cdot}_{\cdot\,i\,\cdot\,\cdot\,\cdot}, \tag{3.3}$$

where $b^{\cdot\,\cdot\,k\,\cdot\,\cdot}_{\cdot\,i\,\cdot\,\cdot\,\cdot}(x, t)$ is any tensor of weight W, v^i is the covariant velocity at x^i and time t, and a comma denotes a covariant derivative. The successive rate of strain components become

$$e^{(N)}_{ik}(x, t) = \tfrac{1}{2}\, d^N g_{ik}(x)/dt^N, \tag{3.4}$$

where $g_{ik}(x)$ is the fixed metric tensor and the first of Oldroyd's equations of state, (3.1), becomes

$$p_{ik}(x, t) = p'_{ik}(x, t) - p''(x, t) g_{ik}(x), \tag{3.5}$$

p_{ik} now being the stress tensor, and p'_{ik} being given by an integro-differential equation. Oldroyd (see Oldroyd 1961) chose as the simplest plausible *model* for elasticoviscous fluids the differential form

$$p'_{ik} + \lambda_1 \mathscr{D}p'_{ik}/\mathscr{D}t + \mu_0 p'^{j}_{j} e^{(1)}_{ik} - \mu_1 (p'^{j}_{i}\, e^{(1)}_{jk} + p'^{j}_{k}\, e^{(1)}_{ij}) + \nu_1 p'^{jl}\, e^{(1)}_{jl} g_{ik}$$
$$= 2\eta_0 (e^{(1)}_{ik} + \lambda_2\, \mathscr{D}e^{(1)}_{ik}/\mathscr{D}t - 2\mu_2 e^{(1)j}_{i} e^{(1)}_{jk} + \nu_2\, \epsilon^{(1)}_{jl}\, e^{(1)jl} g_{ik}), \qquad (3.6)$$

where $\mu_0, \mu_1, \mu_2, \nu_1, \nu_2$ and η_0 are scalar physical constants and $\mathscr{D}/\mathscr{D}t$ is a material (or Jaumann) derivative closely related to $\mathfrak{d}/\mathfrak{d}t$ (see Fredrickson 1964, pp. 293–295, for details).

λ_1 and λ_2 are sometimes known as the retardation and relaxation times. Walters (1960) has extended the model to allow for a distribution of relaxation times, in the form

$$p'^{ik}(x, t) = 2\int_{-\infty}^{t} \Psi(t - t')\frac{\partial x^i}{\partial x'^m}\frac{\partial x^k}{\partial x'^r}\, e^{(1)mr}(x', t')\, dt', \qquad (3.7)$$

where $x'(x, t')$ is the position occupied at time t' by the particles occupying position x at time t, and

$$\Psi(t - t') = \int_0^{\infty} \frac{N(\tau)}{\tau}\, e^{-(t-t')/\tau}\, d\tau, \qquad (3.8)$$

where $N(\tau)$ is the spectrum of relaxation times.

Lodge (1964) has also used Oldroyd's methods to derive an equation for a rubberlike liquid in the form

$$\pi^{ij} + p\gamma^{ij} = \int_{-\infty}^{t} \mu(t - t')\gamma^{ij}(t')\, dt', \qquad \det \gamma^{ij} = \text{const}, \qquad (3.9)$$

and has discussed other extensions.

A parallel development, using different notation and emphasizing different aspects, has been based on the work of Truesdell (1952), Rivlin and Ericksen (1955), Noll (1955) and Coleman and Noll (1961). Noll (1958) introduces the concept of a simple fluid whose constitutive relation can be written

$$\mathbf{S}(t) = \underset{s=0}{\overset{\infty}{\mathbf{H}}}\,(\mathbf{C}_t(t - s);\, \rho(t)) \qquad (3.10)$$

where t is time, ρ the density, $\mathbf{H}$ is a functional, $\mathbf{S}$ the stress and $\mathbf{C}_t(\tau)$

the right Cauchy–Green tensor at time τ relative to the configuration at time t given by

$$\mathbf{C}_t(\tau) = \mathbf{F}_t^{\,T}(\tau)\mathbf{F}_t(\tau), \qquad \mathbf{F}_t(\tau) = \nabla_x \mathbf{X}_t(x, \tau), \tag{3.11}$$

$\mathbf{X}_t(\mathbf{x}, \tau)$ being the displacement function giving the position at time τ of the material point having the position $\mathbf{x}$ at time t. For the case of slow motions, where the slowness is measured in terms of a time-scaling parameter α, Coleman and Noll (1960) introduce the notion of successive approximations, from which they derive a series of models termed *Nth*-order fluids of which the first three are given by

$$\mathbf{S} = -p\mathbf{I} + \eta\mathbf{A}_1, \tag{3.12.1}$$

$$\mathbf{S} = -p\mathbf{I} + \eta\mathbf{A}_1 + \beta\mathbf{A}_1^{\,2} + \gamma\mathbf{A}_2, \tag{3.12.2}$$

$$\mathbf{S} = -p\mathbf{I} + \eta\mathbf{A}_1 + \beta\mathbf{A}_1^{\,2} + \gamma\mathbf{A}_2 + \delta\mathbf{A}_3 + \varepsilon(\mathbf{A}_1\mathbf{A}_2 + \mathbf{A}_2\mathbf{A}_1), \tag{3.12.3}$$

where η, β, ..., ε are constants and

$$\mathbf{A}_k = (d^k\mathbf{C}_t\,(\tau)/d\tau^k)_{\tau=t}\,. \tag{3.13}$$

Much of this is implicit in the work of Rivlin and Ericksen (1955) who considered the general form

$$\mathbf{S} = -p\mathbf{I} + \mathbf{F}(\mathbf{A}_1, \mathbf{A}_2, \cdots, \mathbf{A}_N), \tag{3.14}$$

where $\mathbf{F}$ is a polynomial in the $\mathbf{A}_i$ with coefficients functions of their invariants and joint invariants. The particular model they suggested explicitly was the one relevant for $\mathbf{A}_i = 0$, $i > 2$ (the case corresponding to uniform steady simple shear flow),

$$\begin{aligned}\mathbf{S} = {} & -p\mathbf{I} + 2\alpha_1\mathbf{A}_1 + 4\alpha_2\mathbf{A}_1^{\,2} + 2\alpha_3\mathbf{A}_2 + 4\alpha_4\mathbf{A}_2^{\,2} \\ & + 4\alpha_5(\mathbf{A}_1\mathbf{A}_2 + \mathbf{A}_2\mathbf{A}_1) + 8\alpha_6(\mathbf{A}_1^{\,2}\mathbf{A}_2 + \mathbf{A}_2\mathbf{A}_1^{\,2}) \\ & + 8\alpha_7(\mathbf{A}_1\mathbf{A}_2^{\,2} + \mathbf{A}_2^{\,2}\mathbf{A}_1) + 16\alpha_8(\mathbf{A}_1^{\,2}\mathbf{A}_2^{\,2} + \mathbf{A}_2^{\,2}\mathbf{A}_1^{\,2}),\end{aligned} \tag{3.15}$$

the 8 α's being functions of the ten invariants derived from $\mathbf{A}_1$ and $\mathbf{A}_2$.

We are not here concerned with the detailed arguments that relate to the relative merits of these various models (Oldroyd for example suggests that the explicit appearance of $\mathbf{S}(p_{ik})$ in the general forms (3.10) and (3.14) restricts the class of fluids, or of flows, to which they can be applied, and infers that $\mathbf{S}$ should be an argument of the functional $\mathbf{H}$) but are more concerned with detailing those models that have been, and are likely to be, employed in the analysis of flow problems, so that we can try to incorporate them into our general perturbation or approximation scheme. There is one important simplification that arises for quite general fluids undergoing steady simple shear, i.e., when the deformation tensor can be constructed from a single scalar constant, k say; the stress tensor can then be represented, apart from the isotropic component, by three material functions, scalar functions of k. These are the viscosity and the two normal stress differences, and are the functions that have been most earnestly sought experimentally. We shall see that a knowledge of the viscosity functions alone is sufficient for us to obtain our first-order lubrication approximation. More detailed rheological information is only necessary for higher-order approximations.

4. Perturbation and Iterative Solution Scheme

4.1. Flow Field Geometry

The essential point about the proposed perturbation scheme is that it begins with, and is determined by, the geometry of the flow region. This is in keeping with the original concept of the lubrication approximation; the only difference here is that a class of related flow regions are considered, each with its (related) coordinate system (frame of reference). For comprehensiveness, general coordinate systems will be used and the terminology of co- and contravariant tensors will be employed.

We start with an actual physical flow channel of interest, which is normally the volume between bounded smooth surfaces S_1 and S_2. S_1 and S_2 are to be thought of as "close together" and are such that no part of S_1 is far from S_2 and vice versa, this being the characteristic suggesting a lubrication approximation. We call S_0 the center surface of the flow region, where S_0 is such that perpendiculars erected from any point P on S_0 to S_1 and S_2 are of equal length; this length we shall call $h(P)$ and is half the local 'width' of the channel. The boundary curve of S_0 will be called L_0. In order to attach any precise meaning to the concept of *smooth* surfaces *close* together we need a length scale, l_0 say, with which to compare h. The obvious lengths that suggest themselves are the lateral dimensions of S_0, and its radii of curvature. We shall suppose that the radii of curvature are of at least the same order as the lateral dimensions, and take a representative value of the latter to be l_0. We now interpret the closeness of S_1 and S_2 as meaning that h/l_0 is everywhere $o(1)$, and the smoothness of S_1 and S_2 as further meaning that $|\text{grad}^*h|$ is also $o(1)$ where grad*h represents the local vectorial rate of variation of h on S_0; in simpler terms this means that the change in h over any distance l measured on S_0 is always very much smaller than l.

The next step is to choose a coordinate system for the flow region. On S_0, we choose an orthogonal set of coordinate lines (x^1, x^3) subject only to the requirement that $h_1(x^1, x^3)$ and $h_3(x^1, x^3)$ be everywhere $O(1)$. We assume that we can do this because of the smoothness of S_1. Here h_1 and h_3 are given by the usual metrical relation

$$ds^2 = h_1{}^2(dx^1)^2 + \{h_3{}^2(dx^3)\}^2, \tag{4.1}$$

which is the relevant degenerate form of

$$ds^2 = g_{ij}x^i x^j, \tag{4.2}$$

where g_{ij} is the metric tensor. If ds and dx^1, dx^3 have the dimension of length then h_1, h_3 are dimensionless whereas if dx^1, dx^3 are thought of as dimensionless then h_1, h_3 must have the dimension of length. This rather trivial duality is significant later in alternative visualization of the asymptotic procedure. The coordinate lines x^2 are now chosen to be straight and perpendicular to S_0. To each point P, i.e., $(x_p{}^1, x_p{}^3)$, there corresponds an $h(x^1, x^3)$; we now choose $g_{22}(x^1, x^2, x^3)$ to be

$$g_{22}(x^1, x^2, x^3) = \{h(x^1, x^3)\}^2, \tag{4.3}$$

independent of x^2. This makes S_1 and S_2 the coordinate surfaces $x^2 = \pm 1$. Furthermore, because of the smallness of h compared with the radii of curvature of S_0 (and of S_1 and S_2), we can now associate a unique value $(x_R{}^1, x_R{}^2, x_R{}^3)$ with every point R of the flow region F (which is bounded by the surfaces S_1 and S_2 and the x^2 lines through L_0†) and vice versa

† There is the difficulty that the original physical choice of S_1 and S_2 generally must lead to a surface S_0 and a boundary L_0 such that the x^2 lines through L_0 will not intersect both S_1 and S_2. However, a small reduction in area of the surface S_0 and change in L_0 will remove this difficulty; F will then be contained within the original flow region.

provided that the x's lie within certain obvious limits. Thus we have defined a specific nonorthogonal coordinate system, x say. It will be noted that on S_0, this system is "locally" orthogonal, and that for "small" h, it will be "nearly" orthogonal within F. Furthermore, $g_{ik}(x)$ will be prescribed every where, because of the relation (4.2). If we choose our original length scale to be h_0 (and not some l_0), then we can arrange for all the diagonal elements of g_{ik} to be $O(1)$, by suitable "numbering" of the x^1, x^3 lines on S_0. This makes the derivatives $\partial g_{ik}/\partial x^1$ and $\partial g_{ik}/\partial x^3$ evaluated on S_0 small, i.e., $o(1)$; the vanishing of the Riemann–Christoffel tensor (see e.g., Spain 1960, p. 50) implies that all the derivaties $\partial g_{ik}/\partial x^j$ will be small within F, and hence that the Taylor expansion

$$g_{ik}(x^1, x^2, x^3) = [g_{ik}]_0 + x^2[\partial g_{ik}/\partial x^2]_0 + \tfrac{1}{2}(x^2)^2 [\partial^2 g_{ik}/\partial (x^2)^2]_0 + \cdots \tag{4.4}$$

holds, where $[\ \ ]_0$ refers to the value at $(x^1, 0, x^3)$.

The last step is to consider a set of flow regions F_α and coordinate systems ξ_α, that are obtained by holding the surface S_0 "constant" and considering the set of functions $\alpha h(P)$, where $\{\alpha\}$ is a sequence tending to zero. Because the relevant length scale changes from h_0 to αh_0 the coordinate numbers associated with a point P on S_0 will change from

$$x_p^{\ i} \quad \text{to} \quad \xi^i_{\alpha p} = \alpha^{-1} x_p^{\ i}, \qquad (i = 1, 3) \tag{4.5}$$

if the same network of $(\xi_\alpha^{\ 1}, \xi_\alpha^{\ 3})$ lines is maintained on S_0. By repeating the construction of the x coordinate system, we obtain a ξ_α coordinate system whose metric $g_{\alpha ij}$ has the similarity property

$$g_{\alpha ij}(\xi_\alpha^{\ 1}, 0, \xi_\alpha^{\ 3}) = g_{ij}(\alpha\xi_\alpha^{\ 1}, 0, \alpha\xi_\alpha^{\ 3}) \tag{4.6}$$

or

$$[\partial g_{\alpha ij}/\partial \xi_\alpha^{\ k}]_0 = \alpha[\partial g_{ij}/\partial x^k]_0, \tag{4.7}$$

the $[\ \ ]_0$ referring to the same point on S_0.

The surfaces $S'_{\alpha 2}$ and $S'_{\alpha 1}$ given by $\xi_\alpha^{\ 2} = \pm 1$ will be in every case the boundaries of the flow field. Viewed from the original coordinate system x, with length scale h_0, the pair of surfaces $S'_{\alpha 1}$ and $S'_{\alpha 2}$ coincide in the limit $\alpha \to 0$. However, viewed from the changing coordinate system ξ_α with length scale αh_0 and with $(\xi_\alpha^{\ 1}, \xi_\alpha^{\ 3})$ held constant, the pair of surfaces tend in the limit to a pair of parallel planes two units distance apart. It is the latter point of view which is the relevant one. It will be noted that

Eq. (4.7) suggests that we could have relaxed the initial requirement that $[\partial h/\partial x^j]_0$ be small; the reason that we did not was that it was necessary to defined our coordinate system uniquely and we shall aim to use the asymptotic expansion in α (to be developed later) as the basis for an approximation to the physical flow of interest, i.e., we shall want it to be useful up to α of order unity.

4.2. Governing Equations and Boundary Conditions

The equations to be satisfied by the field variables are (e.g., Aris 1962)

the equation of continuity

$$\partial \rho/\partial t + (\rho v^i),_i = 0, \tag{4.8}$$

the equations of motion

$$\partial v^i/\partial t + v^j v^i,_j = f^i + p^{ij},_j, \tag{4.9}$$

the equations of state (see Section 3),

where, as before, ρ is the fluid density, t is time, v^i is the contravariant velocity vector, p^{ij} is the doubly contravariant stress tensor, a comma denotes covariant differentiation and f^i is the contravariant body force. These are not affected by the transformation $x \to \xi_\alpha$ except by the addition of a subscript α.

The boundary conditions to be satisfied need more careful definition. We shall suppose that in the original, physical problem the velocity of the boundary surfaces S_1 and S_2 is everywhere prescribed, and that by use of the no-slip boundary condition this specifies the boundary values $v^i_{S_1}$, $v^i_{S_2}$ for the fluid. Furthermore, we shall suppose that on L_0, the boundary curve of the surface S_0, we can prescribe either

(i) the volume flux V across L_0 between S_1 and S_2, where†

$$V = h \int_{-1}^{1} v^i n_i \, dx^2, \tag{4.10}$$

n_i being the unit normal to the surface formed by the x^2 lines through L_0, or

† In most cases we have $V = 0$.

(ii) the hydrostatic pressure p above (or below) some ambient pressure p_0,

or some combination of the two.

There is no need at this stage to make these time-independent. As we carry out the α-limit procedure described in Section 4.1 for the *geometry* of the flow field, so we must attempt to impose such boundary conditions as will retain the desired *dynamical* relationships between the set of flows. In particular, because we are concerned with equations of state involving rate-of-deformation histories, we shall aim to keep rates-of-strain constant. Thus for the α-flows, we have $v_{\alpha i}$ prescribed on $S'_{\alpha 1}$ and $S'_{\alpha 2}$, or

$$v_{\alpha i}(\xi_\alpha{}^1, \pm 1, \xi_\alpha{}^3) = v_{S_{1,2}i}(\alpha\xi_\alpha{}^1, \alpha\xi_\alpha{}^3), \tag{4.11†}$$

and on L_0 or

$$g_{L_0}(\alpha\xi_\alpha{}^1, \alpha\xi_\alpha{}^3) = 0, \text{ say}, \tag{4.12}$$

we have either

(i) the 'volume flux' vector prescribed, i.e.,

$$\int_{-1}^{1} (v_\alpha{}^i\, n_{\alpha i})\, d\xi_\alpha{}^2 = V/h = \bar{V}_\alpha \tag{4.13}$$

or

(ii) the "pressure" P_α above some ambient pressure $P_{\alpha 0}$ prescribed, where

$$P_\alpha = \alpha p_\alpha; \tag{4.14}$$

for convenience we have taken $P_{\alpha 0} = p_0$.

The last condition introduces no difficulties if the fluid is incompressible. If it is compressible, then the requisite "similarity" between the α flows can obviously only be achieved if the compressibility decreases linearly with α, and the density $\rho_{\alpha 0}$ at the reference state $P_{\alpha 0}$ remains constant. Most of the equations of state used for non-Newtonian liquids assume incompressibility, so this restriction is not severe in practice. However, even if p did figure in the equations of state, the transformation (4.14) could be accommodated by taking a sequence of equations of state showing

† We take the covariant form so as to be able to put $v_{S_{1,2}2} = 0$ later as the condition for steadiness and fixed geometry.

the necessary α scaling for p. Langlois (1963a) among others has pointed out that normal stress, rather than hydrostatic pressure, is the more sensible variable to prescribe at boundaries—indeed when calculating bearing loads in conventional lubrication theory this is evidently so; however, as we shall see later, the perturbation scheme inevitably makes the hydrostatic component P_α dominant in the limit, as is already suggested by the scaling (4.14). In simple terms we are saying that provided the flow field is such that a lubrication approximation is relevant, then normal forces cannot be significant in a first approximation (except in the rather artificial case where normal forces are much larger than shear forces).

The last point concerns the behavior of $f_\alpha^{\,1}$: this, to be compatible with "similar" dynamical behavior of the α flows, must scale such that

$$f_\alpha^i(\xi_\alpha^{\,1}, 0, \xi_\alpha^{\,3}) = f^i(\alpha\xi_\alpha^{\,1}, 0, \alpha\xi_\alpha^{\,3}) \tag{4.15.1}$$

where f^i is derived from the original physical problem. The dependence on $\xi_\alpha^{\,2}$ is not so crucial.

4.3. Expansions for Field Variables

We look for general asymptotic expansions of the form

$$y = F_0(\alpha)y^{(0)} + F_1(\alpha)y^{(1)} + F_2(\alpha)y^{(2)} + \cdots + o(F_N(\alpha)), \tag{4.16}$$

where y is a typical field variable, and $F_S(\alpha)$ are dimensionless functions with the property

$$\frac{F_{S+1}(\alpha)}{F_S(\alpha)} \to 0 \qquad \text{as} \quad \alpha \to 0 \qquad (S = 0, 1, 2 \ldots) \tag{4.17}$$

and $y^{(i)}$ are functions of η_α (see below). By taking a finite number N of terms of the expansion (4.16) we hope to obtain an approximation over most, if not all, the flow field valid to order $F_N(\alpha)$.

In the first instance, we try the simplest form for $F_S(\alpha)$, notably $(\alpha)^s$, though we must be prepared in special cases to have other functional forms arise. Thus we set

$$v_\alpha^{\,i} = v^{(0)i}(\eta_\alpha) + \alpha v^{(1)i}(\eta_\alpha) + \alpha^2 v^{(2)i}(\eta_\alpha) + \cdots, \tag{4.18.1}$$

$$p_\alpha = \alpha^{-1}p^{(-1)}(\eta_\alpha^*) + p^{(0)}(\eta_\alpha) + \alpha p^{(1)}(\eta_\alpha) + \cdots, \tag{4.18.2}$$

$$g_{ij} = g_{ij}^{(0)}(\eta_\alpha^*) + \alpha g_{ij}^{(1)}(\eta_\alpha) + \alpha^2 g_{ij}^{(2)}(\eta_\alpha) + \cdots, \tag{4.18.3}$$

where $g_{ij}^{(0)}$ is of diagonal form,

$$p'_{ik} = p'^{(0)}_{ik}(\eta_\alpha) + \alpha p'^{(1)}_{ik}(\eta_\alpha) + \cdots, \tag{4.18.4}$$

and where necessary

$$\rho_\alpha = \rho^{(0)}(\eta_\alpha^*) + \alpha\rho^{(1)}(\eta_\alpha) + \alpha^2\rho^{(2)}(\eta_\alpha) - \cdots, \tag{4.18.5}$$

$$f_\alpha^{\ i} = f^{(0)i}(\eta_\alpha) + \alpha f^{(1)i}(\eta_\alpha) + \alpha^2 f^{(2)i}(\eta_\alpha) - \cdots, \tag{4.18.6}$$

where η_α is the stretched variable given by

$$(\eta_\alpha^{\ 1}, \eta_\alpha^{\ 2}, \eta_\alpha^{\ 3}) = (\alpha\xi_\alpha^{\ 1}, \xi_\alpha^{\ 2}, \alpha\xi_\alpha^3) \tag{4.19}$$

and η_α^* is the two-dimensional variable given by $(\eta_\alpha^{\ 1}, \eta_\alpha^{\ 3})$. The leading term in the p-expansion takes the form it does because αp is the variable that would fit the expansion scheme (4.16). The metric tensor g_{ik} can be written in the form (4.18.3) by means of a relatively trivial use of Eq. (4.4). We note that raising and lowering of suffices on any term $a^{(s)\cdot\cdot i\cdot\cdot}_{\cdot\cdot j\cdot\cdot}$ cannot be effected in the usual fashion because the metric tensor is itself an expansion in α. The precise form of the expansions (4.18) has been chosen to correspond to the boundary conditions (4.11)–(4.14), or

$$v_{\alpha 1}(\xi_\alpha^{\ 2} = \pm 1) = v_{S_{1,2}i}(\eta_\alpha^*) \tag{4.20}$$

and either

$$\int_{-1}^{1} v_\alpha^{\ i} n_{\alpha i}\, d\xi_\alpha^{\ 2} = \bar{V}_\alpha(\eta_\alpha^*) \tag{4.21.1}$$

or (ii)

$$\alpha p_\alpha = P_\alpha(\eta_\alpha^*) \tag{4.21.2}$$

on

$$g_{L_0}(\eta_\alpha^*) = 0. \tag{4.22}$$

For the present, these expansions are to be regarded as conjectural. As is usual in similar situations their suitability is to be determined by whether or not a feasible scheme of successive solution for the various terms $v^{(s)i}$, $p^{(s-1)}$, etc. ($S = 0, 1, 2, \ldots$) can be described. We may remark in advance that we expect $v^{(0)2}$ to be zero for steady conditions.

In all the above, we have not shown an explicit dependence upon time.

In what follows we shall for simplicitly assume steady incompressible conditions and then append some further remarks to cover the case of time-varying and compressible flows.

4.4. Solution Scheme

This is based on the straightforward principle of substituting the expansions (4.18) into the Eqs. (4.8), (4.9), and one of the sets of equations of state given in Section 3, and equating terms corresponding to successively higher powers of α. Some care must be exercised in choosing boundary conditions for various terms, whilst it is not immediately obvious in what order the various differential equations obtained should be solved. It must also be shown that substitution of the chosen expansions into the equations yields only powers and not other functions of α.

First we observe that covariant derivatives can be written in the form (e.g., Aris 1962)

$$\phi_{\alpha,i} = \frac{\partial \phi_\alpha}{\partial \xi_\alpha{}^i}, \tag{4.23.1}$$

$$a^i_{\alpha,j} = \frac{\partial a_\alpha{}^i}{\partial \xi_\alpha{}^j} + \left\{ \begin{matrix} i \\ j\,k \end{matrix} \right\}_\alpha a_\alpha{}^k, \tag{4.23.2}$$

$$b^{ik}_{\alpha,j} = \frac{\partial b^{ik}_\alpha}{\partial \xi_\alpha{}^j} + \left\{ \begin{matrix} i \\ p\,j \end{matrix} \right\}_\alpha b^{pk}_\alpha + \left\{ \begin{matrix} k \\ p\,j \end{matrix} \right\}_\alpha b^{ip}_\alpha, \tag{4.23.3}$$

where

$$\left\{ \begin{matrix} i \\ j\,k \end{matrix} \right\}_\alpha = \left\{ \begin{matrix} i \\ k\,j \end{matrix} \right\}_\alpha = \frac{1}{2} g^{il}_\alpha \left[\frac{\partial g_{\alpha lj}}{\partial \xi_\alpha{}^k} + \frac{\partial g_{\alpha lk}}{\partial \xi_\alpha{}^j} - \frac{\partial g_{\alpha jk}}{\partial \xi_\alpha{}^l} \right]. \tag{4.24}$$

We also recall that

$$a^i = g^{ij} a_j \tag{4.25}$$

is the typical way in which indices can be raised or lowered. By use of the expression (4.18.3) and transforming to the coordinate system η_α, we see that all the terms in (4.24) are of order α; indeed we can write

$$\left\{ \begin{matrix} i \\ j\,k \end{matrix} \right\}_\alpha = \alpha \left\{ \begin{matrix} i \\ j\,k \end{matrix} \right\}^{(1)} + \alpha^2 \left\{ \begin{matrix} i \\ j\,k \end{matrix} \right\}^{(2)} + \cdots . \tag{4.26}$$

Thus (4.23) can be rewritten

$$\phi_{\alpha,i} = \delta_i^{\,2}\frac{\partial\phi_\alpha}{\partial{\eta_\alpha}^2} + \alpha\left(\frac{\partial\phi_\alpha}{\partial{\eta_\alpha}^i} - \delta_i^{\,2}\frac{\partial\phi_\alpha}{\partial{\eta_\alpha}^2}\right) + \cdots, \tag{4.27.1}$$

$$a^i_{\alpha,j} = \delta_j^{\,2}\frac{\partial {a_\alpha}^i}{\partial{\eta_\alpha}^2} + \alpha\left[\left(\frac{\partial {a_\alpha}^i}{\partial{\eta_\alpha}^j} - \delta_j^{\,2}\frac{\partial {a_\alpha}^i}{\partial{\eta_\alpha}^2}\right) + \left\{ {i \atop j\,k} \right\}^{(1)} {a_\alpha}^k\right] + \cdots, \tag{4.27.2}$$

$$b^{ik}_{\alpha,j} = \delta_j^{\,2}\frac{\partial b^{ik}_\alpha}{\partial{\eta_\alpha}^2} + \alpha\left[\left(\frac{\partial b^{ik}_\alpha}{\partial{\eta_\alpha}^j} - \delta_j^{\,2}\frac{\partial b^{ik}}{\partial{\eta_\alpha}^2}\right) + \left\{ {i \atop p\,j} \right\}^{(1)} b^{pk}_\alpha + \left\{ {k \atop p\,j} \right\}^{(1)} b^{ip}_\alpha\right] + \cdots. \tag{4.27.3}$$

Next, we observe that the rate-of-strain tensor

$$e^{(1)}_{\alpha ij} = \tfrac{1}{2}(v_{\alpha i,j} + v_{\alpha j,i}), \tag{4.28}$$

which was first introduced in Section 3, Eq. (3.4), and is better known in the form (4.28) (see, for example, Fredrickson 1964), can be written

$$\begin{aligned} e^{(1)}_{\alpha ij} = \frac{1}{2}\left[\frac{\partial v_i^{(0)}}{\partial{\eta_\alpha}^2}\delta_j^{\,2} + \frac{\partial v_j^{(0)}}{\partial{\eta_\alpha}^2}\delta_i^{\,2}\right] + \frac{1}{2}\alpha\Bigg[\left(\frac{\partial v_i^{(0)}}{\partial{\eta_\alpha}^j} - \delta_j^{\,2}\frac{\partial v_i^{(0)}}{\partial{\eta_\alpha}^2}\right) \\ + \left(\frac{\partial v_j^{(0)}}{\partial{\eta_\alpha}^i} - \delta_i^{\,2}\frac{\partial v_j^{(0)}}{\partial{\eta_\alpha}^2}\right) - 2\left\{ {k \atop i\,j} \right\}^{(1)} v_k^{(0)} \\ + \left(\frac{\partial v_i^{(1)}}{\partial{\eta_\alpha}^2}\delta_j^{\,2} + \frac{\partial v_j^{(1)}}{\partial{\eta_\alpha}^2}\delta_i^{\,2}\right)\Bigg] + O(\alpha^2); \end{aligned} \tag{4.29}$$

similarly the vorticity tensor

$$w_{\alpha ij} = \tfrac{1}{2}(v_{i,j} - v_{j,i}) \tag{4.30}$$

can be expressed as a series of terms multiplying successively higher powers of α. It further follows from relations (4.26) and (4.27) that the derivatives $\mathfrak{d}/\mathfrak{d}t$ (see Eq. (33)) and $\mathcal{D}/\mathcal{D}t$ operating on the variables (4.18) or (4.28) also yield a series of terms multiplying successively higher powers of α, which can be truncated as in the expansions (4.18).

In order to deal with equations of state like (3.7) or (3.10), it is necessary to evaluate terms like $\delta\xi'^i_\alpha/\delta{\xi_\alpha}^j$, where ${\xi_\alpha}'$ is the position occupied at time t' by the fluid particle occupying the position ξ_α at time t. To do this we must use the equations

$$d\xi'^i_\alpha/ds = -v'^i_\alpha({\xi_\alpha}'), \tag{4.31}$$

where $s = t - t'$, and positive values of s will be relevant. We shall suppose that

$$\xi_\alpha'^i = \xi_\alpha{}^i + \xi'^{(0)i}(\xi_\alpha, s) + \alpha\xi'^{(1)i}(\xi_\alpha, s) + \cdots + o(\alpha^N) \tag{4.32}$$

and we shall use the Taylor expansion

$$v_\alpha'^i = v_\alpha{}^i + (\xi_\alpha'^j - \xi_\alpha{}^j)\frac{\partial v_\alpha{}^i}{\partial \xi_\alpha{}^j} + \frac{1}{2}(\xi_\alpha'^j - \xi_\alpha{}^j)(\xi_\alpha'^k - \xi_\alpha{}^k)\frac{\partial^2 v_\alpha{}^i}{\partial \xi_\alpha{}^j\,\partial \xi_\alpha{}^k} + \cdots \tag{4.33.1}$$

and (4.18.1) to get the elaborate expression

$$\begin{aligned} v_\alpha'^i = {} & v^{(0)i} + \xi'^{(0)2}\frac{\partial v^{(0)i}}{\partial \eta_\alpha{}^2} + \frac{1}{2}(\xi'^{(0)2})^2\frac{\partial^2 v^{(0)i}}{(\partial \eta_\alpha{}^2)^2} + \cdots \\ & + \alpha\Bigg[v^{(1)i} + \xi'^{(0)2}\frac{\partial v^{(1)i}}{\partial \eta_\alpha{}^2} + \frac{1}{2}(\xi'^{(0)2})^2\frac{\partial^2 v^{(1)i}}{(\partial \eta_\alpha{}^2)^2} + \cdots \\ & + \xi'^{(0)j}\left(\frac{\partial v^{(0)i}}{\partial \eta_\alpha{}^j} - \delta_j{}^2\frac{\partial v^{(0)i}}{\partial \eta_\alpha{}^2}\right) + \xi'^{(0)2}\xi'^{(0)j} \\ & \times\left(\frac{\partial^2 v^{(0)i}}{\partial \eta_\alpha{}^2\,\partial \eta_\alpha{}^j} - \delta_j{}^2\frac{\partial^2 v^{(0)2}}{(\partial \eta_\alpha{}^2)^2}\right) + \cdots \\ & + \xi'^{(1)2}\frac{\partial v^{(0)i}}{\partial \eta_\alpha{}^2} + \xi'^{(1)2}\xi'^{(0)2}\frac{\partial^2 v^{(0)i}}{(\partial \eta_\alpha{}^2)^2} + \cdots\Bigg] + O(\alpha^2), \end{aligned} \tag{4.33.2}$$

where all terms are evaluated at ξ_α. The next stage is to substitute Eqs. (4.32) and (4.33.2) into (4.31) to obtain a set of first-order differential equations for the $\xi'^{(M)i}$, of which the first three are given by

$$\begin{aligned} \frac{d\xi'^{(0)i}}{ds} &= -\left[v^{(0)i} + \xi^{1(0)2}\frac{\partial v^{(0)i}}{\partial \eta_\alpha{}^2} + \frac{1}{2}(\xi'^{(0)2})^2\frac{\partial^2 v^{(0)i}}{(\partial \eta_\alpha{}^2)^2} + \cdots\right] \\ &= v^{(0)i}(\eta_\alpha{}^1, \eta_\alpha{}^2 + \eta'^{(0)2}, \eta_\alpha{}^3). \end{aligned} \tag{4.34.1}$$

At first sight, $\xi'^{(0)2}$ seems to be involved implicitly in the Eqs. (4.34.1). However, we shall see (Eq. (4.40)) that $v^{(0)2} \equiv 0$ and so, because $\xi'^{(1)2}(s = 0) = 0$, $\xi'^{(0)2} \equiv 0$. This allows (4.34.1) to be integrated directly to give

$$\xi'^{(0)i} = -v^{(0)i}s. \tag{4.35.1}$$

The next three equations are given by

$$\frac{d\xi'^{(1)i}}{ds} = -\left[v^{(1)i} + \xi'^{(1)2}\frac{\partial v^{(0)i}}{\partial \eta_\alpha{}^2} + \xi'^{(0)j}\left(\frac{\partial v^{(0)i}}{\partial \eta_\alpha{}^j} - \delta_j{}^2\frac{\partial v^{(0)i}}{\partial \eta_\alpha{}^2}\right)\right], \tag{4.34.2}$$

which can be integrated to give

$$\xi'^{(1)i} = \left[-v^{(1)i}s + \frac{1}{2}s^2\left\{v^{(1)2}\frac{\partial v^{(0)i}}{\partial \eta_\alpha{}^2} + v^{(0)j}\left(\frac{\partial v^{(0)i}}{\partial \eta_\alpha{}^j} - \delta_j{}^2\frac{\partial v^{(0)i}}{\partial \eta_\alpha{}^2}\right)\right\}\right]. \tag{4.35.2}$$

Clearly integrals for higher-order terms can similarly be obtained. From (4.35), we obtain directly functions like $\delta\xi_\alpha'^i/\delta\xi_\alpha{}^j$ (i.e., $\mathbf{F}_i(\tau)$ in (3.11)) in the form of expansions in α, and hence the $\mathbf{C}_t(\tau)$ of Eq. (3.11).

Turning to the various equations of state given in Section 3, we see that for Oldroyd's model (3.6), using the expansion (4.18.4) and the relations developed above, the various $p_{ik}^{(N)}(\eta_\alpha)$ can be expressed in terms of the $v^{(M)i}(\eta_\alpha)$, $M = 0, 1, 2, \ldots, N$; similarly for the Walters model (3.7) using (4.35) and (4.29), since $e^{(1)mt}$ (ξ_α', t') can be written as an expansion of the form (4.33). The Coleman and Noll fluids given by (3.12), (3.13) are immediately in the right form as is the Rivlin and Ericksen fluid given by (3.14) or its special case (3.15). It is however not immediately clear that the general relation (3.10) for a simple fluid, even with ρ constant, leads to such a relation: for it to be so our functional $\mathbf{H}$ must be such that

$$\underset{s=0}{\overset{\infty}{\mathbf{H}}}(\mathbf{g}^{(0)} + \alpha\mathbf{g}^{(1)} + O(\alpha^2)) = \underset{s=0}{\overset{\infty}{\mathbf{H}}}(\mathbf{g}^{(0)}) + \alpha\underset{s=0}{\overset{\infty}{\overline{\mathbf{H}}}}(\mathbf{g}^{(0)}, \mathbf{g}^{(1)}) + O(\alpha^2), \tag{4.36}$$

and so on for higher approximations. We shall henceforth assume that this is so. We observe finally that raising and lowering of suffices does not affect the general argument.

Armed with the above results we can now convert our equations (4.8), (4.9), and the relevant equations of state into a series of equations, each independent of α. From the continuity equation (4.8) written in the form

$$v_{\alpha,i}^i = 0 \tag{4.37}$$

to correspond to the case of steady incompressible flow, we get

$$\frac{\partial v^{(0)2}}{\partial \eta_\alpha{}^2} = 0, \tag{4.38.0}$$

$$\frac{\partial v^{(1)2}}{\partial \eta_\alpha{}^2} + \frac{\partial v^{(0)1}}{\partial \eta_\alpha{}^1} + \frac{\partial v^{(0)3}}{\partial \eta_\alpha{}^3} + \begin{Bmatrix} i \\ i\ k \end{Bmatrix}^{(1)} v^{(0)k} = 0, \tag{4.38.1}$$

$$\vdots$$

$$\frac{\partial v^{(M)2}}{\partial \eta_\alpha{}^2} + \frac{\partial v^{(M-1)1}}{\partial \eta_\alpha{}^1} + \frac{\partial v^{(M-1)3}}{\partial \eta_\alpha{}^3} + \left(\begin{Bmatrix} i \\ i\ k \end{Bmatrix}^{(M)} v^{(0)k} + \begin{Bmatrix} i \\ i\ k \end{Bmatrix}^{(M-1)} v^{(1)k} + \cdots + \begin{Bmatrix} i \\ i\ k \end{Bmatrix}^{(1)} v^{(M-1)k} \right) = 0. \tag{4.38.M}$$

The relevant boundary conditions for these will be, again appealing to the steady flow condition,

$$v_i^{(M)}(\eta_\alpha{}^2 = \pm 1) = 0 \qquad (M = 1, 2 \cdots),$$

with

$$\begin{aligned} v_i^{(0)}(\eta_\alpha{}^2 = \pm 1) &= v_{S_{1,2}i} \qquad (i = 1, 3); \\ v_2^{(0)}(\eta_\alpha{}^2 = \pm 1) &= 0. \end{aligned} \tag{4.39}$$

From (4.39) and (4.18.3) the conditions on $v^{(M)i}$ can be obtained. From (4.38.0) and (4.39) we can immediately deduce that

$$v^{(0)2} \equiv 0, \tag{4.40†}$$

a familiar and significant property of lubrication flows, and an essential feature in our solution scheme. Using (4.40) and (4.29), we find that, in matrix form,

$$e_{ij}^{(1)(0)} = \begin{bmatrix} 0 & \dfrac{1}{2}\dfrac{\partial v_1^{(0)}}{\partial \eta_\alpha{}^2} & 0 \\ \dfrac{1}{2}\dfrac{\partial v_1^{(0)}}{\partial \eta_\alpha{}^2} & 0 & \dfrac{1}{2}\dfrac{\partial v_3^{(0)}}{\partial \eta_\alpha{}^2} \\ 0 & \dfrac{1}{2}\dfrac{\partial v_3^{(0)}}{\partial \eta_\alpha{}^2} & 0 \end{bmatrix} \tag{4.41.0}$$

† This follows because $g_{ij}^{(0)}$ is diagonal.

$$e_{ij}^{(1)(1)} = \begin{bmatrix} 0 & \frac{1}{2}\frac{\partial v_1^{(1)}}{\partial \eta_\alpha{}^2} & 0 \\ \frac{1}{2}\frac{\partial v_1^{(1)}}{\partial \eta_\alpha{}^2} & 0 & \frac{1}{2}\frac{\partial v_3^{(1)}}{\partial \eta_\alpha{}^2} \\ 0 & \frac{1}{2}\frac{\partial v_3^{(1)}}{\partial \eta_\alpha{}^2} & 0 \end{bmatrix} + \tilde{e}_{ij}^{(1)(1)}, \tag{4.41.1}$$

where

$$\begin{aligned}
\tilde{e}_{11}^{(1)(1)} &= \frac{\partial v_1^{(0)}}{\partial \eta_\alpha{}^1} - \begin{Bmatrix} 1 \\ 1\ 1 \end{Bmatrix}^{(1)} v_1^{(0)} - \begin{Bmatrix} 3 \\ 1\ 1 \end{Bmatrix}^{(1)} v_3^{(0)}, \\
\tilde{e}_{21}^{(1)(1)} = \tilde{e}_{12}^{(1)(1)} &= -\left[\begin{Bmatrix} 1 \\ 1\ 2 \end{Bmatrix}^{(1)} v_1^{(0)} + \begin{Bmatrix} 3 \\ 1\ 2 \end{Bmatrix}^{(1)} v_3^{(0)}\right], \\
\tilde{e}_{32}^{(1)(1)} = \tilde{e}_{23}^{(1)(1)} &= -\left[\begin{Bmatrix} 1 \\ 3\ 2 \end{Bmatrix}^{(1)} v_1^{(0)} + \begin{Bmatrix} 3 \\ 3\ 2 \end{Bmatrix}^{(1)} v_3^{(0)}\right], \\
\tilde{e}_{22}^{(1)(1)} &= -\begin{Bmatrix} 1 \\ 2\ 2 \end{Bmatrix}^{(1)} v_1^{(0)} - \begin{Bmatrix} 3 \\ 2\ 2 \end{Bmatrix}^{(1)} v_3^{(0)} + \frac{\partial v_2^{(1)}}{\partial \eta_\alpha{}^2}, \\
\tilde{e}_{33}^{(1)(1)} &= \frac{\partial v_3^{(0)}}{\partial \eta_\alpha{}^3} - \begin{Bmatrix} 1 \\ 3\ 3 \end{Bmatrix}^{(1)} v_1^{(0)} - \begin{Bmatrix} 3 \\ 3\ 3 \end{Bmatrix}^{(1)} v_3^{(0)}, \\
\tilde{e}_{13}^{(1)(1)} = \tilde{e}_{31}^{(1)(1)} &= \frac{1}{2}\left(\frac{\partial v_1^{(0)}}{\partial \eta_\alpha{}^3} + \frac{\partial v_3^{(0)}}{\partial \eta_\alpha{}^1}\right) - \begin{Bmatrix} 3 \\ 1\ 3 \end{Bmatrix}^{(1)} v_3^{(0)} - \begin{Bmatrix} 1 \\ 1\ 3 \end{Bmatrix}^{(1)} v_1^{(0)},
\end{aligned} \tag{4.42.1}$$

and

$$e_{ij}^{(1)(M)} = \begin{bmatrix} 0 & \frac{1}{2}\frac{\partial v_1^{(M)}}{\partial \eta_\alpha{}^2} & 0 \\ \frac{1}{2}\frac{\partial v_1^{(M)}}{\partial \eta_\alpha{}^2} & 0 & \frac{1}{2}\frac{\partial v_3^{(M)}}{\partial \eta_\alpha{}^2} \\ 0 & \frac{1}{2}\frac{\partial v_3^{(M)}}{\partial \eta_\alpha{}^2} & 0 \end{bmatrix} + \tilde{e}_{ij}^{(1)(M)}, \tag{4.41.M}$$

where $\tilde{e}_{ij}^{(1)(M)}$ involves the variables $v_1^{(P)}, v_3^{(P)}, v_2^{(P+1)}$ $(P = 0, 1, \ldots, M-1)$, the known $g_{ij}^{(P)}$ and their various derivatives. We observe at once that $e_{ij}^{(1)(0)}$ and $[e_{ij}^{(1)(M)} - \tilde{e}_{ij}^{(1)(M)}]$ represent a state of simple shear, the shear planes being ${\eta_\alpha}^2$ = constant. To prove this, we observe that, locally, the axes are essentially Cartesian rectangular, and so a simple rotation of the 1 and 3 axes about the 2 axis can transform (4.41.0), etc, into the more usual form for simple shear with

$$\bar{e}^{(1)(0)}_{(12)} = \bar{e}^{(1)(0)}_{(21)} = \tfrac{1}{2}K^{(0)}, \quad \text{all other} \quad \bar{e}^{(1)(0)}_{(ij)} = 0, \tag{4.43}$$

where

$$K^{(0)} = 2\{[e^{(0)}_{(12)}]^2 + [e^{(0)}_{(23)}]^2\}^{1/2}. \tag{4.44}$$

This rather elementary property of the zero-order (or Reynolds) approximation to a lubrication flow is of great significance in our extension of the Reynolds approximation to simple fluids, as we shall now see. For, using (4.40), (4.41.0), (4.41.1), etc together with (4.43), (4.44) in the various equations of state, we can obtain functional relations for $p'^{(0)}_{ik}$, $p'^{(1)}_{ik}$, ..., $p'^{(M)}_{ik}$ or for their contravariant forms. In particular for the Oldroyd model (3.6) we find that $p'^{(0)}_{ik}$ is a function of $e^{(1)(0)}_{ik}$ only and is thus given by the solution for the *steady* uniform shearing problem. Similarly for the Walters model (3.7), we have from (4.34) that

$$\frac{\partial \xi'^i_\alpha}{\partial {\xi_\alpha}^j} = \begin{bmatrix} 1 & -\dfrac{\partial v^{(0)1}}{\partial {\eta_\alpha}^2} s & 0 \\ 0 & 1 & 0 \\ 0 & -\dfrac{\partial v^{(0)3}}{\partial {\eta_\alpha}^2} s & 1 \end{bmatrix} + O(\alpha)$$

and that

$$e^{(1)ij}({\xi_\alpha}') = e^{(1)(0)ij}(\xi_\alpha) + O(\alpha),$$

so that once again $p'^{(0)}_{ik}$ is a function of $e^{(1)(0)}_{ik}$ only. Precisely the same arguments can be used for the other models, including the simple fluid if (4.36) is assumed. But we have earlier referred, in Section 3, to the fact that in steady simple shear, characterized by $K^{(0)}$ in (4.43), (4.44), the stress tensor is given by 3 material functions (Coleman and Noll 1961),

τ, σ_1, and σ_2, themselves given by

$$\bar{p}'_{(12)} = \tau(K^{(0)}) = \eta(K^{(0)})K^{(0)}, \tag{4.45.1}$$

$$\bar{p}'_{(11)} - \bar{p}'_{(33)} = \sigma_2(K^{(0)}), \tag{4.45.2}$$

$$\bar{p}'_{(22)} - \bar{p}'_{(33)} = \sigma_1(K^{(0)}), \tag{4.45.3}$$

with $\bar{p}'_{(13)} = \bar{p}'_{(23)} = 0$. Thus, in our case we have

$$p'^{(0)}_{12} = \eta(K^{(0)})\, \partial v_1^{(0)}/\partial \eta_\alpha^{\ 2}, \tag{4.46.1}$$

$$p'^{(0)}_{13} = \eta(K^{(0)})\, \partial v_3^{(0)}/\partial \eta_\alpha^{\ 2}, \tag{4.46.2}$$

$$p^{(0)}_{11} - p^{(0)}_{22} = F_1(\sigma_1, \sigma_2, \beta), \tag{4.46.3}$$

$$p^{(0)}_{33} - p^{(0)}_{22} = F_3(\sigma_1, \sigma_2, \beta), \tag{4.46.4}$$

where F_1, F_3 are known functions of $K^{(0)}$ through σ_1 and σ_2 and of

$$\beta = \tan^{-1}\left[\frac{\partial v_1^{(0)}/\partial \eta_\alpha^{\ 2}}{\partial v_3^{(0)}/\partial \eta_\alpha^{\ 2}}\right].$$

We may, for specified model fluids, write out relations for $p'^{(1)}_{ik}, \cdots, p'^{(M)}_{ik}$. These will be such that the $p'^{(M)}_{ik}$ involve the "unknown" functions $\partial v_1^{(M)}/\partial \eta_\alpha^{\ 2}$ and $\partial v_3^{(M)}/\partial \eta_\alpha^{\ 2}$, largely through $e^{(1)(M)}_{ij}$, which can be written as (4.41.M), and the "known" functions $v_1^{(P)}$, $v_3^{(P)}$, $v_2^{(P+1)}$ $(P = 0, 1, \ldots, M-1)$ and their derivatives—the distinction between "known" and "unknown" functions will become clear when the iterative steps of the solution scheme have been explained. Of course, the dependence on $\partial v_1^{(M)}/\partial \eta_\alpha^{\ 2}$ and $\partial v_3^{(M)}/\partial \eta_\alpha^{\ 2}$ will not be so simple in the general case as it is for $M = 0$, and so the use of material functions alone is limited to the Reynolds approximation.

We are now in a position to make use of the momentum equations in the form

$$v_\alpha^{\ j}\, v^{\ i}_{\alpha,j} = f_\alpha^{\ i} + p^{ij}_{\alpha,j}. \tag{4.47}$$

Written in full, after separation into powers of α, and using (4.46) we have

$$\frac{\partial p^{(-1)}}{\partial \eta_\alpha^{\ 1}} = f^{(0)1} + \frac{\partial}{\partial \eta_\alpha^{\ 2}}\left(\eta(K^{(0)})\frac{\partial v^{(0)1}}{\partial \eta_\alpha^{\ 2}}\right) + O(\alpha), \tag{4.48.1}$$

$$\frac{\partial p^{(0)}}{\partial \eta_\alpha^{\ 2}} = f^{(0)2} + \frac{\partial}{\partial \eta_\alpha^{\ 2}}\left(F_2(K^{(0)}, \beta)\right) + O(\alpha), \tag{4.48.2}$$

$$\frac{\partial p^{(-1)}}{\partial \eta_\alpha^{\ 3}} = f^{(0)3} + \frac{\partial}{\partial \eta_\alpha^{\ 2}}\left(\eta(K^{(0)})\frac{\partial v^{(0)3}}{\partial \eta_\alpha^{\ 3}}\right) + O(\alpha), \tag{4.48.3}$$

for the lowest order terms, with later relations of the form

$$\frac{\partial p^{(M-1)}}{\partial \eta_\alpha{}^1} = f^{(M)1} + \frac{\partial p'^{(M)12}}{\partial \eta_\alpha{}^2} + \frac{\partial p'^{(M-1)11}}{\partial \eta_\alpha{}^1} + \frac{\partial p'^{(M-1)13}}{\partial \eta_\alpha{}^3} - \sum_{R=0}^{M-1} \left(v^{(R)1} \frac{\partial v^{(M-R-1)1}}{\partial \eta_\alpha{}^1} + v^{(R+1)2} \frac{\partial v^{(M-R-1)1}}{\partial \eta_\alpha{}^2} + v^{(R)3} \frac{\partial v^{(M-R-1)1}}{\partial \eta_\alpha{}^3} \right), \tag{4.49.1}$$

$$\frac{\partial p^{(M)}}{\partial \eta_\alpha{}^2} = f^{(M)2} + \frac{\partial p'^{(M)22R}}{\partial \eta_\alpha{}^2} + \frac{\partial p'^{(M-1)12}}{\partial \eta_\alpha{}^1} + \frac{\partial p'^{(M-1)23}}{\partial \eta_\alpha{}^3} - \sum_{R=0}^{M-2} \left(v^{(R)1} \frac{\partial v^{(M-R-1)2}}{\partial \eta_\alpha{}^1} + v^{(R+1)2} \frac{\partial v^{(M-R-1)2}}{\partial \eta_\alpha{}^2} + v^{(R)3} \frac{\partial v^{(M-R-1)2}}{\partial \eta_\alpha{}^3} \right), \tag{4.49.2}$$

$$\frac{\partial p^{(M-1)}}{\partial \eta_\alpha{}^3} = f^{(M)3} + \frac{\partial p'^{(M)23}}{\partial \eta_\alpha{}^2} + \frac{\partial p'^{(M-1)13}}{\partial \eta_\alpha{}^1} + \frac{\partial p'^{(M-1)33}}{\partial \eta_\alpha{}^3} - \sum_{R=0}^{M-1} \left(v^{(R)1} \frac{\partial v^{(M-R-1)3}}{\partial \eta_\alpha{}^1} + v^{(R+1)2} \frac{\partial v^{(M-R-1)3}}{\partial \eta_\alpha{}^2} + v^{(R)3} \frac{\partial v^{(M-R-1)3}}{\partial \eta_\alpha{}^3} \right). \tag{4.49.3}$$

The $O(\alpha)$ terms have been included in Eqs. (4.48) to allow for the effect of raising suffices. In practice they must appear as additional terms in Eqs. (4.49), but this has not been done here to avoid extensive detail which adds nothing to the general argument. We select the pair of Eqs. (4.48.1) and (4.48.3) to work with.

Because $p^{(-1)}$ is a function of $\eta_\alpha{}^*$ only (independent of $\eta_\alpha{}^2$)† we can treat $\eta_\alpha{}^*$ (i.e., $\eta_\alpha{}^1$ and $\eta_\alpha{}^3$) and hence $p^{(-1)}$ as parameters, with $v^{(0)1}$ and $v^{(0)3}$ regarded as functions of a single independent variable $\eta_\alpha{}^2$. It will readily be seen that only second derivatives of $v^{(0)1}$ and $v^{(0)3}$ arise and so

† This result would have followed in any case, even if we had started by regarding $p^{(-1)}$ as a function of η_α. The lowest-order equation of motion for the 2-direction would have yielded $\partial p^{(-1)}/\partial \eta_\alpha{}^2 = 0$.

the boundary conditions (4.11) written in the form

$$v^{(0)i}\,(\eta_\alpha{}^2 = \pm 1) = v_{S_{1,2}}^{(0)i} \qquad \text{for } i = 1, 3 \tag{4.50}$$

allow in principle of a unique solution. Note that this will be of the form

$$v^{(0)i} = v^{(0)i}(p^{(-1)}, f^{(0)i}, \eta, \eta_\alpha{}^2), \tag{4.51}$$

where $f^{(0)j}$ is a prescribed function of η_α and $p^{(-1)}$ is an unknown function of $\eta_\alpha{}^*$.

Now we return to the continuity condition and take Eq. (4.38.1) and integrate it between $\eta_\alpha{}^2 = \pm 1$. Using the steady boundary condition (4.39) with $M = 1$, we have then that

$$\int_{-1}^{1} \left[\frac{\partial v^{(0)1}}{\partial \eta_\alpha{}^1} + \frac{\partial v^{(0)3}}{\partial \eta_\alpha{}^3} + \left\{ \begin{matrix} i \\ i\;1 \end{matrix} \right\}^{(1)} v^{(0)1} + \left\{ \begin{matrix} i \\ i\;3 \end{matrix} \right\}^{(1)} v^{(0)3} \right] \partial \eta_\alpha{}^2 = 0. \tag{4.52}$$

This, because of (4.51), becomes an equation for $p^{(-1)}\,(\eta_\alpha{}^*)$ in terms of the independent variables $\eta_\alpha{}^1$, $\eta_\alpha{}^3$. Once again we see that up to second derivatives of $p^{(-1)}$ are involved—and we expect the equation to be of elliptic type. The boundary conditions, along the curve given by $g_{L_0}(\eta_\alpha{}^*) = 0$, on $p^{(-1)}$ can be obtained from (4.14) written as

$$p^{(-1)} = P_\alpha \tag{4.53}$$

and (4.13) written as

$$\int_{-1}^{1} v^{(0)i}\, n_i d\eta_\alpha{}^2 = \bar{V}. \tag{4.54}$$

Equation (4.54) is in effect a relation between the first derivatives of $p^{(-1)}$—for the case $V = 0$ it reduces to the simple form $\mathbf{n} \cdot \nabla p^{(-1)} = 0$. Again we argue that in principle a unique solution for $p^{(-1)}$ can be obtained, because we have prescribed just the right number of boundary conditions. This completes the Reynolds approximation, i.e., the zero-order solution to the lubrication problem.

To continue the solution to first-order terms we take Eq. (4.38.1) to give $v^{(1)2}$, by simple integration of known functions using the boundary condition (4.39). Then we can use the momentum equation (4.48.2) to give $p^{(0)}$ in the form

$$p^{(0)} = \bar{p}^{(0)}(\eta_\alpha) + p^{(0)*}(\eta_\alpha{}^*), \tag{4.55}$$

where $\bar{p}^{(0)}$ is the particular integral obtained by direct integration and $p^{(0)*}$ is an as-yet-undertermined "parameter": knowing $v^{(0)1}$, $v^{(0)3}$, and $v^{(1)0}$, we can now, as explained earlier, use the equations of state to evaluate $p'^{(1)ik}$ in terms of $v^{(1)1}$, $v^{(1)3}$, or their derivatives, other terms involved being already known. The Eqs. (4.49.1) and (4.49.3) can now be treated as a pair of equations in $v^{(1)1}$ and $v^{(1)3}$ with $\eta_\alpha{}^2$ as the only independent variable, and with $p^{(0)*}$ as an unknown parameter. These equations, as with (4.48.1) and (4.48.2) will involve only the second derivatives of $v^{(1)1}$ and $v^{(1)3}$ and a unique solution follows by the choice of boundary conditions

$$v^{(1)i}\,(\eta_\alpha{}^2 = \pm 1) = v_{S_{1,2}}^{(1)i} \qquad (i = 1, 3). \tag{4.56}$$

To complete this first-order approximation, we turn again to the continuity equation, using it in the form

$$\int_{-1}^{1} \left[\frac{\partial v^{(1)1}}{\partial \eta_\alpha{}^1} + \frac{\partial v^{(1)3}}{\partial \eta_\alpha{}^3} + \left(\text{terms in} \left\{\ \right\}^{(1)}, \left\{\ \right\}^{(2)}, v^{(0)1}, v^{(0)3}, v^{(1)1}, v^{(1)3}\right)\right] \partial \eta_\alpha{}^2 = 0. \tag{4.57}$$

This provides an equation for $p^{(0)*}$, for which we shall take the necessary boundary conditions on $g_{L_0}(\eta^*) = 0$ to be either

$$p^{(0)*} = 0 \tag{4.58}$$

or

$$\int_{-1}^{1} v^{(1)i} n_i \, d\eta_\alpha{}^2 = 0. \tag{4.59}$$

Other equally plausible boundary conditions could be used to take account of the mean effect of $\bar{p}^{(0)}$ at the boundary. However at this point, we are most interested in demonstrating that an iterative method of solution is possible. The question of what set of boundary conditions would be most likely to correspond to those relevant to a physical flow will be taken up in a later section.

The sequence of equations to be used in an iterative solution is now clear. For the Mth approximation we use (4.38.M) to give $v^{(M)2}$, then the 2-momentum equation to give a particular integral for $\bar{p}^{(M-1)}$, then evaluate $p'^{(M)}{}_i^k$ in terms of $v^{(M)1}$ and $v^{(M)3}$, substitute into the 1- and 3-momentum equations involving the derivatives of $p^{(M-1)*}$ to give solutions for $v^{(M)1}$ and $v^{(M)3}$ in terms of $p^{(M-1)*}$, and then finally into an integrated form of (4.38.$M + 1$) to give the required equation for $p^{(M-1)*}$, which when solved provides the Mth approximation.

What is important to note is that the process can be satisfactorily terminated after a finite number, M say, of iterations. All the terms neglected are $o(\alpha^M)$ uniformly within the flow region provided the original prescribed geometrical flow field parameters and the prescribed functions arising in the boundary conditions are sufficiently differentiable. By virtue of this we can expect that the solution is a uniformly valid approximation to the mathematical problem posed. The question of how far this is a uniformly valid approximation to any physical flow is quite another matter and will be discussed in Section 6. Before that, we shall describe how the scheme could be extended to treat unsteadiness, compressibility, and to include temperature effects through an energy equation.

5. Extension to Include Unsteadiness, Compressibility, and Heat Effects

5.1. Unsteadiness

In Section 4.4, we assumed that the boundary conditions on the flow were independent of time. This restriction can be relaxed in two ways:

(i) by holding the geometry, i.e., $h(x)$, constant and letting the conditions, V and P, along L_0 vary with time;

(ii) by allowing $v_{S_{1,2}2}$ to be nonzero, i.e., allowing $h(x^*)$ to vary with time (see for example Langlois 1962). In both cases, the effects can be readily accommodated in our perturbation scheme if the terms $\partial(\ )/\partial t$ in Eqs. (4.8), (4.9) and the equations of state are replaced by terms $\alpha\partial(\ )/\partial t$ after the α-coordinate transformation, i.e., if the coordinate transformation (4.5) is extended by the relation

$$t_\alpha = \alpha^{-1} t, \tag{5.1}$$

so that

$$P_\alpha(\eta_\alpha^*; t_\alpha) = P(x^*; \alpha t) \qquad \text{on} \quad L_0 \tag{5.2}$$

and

$$v_{\alpha 2}(\xi_\alpha{}^1, \pm 1, \xi_\alpha{}^3; t_\alpha) = v_{S_1,22}(\alpha\xi^1, \alpha\xi^3, \alpha t). \tag{5.3}$$

This is clearly consistent with the scaling already imposed on $v_{\alpha 1}$ and $v_{\alpha 3}$ at the boundary S_0. It corresponds physically to "slowly-varying" flows. The details of the extra terms involved need not be rehearsed here; the important point is that the zero-order approximation remains unaltered.

5.2. Compressibility

The expansion (4.18.5) has already been chosen to allow for compressibility, but it has not so far been used. To do so we must postulate a conventional equation of state

$$\rho = \rho(p, T), \qquad (5.4)\dagger$$

where p is now the thermodynamic pressure, and T the temperature. This introduces two complications: (a) that of temperature, an often significant thermodynamic variable, which for simplicity is taken to be constant throughout the flow (the extension to include variable T is discussed in Section 5.3 below); (b) that of distinguishing between p and the mean isotropic component $\bar{p}$ of $p_k{}^i$, given by

$$p_k{}^i = \bar{p}\,\delta_k{}^i + p_k'^i, \qquad p_i'^i = 0. \tag{5.5}$$

The latter problem has been discussed by many authors, and arises even for a compressible linear viscous fluid (see for example Lamb 1953, §§325, 358); it is however part of the specification problem for the fluid and is not a result of our perturbation scheme.

If we take the simplest case, that of T constant, then the terms $\rho^{(0)}$, $\rho^{(1)}$, etc., can be regarded as known functions of $p^{(-1)}$, $p^{(0)}$, etc. In particular $\rho^{(0)}$ is a function of $p^{(-1)}$ only. We find, on carrying through the procedure given in section 4.4, that the result (4.40) still holds, i.e., $v^{(2)2} \equiv 0$; next that the relation (4.41.0) still holds, and so Eqs. (4.46) follow; when using (4.47), $\rho^{(0)}$ and hence $p^{(-1)}$ can enter through $f^{(0)1}$. Thus solution for $v^{(0)i}$ in terms of $p^{(-1)}$ follows as before. It is at the next

---

† In some systems, we might have to regard p_i^k and not just p as the relevant thermodynamic variable. Also for the α-flow we must replace (5.4) by $\rho_\alpha = \rho(\alpha p_\alpha, T_\alpha)$.

stage that the variation of $\rho^{(0)}$ with $\eta_\alpha{}^*$ becomes relevant: equation (4.38.1) has to be replaced by the equation

$$\rho^{(0)}\left(\frac{\partial v^{(1)2}}{\partial \eta_\alpha{}^2} + \frac{\partial v^{(0)1}}{\partial \eta_\alpha{}^1} + \frac{\partial v^{(0)3}}{\partial \eta_\alpha{}^2} + \begin{Bmatrix} i \\ i\ k \end{Bmatrix}^{(1)} v^{(0)k}\right)$$
$$+ v^{(0)1}\frac{\partial \rho^{(0)}}{\partial \eta_\alpha{}^1} + v^{(0)3}\frac{\partial \rho^{(0)}}{\partial \eta_\alpha{}^3} = 0. \tag{5.6}$$

Integration between $\eta_\alpha{}^2 = \pm 1$ is still possible, and so the effect of compressibility is to alter the equation for $p^{(-1)}(\eta^*)$. Boundary condition (4.53) is unaltered, but care has to be exercised in interpreting (4.54) unless the special case $V = 0$ only is employed. In principle however solution is still possible, though the zero-order solution *is* substantially affected by compressibility.

The general development of the iteration procedure for first- and higher-order terms runs analogously to that given in Section 4.4, though extra terms arise in all the relations.

5.3. Heat Effects

Equation (5.4) above introduced the temperature T as a thermodynamic variable. In general, it should appear also in the other, rheological, equations of state described in Section 3. In order to include T as a field variable, we need an energy equation of the type (Frederickson 1964, p. 60)†

$$\rho c_v\left(\frac{\partial T}{\partial t} + v^k T_{,k}\right) = -T\left(\frac{\partial p}{\partial T}\right)_\rho v^k{}_{,k} + (k g^{ik} T_{,i})_{,k} + p^{ik} v_{i,k}, \tag{5.7}$$

where c_v is the specific heat at constant volume and k is the thermal conductivity. Here the last term on the right-hand side is the viscous dissipation and the last-but-one the thermal diffusion. We suppose there to be an expansion of the form

$$T_\alpha = T^{(0)}(\eta_\alpha) + \alpha T^{(1)}(\eta_\alpha) + \cdots, \tag{5.8}$$

---

† Equation (5.7) can really only be regarded as an approximation, for it neglects any dependence of the internal energy on the strain history of the material. However, the arguments used for dealing with the rheological equations of state could similarly be applied here, if 'fading-memory' functionals were used.

and we suppose that temperature boundary conditions

$$T(\eta_\alpha{}^1, \pm 1, \eta_\alpha{}^3) = T_{S_{1,2}} \quad \text{on} \quad S_{1,2} \tag{5.9†}$$

and others on L_0 are prescribed.

The lowest-order equation, obtained by substitution of the expansions (4.18) and (5.8) into (5.7), is

$$h^2 \frac{\partial}{\partial \eta_\alpha{}^2}\left[k(T^{(0)})\frac{\partial T^{(0)}}{\partial \eta_\alpha{}^2}\right] + p^{(0)12}\frac{\partial v_1^{(0)}}{\partial \eta_\alpha{}^2} + p^{(0)13}\frac{\partial v_3^{(0)}}{\partial \eta_\alpha{}^2} = 0, \tag{5.10}$$

which, like (4.48.1) and (4.48.3), is an equation involving $T^{(0)}$, $v_3^{(0)}$ and $v_1^{(0)}$ as functions of the single variable $\eta_\alpha{}^2$,$\eta_\alpha{}^1$ and $\eta_\alpha{}^3$ being independent parameters. (5.9) with $T^{(0)}$ written for T on the left-hand side provides requisite boundary conditions. Thus the zero-order approximation now involves a simultaneous solution of the three equations (4.41.1), (4.48.3), and (5.10) in terms of the as yet-unknown parameter $p^{(-1)}$ and its derivatives. In principle this can be achieved, and once again the formal iterative method of solution can be recovered. As in the case of compressibility, the zero-order approximation *is* substantially affected by temperature variations.

6. Discussion

At first sight, the perturbation and solution scheme proposed above seems entirely satisfactory. Reynolds' original approximate equation, or an equally manageable set of zero-order equations, are obtained from an expansion scheme based on a single dimensionless variable. Because, in the physical problem of interest, the quantities $\partial h/\partial x^1$ and $\partial h/\partial x^3$ are usually small‡, the zero-order approximation can be expected to apply up to values of α comparable with unity, at least as far as geometrical factors in the equations are concerned. However, when considering the other (dynamical, rheological, and thermodynamic) factors that arise in the first and higher-order equations, we find that the absolute magnitude of many other dimensionless quantities becomes relevant when we try to

† Alternative conditions on the heat flux and temperature taken together might be prescribed.

‡ And consequently, the length scale h_0 is usually small compared with the length scale l_0 associated with L_0, e.g., its perimeter.

apply our perturbation solution to a physical problem—in particular for the obvious case of $\alpha = 1$.

If we write h_0 and l_0 for the length scales in the 2 and 1, 3 directions respectively, u for the velocity scale (obtained from either $v_{S_{1,2}}$, V or P), ν for the kinematic viscosity scale, λ for the generalised "memory" time scale of the fluid, and κ for the thermometric conductivity scale, then the order of magnitude of the various first-order effects will be as follows:

purely geometrical terms, including curvature and variation of channel depth, $O(h_0/l_0) = O(\varepsilon)$;
inertia terms, $O(uh_0{}^2/l_0\nu) = O(R\varepsilon)$;
fluid-memory (elastic)effects, $O(\lambda u/l_0) = O(\varepsilon\lambda u/h_0)$;
heat effects, $O(uh_0{}^2/l_0\kappa) = O(\varepsilon uh_0/\kappa)$.

Thus we see that not only must ε be small, but also $R\varepsilon$, where R is the characteristic Reynolds number of the local flow; $\Lambda\varepsilon$, where Λ is a dimensionless local memory time; and $K\varepsilon$, where $K = R\,Pr$, Pr being the Prandtl number. In our limiting process, we effectively let $\varepsilon \to 0$, thus achieving the required result for any fixed R, Λ, and K; in the physical problem, all are given, and so the validity of the zero-order approximation requires that all of ε, $R\varepsilon$, $\Lambda\varepsilon$, and $K\varepsilon$ be small. We note in passing that one (or more) of these four may be much larger than the others, even though they may all be small compared with unity; in such a case, a first-order correction might be calculated taking account only of the terms corresponding to the largest dimensionless number(s). Indeed this is precisely what other workers have done, at the same time avoiding the labor of setting out the problem in perturbation form and instead relying on heuristic or order-of-magnitude arguments.†

An equally important, but more subtle, restriction arises when the boundary conditions are considered in detail. Those specified above have been carefully chosen to provide just sufficient boundary conditions for the set of differential equations obtained in our solution scheme. Unfortunately, this mathematical sufficiency is clouded by a physical inadequacy. The success of the mathematics is achieved by making the problem depend only on the flow within the chosen flow region, F, i.e., the volume bounded by S_1, S_2, and the x^2 coordinate lines passing through L_0; what happens

† The reader may well feel, by this stage, how wise they were!

to the fluid outside F does not *appear* to affect the solution obtained. However the fact that all field variables (including h) are assumed to be N-times differentiable up to *and on* L_0 really imposes very considerable restrictions on the "outer" flow. Yet, in most physical situations to which lubrication theory is applied, the boundary L_0 usually represents a boundary of discontinuity in h (or one of its derivatives) and so a breakdown in the approximation must be expected at least in the neighborhood of the boundary L_0. (This may be readily understood in a physical sense by considering the viscous behavior of a fluid with memory flowing over the boundary *from* the outer region into the flow region F. Clearly this behavior must be determined by the deformation history of the fluid when outside the boundary, and so the flow pattern outside the flow region F must be relevant.) Alternatively, if boundary condition (4.13) or (4.21.1) with $V = 0$ is prescribed along part of L_0, this may have been chosen to correspond with a fixed solid boundary for which the no-slip boundary condition $v_\alpha^{\ i} \equiv 0$ is really relevant. Our solution will however yield a nonzero velocity along the boundary, and so will be in error to order unity near a solid boundary. It is not proposed to undertake a careful investigation of this situation here: in a rather general way, we may argue that some analogue of St. Venant's principle will be applicable and that boundary effects will decay "rapidly" towards the interior of the region F; in a more formal fashion, we might attempt expansions, in the neighborhood of the boundary, of the form

$$z = G_0(\alpha)z^{(0)} + G_1(\alpha)z^{(1)} + \cdots \qquad (6.1)$$

by analogy with (4.16), where now the $z^{(i)}$ are functions of a coordinate system $\zeta_\alpha^{\ i}$ with origin on the line L_0 such that L_0 is a ζ^1-line, $\eta_\alpha^{\ 2} = \xi_\alpha^{\ 2} = \zeta_\alpha^{\ 2}$, and $\zeta_\alpha^{\ 3}$ is positive in F. The 33 component of the metric tensor for the coordinate system ζ is to be of the same order as the 22 component. The governing equations for this problem would still be those of Section 4.2, but the "boundary" conditions would be those relevant to an entry, exit or "edge" problem. In particular this would involve matching the expansions of type (6.1) for $\zeta_\alpha^{\ 3} \to \infty$ to those of type (4.16) (or rather (4.18)) with $\eta_\alpha{}^*$ given by $g_{L_0}(\eta_\alpha{}^*) = 0$ (from (4.22)). The important point to note is that this extra complication arises in the zero-order or Reynolds approximation, and is not peculiar to non-Newtonian fluids. In practice, the difficulty of solving entry or exit problems for any but very simple geometries is so great that uniform zero-order approximations of the now singular perturbation problem have not been sought.

For the reasons given above, it is clear that solution of the equations for first-order terms should only be undertaken with discretion. For all but rather special cases, the system of equations involved even for the zero-order approximation would require numerical solution, and so the labor involved in obtaining higher approximations would be considerable, though the equations for these will be mostly linear. As mentioned earlier, some of the first-order correction terms will be more important than others in particular cases, and so separate evaluation of these may be justifiable.

Several authors have treated various special problems in ways that are comparable with our proposed scheme. Elrod (1960) adopted a perturbation approach for steady Newtonian-fluid flow in a journal bearing based on the ratio of film thickness to bearing length as the small parameter, and thus anticipated the basic idea used here. Tanner (1963) used the viscometric material function for a non-Newtonian fluid to extend the Reynolds lubrication solution to non-Newtonian fluids. The method of evaluation of the shear history of a particle in a slowly-changing deformation gradient (used in Section 4.4) was used by Thomas and Walters (1963, 1964) in other perturbation approaches to non-Newtonian flow. Pearson (1962, 1963, 1964) has used the zero-order equations for power-law fluids to obtain flow patterns in extrusion dies, while Zamodits (1964) solved for flow in a constant-depth (h) screw-extruder, making allowance for curvature effects (i.e., for $g_{ij}^{(1)}$). Langlois has contributed most to this field, however. In 1958, he obtained higher-order corrections to Reynolds two-dimensional lubrication solution taking account of inertia terms, with the various (total) derivatives of h as separate expanding parameters. In 1962, he considered unsteady effects in gas-squeeze films, while in a series of papers (1963 a, b, 1964a) he used a recursive approach to slow viscous flow of viscoelastic materials, and discussed the relevance of his work to lubrication theory for such materials. A brief general account is given in Chapter 9 of his monograph on slow viscous flow (1964*b*). A later paper (1965) adopts as the zero-order solution for a classical lubrication problem the Hamel solution for flow between divergent plates rather than the Poiseuille flow between parallel plates.

In conclusion, we see that the advantage of the approach used here is that it allows of expansion in terms of a single parameter and can treat a wide variety of geometries, whereas alternative specialized approaches have used multiple-expansions on particular geometries. The disadvantage is that because it is comprehensive, the scheme is very complicated and is not immediately amenable to simple calculatory or analytic methods. Its significance must be judged in a fundamental rather than a practical sense.

Acknowledgments

It is a pleasure to acknowledge the considerable help provided by my colleagues, Dr. C. J. S. Petrie, Mr. S. Richardson, and Mr. M. A. Matovich in discussing many aspects of this work, and the encouragement afforded by the invitation extended to me by the National Science Foundation and the University of Delaware to present this paper.

References

ARIS, R., 1962, "Vectors, Tensors, and the Basic Equations of Fluid Mechanics." Prentice-Hall, Englewood Cliffs, New Jersey.

COLEMAN, B. D., and NOLL, W., 1961, *Ann. N.Y. Acad. Sci.* **89**, 672, 1960, *Arch. Rat. Mech. Anal.* **6**, 355.

ELROD, H. G., 1960, *Quart. Appl. Math.* **17**, 349.

FREDERICKSON, A. G., 1964, "Principles and Applications of Rheology." Prentice-Hall, Englewood Cliffs, New Jersey.

FULLER, D. D. 1961, Lubrication Mechanics, *in* "Handbook of Fluid Mechanics,"

LAMB, H., 1953, "Hydrodynamics." Cambridge Univ. Press, London and New York.

LANGLOIS, W. E., 1958, *Proc. 3rd. U.S. Nat. Congr. Appl. Mech.*, 777; 1962, *Quart. Appl. Math.* **20**, 132; 1963*a*, *Trans. Soc. Rheol.* **7**, 75; 1963b, *Rendi. Mat. di Roma*, **22**, 169; 1964a, *Trans. Soc. Rheol.* **8**, 33; 1964b; "Slow Viscous Flow." Macmillan, New York; 1965, *Quart. Appl. Math.* **23**, 39.

LODGE, A. S., 1964, "Elastic Liquids." Academic Press, New York.

NOLL, W., 1955, *J. Rat. Mech. Anal.* **4**, 3, 1958, *Arch. Rat. Mech. Anal.* **2**, 197.

OLDROYD, J. G., 1950, *Proc. Roy, Soc. A*, **200**, 533, 1961 *Rheol. Acta.* **1**, 337.

PEARSON, J. R. A., 1962, *Trans. Plast. Inst.* **30**, 230; 1963, *Trans. Plast. Inst.* **31**, 125; 1964, *Trans. Plast. Inst.* **32**, 239.

PINKUS, O., and STERNLICHT, B., 1961, "Theory of Hydrodynamic Lubrication." McGraw-Hill, New York.

RIVLIN, R. S., and ERICKSEN, J. L., 1955, *J. Rat. Mech. Anal.* **4**, 323.

SPAIN, B., 1960, "Tensor Calculus." Oliver and Boyd, Edinburgh and London.

STREETER, V. L. (ed.). McGraw-Hill, New York.

TANNER, R. I., 1963, *Aust. J. Appl. Sci.* **14**, 129.

THOMAS, R. H., and WALTERS, K., 1963, *J. Fluid Mech.* **16**, 228; 1964, *J. Fluid Mech.* **18**, 33; **19**, 557.

Tipei, N. (ed. Cross, W. A.), 1962, "Theory of Lubrication." Stanford Univ. Press, Stanford, California.

Truesdell, C., 1952, *J. Rat. Mech. Anal.* **1**, 125.

Van Dyke, M., 1964, "Perturbation Methods in Fluid Mechanics." Academic Press, New York.

Walters, K., 1960, *Quart. J. Mech. Appl. Math.* **13**, 444.

Zamodits, H. J., 1964, Ph. D. Dissertation, Univ. Cambridge.

The Computation of Compressible Boundary-Layer Flow

I. FLÜGGE-LOTZ

Division of Engineering Mechanics
Stanford University
Stanford, California

This paper will be restricted to steady *laminar* viscous flow along bodies and, in particular, to two-dimensional or to axi-symmetrical three-dimensional flow. The Mach number of the basic flow in which the obstacle stands, may be larger than 1 but perhaps not larger than 10, because we do consider equilibrium flow only, that means dissociation, etc. of the medium are not taken into account. Laminar flow prevails under these conditions for a good portion along the wall starting at the nose. The transition to turbulent flow takes place later than at low Mach numbers.

The Navier–Stokes equations are used to describe this flow. The viscosity coefficient μ is assumed to depend on the absolute temperature T only. Since only media in temperature regions with constant c_p (specific heat at constant pressure) are considered, one can also say that the viscosity is a function of the enthalpy $i = c_pT$. The state law is given by $p = R\rho T$. (p = pressure, ρ = density, R = gas constant.) The Prandtl number $(\mu c_p/k)$ (k = heat conductivity) is assumed to be constant, but not assumed to be 1. The complete energy equation is employed, that means, frictional heat is not neglected.

The problem is described by six equations, namely, four nonlinear partial differential equations (two momentum equations, one energy equation, one continuity equation), by the state law and by the viscosity law. The last two equations are algebraic equations. These six equations determine the unknown which are: two velocity components u, v (parallel and normal to the wall), the pressure p, the density ρ, the absolute temperature $T = i/c_p$ and the viscosity coefficient μ.

At this point, one often follows L. Prandtl's (1904, see Schlichting [1]) observation of the magnitude of terms in two momentum equations and the energy equation. For large Reynolds numbers

$$\mathrm{Re} = \left(\frac{\rho_0 U L}{\mu_0}\right)$$

where U is a characteristic velocity and L a characteristic length of the problem, ρ_0 and μ_0 are reference or average values of density and viscosity coefficient, the strongest action of viscosity is restricted to a thin layer along the wall. In this layer certain terms become very small compared to others. This leads to:

(1) Neglecting longitudinal and transversal curvature of the wall.

(2) Abandoning a u_{xx}-term in the first momentum equation (momentum in direction parallel to the wall), which makes the first momentum equation a parabolic instead of an elliptic differential equation.

(3) Replacing the second momentum equation by a statement that the pressure at the wall can be obtained from the computation of an inviscid flow along the body.

(4) Simplifying the energy equation by only considering energy changes due to heat flux in the direction normal to the wall and to shear forces connected with the changes of the velocity u parallel to the wall.

This procedure then reduces the problem to solving first the inviscid problem of the flow along the body which is often a potential flow. Secondly, one has to solve only a system of three partial differential equations and two algebraic equations for u, v, T, μ, and ρ in the region of the boundary layer. The thickness of this layer is determined by applying rather simple boundary conditions to the differential equations.

With increasing speed of flight, Prandlt's assumptions needed re-examination. In particular, one had to consider the influence of the curvature of the wall, of vorticity of the outer flow† and of slip or nonslip of particles at the surface and of a temperature jump at the same place. A new effort to put Prandtl's idea in a more rigorous mathematical formulation began. Lagerstrom's and Van Dyke's use of the "asymptotic expansions" [2] entered the analytical schemes of solution. Prandtl's suggestion delivered the lowest terms in these expansions which will be described now. The main idea ist still that viscous effects will be strongest

† Mostly caused by a detached shock in front of a blunt body.

in the neighborhood of walls. Therefore, Lagerstrøm and Van Dyke assume that the solution of the full compressible viscous flow problem can be solved with the help of two expansions: one which is based on the wall situation and one which is based on the conditions far away from the obstacle. The expansion parameter is $\varepsilon = 1/\sqrt{R_e}$. Both expansions should be convergent and have a region of overlap; that means near the wall only a few terms of the inner expansion should give practically the correct solution. Far away from the body a few terms of the outer expansion should suffice. In the region of overlap the results obtained from both expansions have to be matched (see Fig. 1). Prandtl's simple "outer edge conditions"

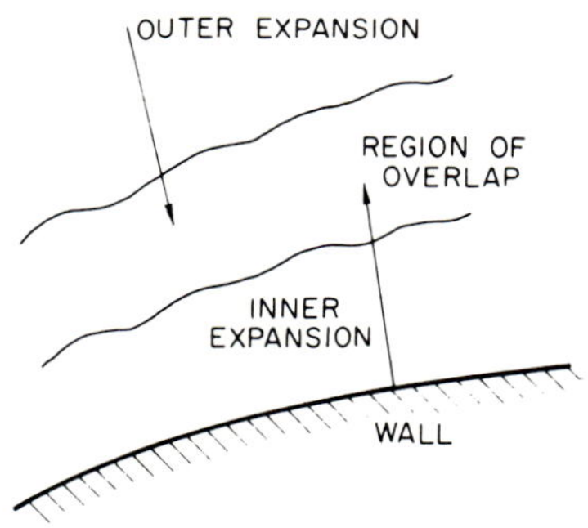

FIG. 1.

are replaced by a mathematically better founded statement based on the comparison of expansion terms. There are still no proofs that the matching (see Van Dyke [2] and Erdelyi [3]) of such two series can always be done. The present state is perhaps best described by comparing theoretical solutions of second-order theory with experiments. If one calls Prandtl's theory "a theory of first order", then Lagerstrøm and Van Dyke have computed certain special cases and others (Davis and Fannelöp) have computed more general cases with a "second-order theory." Interesting results concerning the influence of longitudinal and transversal curvature, of outer vorticity, of slip of velocity and temperature at the wall and of displacement thickness are known, but there becomes evident also a strange behavior of the so-called boundary-layer thickness dependent on the pressure gradient for high Mach numbers (Fannelöp and Flügge-Lotz [16]).

Many years have been spent by solving the first-order problem. The flow along the flat semi-infinite plate has attracted many investigators. Crocco's [4] solution seems to be the most complete one for practical engineering because of its wide variation of Pr, M, c_p/c_v. The flow along the plate is one of the few cases in which one can show that in well chosen coordinates the velocity and temperature profiles along the wall are similar. Another example are the similar solutions for the flow along wedge-shaped cylinders. There are also studies of the compressible laminar boundary flow at nodal or saddle points of attachment† on surfaces. These studies, however, are made with the simplifications Pr = 1 and viscosity coefficient μ proportional to the absolute temperature T; thus similarity solutions for the change normal to the wall of velocity components parallel to the surface and of the temperature are easily obtained (Poots [5]).

It is not the goal of this paper to spend much time on this class of solutions. These examples have given much insight but they do not fulfill the designer's requirements: "Here is the shape of the body, give me shear and heat transfer at the wall for these desired flight conditions or suggest a change of the form such that the heat transfer is kept in certain limits." In order to solve that general problem one has to renounce analytical methods and employ numerical ones.

In the beginning numerical methods were often based on von Kármán's integral equation (see Schlichting [1]). Von Kármán suggested satisfying the momentum equation (direction parallel to the wall) only in the average and to reduce the incompressible boundary-layer problem to the solution of an ordinary differential equation. His idea was extended to comprise the compressible case, but the difference in velocity-layer and thermal-layer thicknesses caused some trouble. References given in Baxter and Flügge-Lotz [6] allow any reader to get acquainted with these procedures which, in the opinion of the author of this survey, have outlived their time since the arrival of high-speed automatic-computing equipment.

Certainly there is still the possibility of replacing differential quotients in x (wall) direction by difference quotients and then solving the ordinary differential equations with y (normal to the wall) as independent variable by computers (Smith and Clutter, [7]). These computations have shown good results in the cases where comparisons could be made. However, it

† In an inviscid flow, those points would be called stagnation points.

is not clear whether the extension to a second-order theory for the various problems of interest will always be easily possible.

The radical decision to replace the partial differential equations by difference equations opens enormous possibilities on one side, but certainly shows also the difficult situation one has to face.

First, the reader should be reminded that intensive studies by mathematicans of linear second-order partial differential equations of parabolic, elliptic, and hyperbolic type have given an insight into two essential computational problems (Courant *et al.* [8]).

A. Does the solution of a difference equation approach the solution of the differential equation for diminishing mesh width?
B. Is the numerical difference scheme stable? That means, does a disturbance or an error vanish with the increase of the independent variable (this may be "time") or does the error increase?

Studying the mathematical literature (for instance Richtmyer [9]) one finds most papers are dealing with stability and convergence problems of one single equation, not of systems of equations—and in most cases, it has been assumed additionally that the coefficients in the single equation are constant.

In general, one found out that, if a difference procedure was stable, one can prove that convergence of the solution of the difference equation toward that of the differential equation existed also.

The difference procedures depend very strongly on the type of differences used for replacing the differential quotients. This is particularly obvious when one investigates parabolic differential equations which are particularly important in the present problem.

Let us consider $\partial q/\partial x = a^2(\partial^2 q/\partial y^2)$; ... $q(x_0, y)$ is given at x_0. Then

$$q[x_0 + (m+1)\,\Delta x] - q[x_0 + m\,\Delta x]$$

$$= a^2\left(\text{difference expression for } \left.\frac{\partial^2 q}{\partial y^2}\right|_{x_0 + n\,\Delta x,\ y}\right)$$

One computes with the help of finite differences all unknowns at station $x_0 + (m+1)\,\Delta x$ on the base of the known values at $(x_0 + m\,\Delta x)$ and if higher accuracy is desired, one may use values at $x_0 + k\ \Delta x$ with $k \leq m - 1$. This means that each single value $q[x_0 + (m+1)\,\Delta x, n\,\Delta y]$

is obtained directly, that no systems of linear equations have to be solved. (Only forward difference quotients are used.) Naturally the new values at station $x_0 + (m + 1)\,\Delta x$ are first approximations of the values at this station and, in general, no check is made, if the differential equation is satisfied to that degree of accuracy which one desires at $[x_0 + (m + 1)\,\Delta x]$. This procedure is stable only if a condition on the ratio $(\Delta y^2/\Delta x)$ is satisfied.

An implicit procedure which expresses the known $q(x_0 + m\,\Delta x, n\,\Delta y)$ as function of the unknown $q(x_0 + (m + 1)\,\Delta x, \text{n}\,\Delta y)$ is more demanding; it requires the solution of a system of N linear equations (one uses backward difference quotients). N is the maximum number of meshes in y direction. Numerous investigations show that the implicit procedure is preferable because there are often no conditions on the mesh width for insuring stability of the numerical procedure. Naturally the difference solution may not coincide with an exact analytic solution (if available) if the mesh width is chosen too large. (Courant *et al.* [8], O'Brien *et al.* [10].)

Elliptic differential equations pose other problems. If one wants to avoid the solutions of a high number of linear equations, one will have to use the well-known relaxation method.

Hyperbolic differential equations have real characteristics. Their diference treatment is well investigated.

This seems to be enough of an introduction for consideration of the problem of the solution of a *system* of partial differential equations, which is the problem to be faced, when the compressible boundary-layer flow problem shall be solved. The following points have to be considered:

I. Are these differential equations linear or nonlinear?
II. Is the coupling strong, weak, or one sided? This may still require class A and B investigations for constant or variable coefficients; that means investigations of convergence and stability.
III. Are the differential equations all of one type or of different types? (See discussion of characteristics in Petrovsky's "Partial Differential Equations" [11]).
IV. What are the boundaries and the conditions on these boundaries?

If the equations are all linear, an implicit method will lead to a system of linear equations. If some or all equations are nonlinear, an implicit scheme will lead to a system of nonlinear equations. This seems to require going back to explicit methods. However, a careful evaluation of the change of nonlinear terms from one mesh to the next one (inclusion of prediction terms), can lead to a valid linearization in the immediate neighborhood of a point. This latter problem needs much experimentation with various difference quotients.

The type of coupling must be carefully considered. It will influence also the linearization scheme.

The equations are not all of the same type [12]. If one can estimate that stability of the numerical scheme is assured for all occuring types for one mesh work (the computer allows only the use of one grid work for one program), one may safely proceed. However, the error propagation may still be different for the different equations.

Point IV of the above-mentioned list seems to me the most difficult one. If the equations are of parabolic or of hyperbolic character one needs initial values to start the computing procedure. The following sketch (Fig. 2) is made for a parabolic form. There is the need to have the values of all unknown on the line x_0, y; the values of unknown or their normal

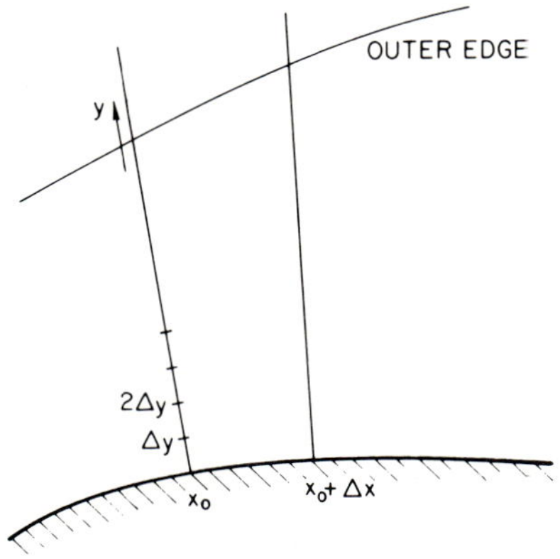

FIG. 2. Coordinates along and normal to the wall.

derivatives on $[x > x_0, y = 0]$, that means at the wall. One expects also to have some statement at what is called "outer edge" in the first-order boundary layer theory. Unfortunately, many difficulties arise at this "outer edge." Firstly, one really does not know in advance where the "outer edge" is located; secondly, the conditions at the "outer edge" will depend on the degree of accuracy of the basic analytic concept of the physical phenomenon. If one merely asks for a smooth transition into the outer inviscid flow and defines the outer edge as that locus where each dependent variable is equal to 99% of its outer flow value, this may still be considered to be a simple condition. It is well known that this trouble was entirely eliminated by the Crocco transformation, which led from independent coordinates x, y and dependent coordinates u, i, to independent coordinates x, u and dependent coordinates $\tau = \mu\, \partial u/\partial y$ and i. Then the problem has a well defined outer edge. With this transformation applied, (Baxter and Flügge-Lotz [6]) solved succesfully a number of problems and showed why simple integral methods had failed. But Crocco's transformation did not allow the handling of velocity profiles $u(x_0, y)$ with overshoot (Fig. 3). In addition, the transformation makes it very difficult to use a second-order theory for solving boundary-layer problems more

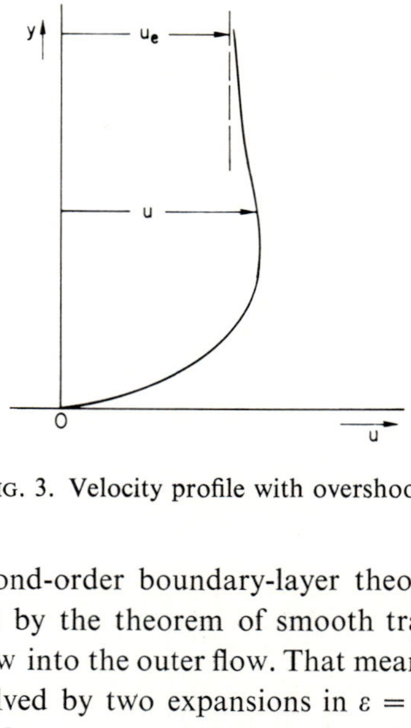

FIG. 3. Velocity profile with overshoot.

accurately. In a second-order boundary-layer theory, the "outer edge" condition is replaced by the theorem of smooth transition of the boundary-layer-proper flow into the outer flow. That means, the entire problem is supposed to be solved by two expansions in $\varepsilon = 1/\sqrt{R_e}$; one solution which is particularly fit to represent the flow close the surface, and another solution which is the expansion of the solution which is valid far away from the surface [2]. These two solutions must meet in the region of what we call neighborhood of the "outer edge" (see Fig. 1) and must give identical results in this region. It is obvious that it is relatively easy to handle this matching of two expansions in the physical plane. How one would handle the case in Crocco's coordinates has never been investigated, and every-one who has studied the singular behavior of $\tau(u)$ near the outer edge, will discard the use of this transformation for a higher-order theory.

Next, the difficulties of a first-order theory will be discussed. Let us assume we stay in the physical plane, that is, use x, y as independent coordinates; then the outer edge condition in a first-order theory means that at a certain $y = N\Delta y$ the velocity $u(x, N\Delta y) = u_e(x)$ and $i(x, N\Delta y) = i_e(x)$. Such boundary conditions pose a programming problem. N is not known, can only be estimated by the value N at $(x - \Delta x)$. Therefore, the above mentioned relations have to be expressed in inequalities in computer programs. That means, one tells the computer that, if $u(x, N\Delta y) - u_e(x) < \bar{\varepsilon}$ the computation at x should stop and proceed to solve the system of linear equations at $(x + \Delta x)$. Since there is no damage in taking N somewhat larger than $N_{\min}(x)$† one expects no computing difficulties, if one started in a correct manner (see Rheinboldt [13] and Blottner and Flügge-Lotz [14]), but unfortunately, there is, in the compressible system, more than one such "floating" condition, namely $|u - u_e| < \bar{\varepsilon}_1$ and $|i - i_e| < \bar{\varepsilon}_2$. Naturally, the larger N will be chosen. However, what happens if, for computertype reasons, one has limited N by $N_0 = 40$ and the layer width grows faster than one expected? Then the computer must deliver a wrong result. See Fig. 4.

Certainly one can build in the computer program an observation of $\partial u/\partial y$ and $\partial T/\partial y$ near the "outer edge" and avoid such bad situations as described in Fig. 4. If one wants to avoid working with an unnecessarily large limiting special number N_0 from the beginning on, one will have to change from time to time (that means with increasing x) the magnitude of Δy. But then the chosen type of difference quotients used for replacing differential quotients will have a great influence on the programming of such a change. Another much worse event is the question of how to choose $\bar{\varepsilon}$; so that one does not merely dance around in the round-offs instead of applying a serious end-condition. It seems therefore that "floating" conditions at one side of the region of solution need particular care.

---

† One takes a grid of width, $N > N_{\min}$, and applies the inequality $|u(N-6)\,\Delta y - u(N-5)\,\Delta y| < \bar{\varepsilon}_1$ for example. On the other hand, if $|u(N-6)\,\Delta y - u(N-5)\,\Delta y| > \bar{\varepsilon}_1$, one has to accommodate for widening.

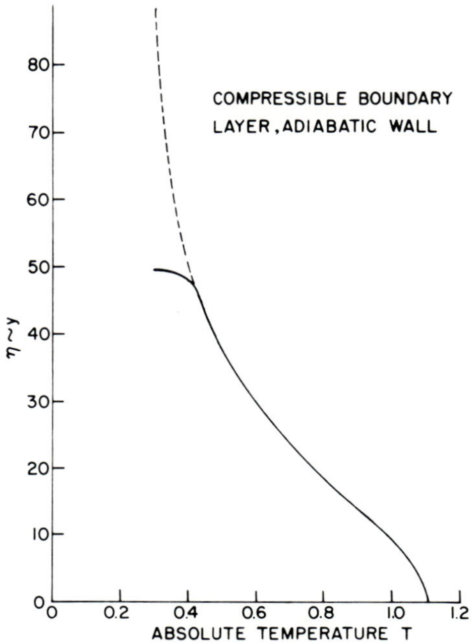

FIG. 4. Incorrect temperature profile.

This difficulty combines with another at the start of the computation. Any finite difference method needs starting values.†

In the compressible problem, all starting profiles are computed under some assumptions. In particular, the viscosity law $\mu(T)$ is preferred in the form of a power law; or there is the idea of setting the Prandtl number 1, etc. Since one needs starting profiles, one takes these solutions, let us say, near the stagnation point of inviscid flow (point of attachment). Naturally, one cannot expect a perfectly smooth transition from these values into the values of the solution of the difference equation which has been set up with realistic Prandtl number and viscosity law. In fact, one will have to

† Only J. C. Wu, (On the finite difference solution of laminar boundary layer problems, in the preprints of *Proc. of the 1961 Heat Transfer and Fluid Mechanics Inst.* (R. C. Binder, M. Epstein, R. L. Mannes, and H. T. Yang, eds.) pp. 55-69, Stanford Univ. Press, 1961) starts with very primitive profiles, which do not always lead to a successful computation of the layer.

investigate whether the disturbance due to "patching" two μ-laws, e.g., will soon die out.†

A number of the points cited here, are described by comparing results of different procedures in the following papers: Blottner and Flügge-Lotz [14], Davis and Flügge-Lotz [12, 15, 17], Fannelöp and Flügge-Lotz [16], and Flügge-Lotz [18].

The difficulties of the viscous flow problem have thus far been discussed only by considering systems of partial differential equations which are, in fact, equations of a first-order theory. Naturally, if the Mach number of the free stream is high and the Reynolds number is relatively low, one will have to look for a second-order theory as was mentioned earlier. This means, a larger set of partial differential equations has to be solved. Principally, one still can proceed by solving first the system of the first-order nonlinear differential equations and then the system of the second-order which are linear differential equations. However, the necessity of avoiding tabulating many functions and providing interpolation leads to solving the two difference-equation systems simultaneously.

At this point, one naturally asks the question if there is not something wrong in the philosophy of using the finite-difference method if one splits in first- and second-order systems. However, this split gives, in our special case of viscous flow, a physical insight into the influence of longitudinal and transversal curvature, of displacement thickness, of vorticity of the exterior flow (due to a curved detached shock), into slip effects of velocity and temperature at the wall. But there is no doubt that probably most people would be reluctant to go into a third-order theory on the lines indicated by Van Dyke. The computing equipment of today would not have enough storage (except if storage tapes are used). But it seems to me also to be a way of diminishing return. Therefore, we are trying now to handle the entire region between body and shock at once, but we do not assume the simplification that shock and boundary-layer are of equal magnitude (see Fig. 5).

Since it is our aim to begin with the body shape, we will now face the problem of determining the shock form, which means we have again to satisfy a known condition in an unknown location (floating boundary condition, expressed by inequalities in the computer program).

---

† This difficulty can be removed by using iteration at each step at the beginning.

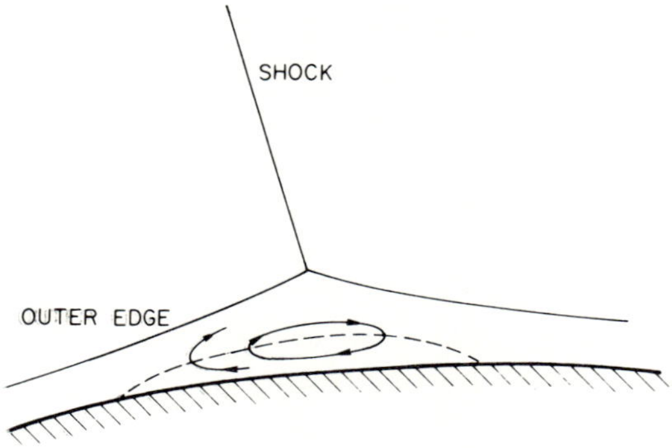

FIG. 5. Blunt body with detached shock and boundary layer.

The inverse problem, assuming the shock form and determining the body shape, has been considered a difficult problem of the inviscid theory, and has been finally solved by Lomax and Inouye [19]. Their work would still be a guide as to the first approximate location of the detached bow shock. Such experience will help in choosing the magnitude of the grid work surrounding the body shape. Naturally, the size of the meshes has to be determined by considering the viscous problem. Work in this direction is in progress. Davis [24], has treated the surrounding of the stagnation point of the inviscid flow with assuming constant density and viscosity coefficient. Kaiser found these results not sufficiently accurate for starting his procedure to investigate the viscous flow in the entire region between shock and body. Therefore, he refined the method by using variable density and viscosity. At present, we are still facing difficulties in determining the shock location further down stream, where shock curvature and body-nose curvature are no longer nearly similar. The idea is still to proceed in a manner proposed for parabolic equations, inspite of being aware that the system of differential equations cannot be considered to be purely parabolic.

Several authors, Thommen [20] and Crocco [21] have made successful attempts to assure the parabolic character of the problem of compressible flow by principally considering the steady-state problem as the final state of an unsteady problem. This means, they consider the viscous flow around a body which proceeds from rest by acceleration toward a motion with constant speed in one fixed direction. In this case, each momentum equation and the energy equation contain a first-order partial derivative with respect to time, and higher-order derivatives of the velocity components or temperature. Naturally, the computing effort is increased, but one avoids the difficulty which causes so much trouble—having differential equations of different types. In using this scheme of unsteady toward steady motion, one faces the difficulty of a not very accurately determined bow shock location (see Thommens' figures), at least if one is not willing to go to extremely small meshes.

In the last months at Stanford, we have experimented with using difference schemes which take into account the varying character of the different equations of a system. In particular, we were interested in those cases where we cannot afford to remove u_{xx} from the first momentum equation, because we would remove, at the same time, the spreading of down-stream flow effects on up-stream behavior. Examples of this type are the boundary layer in the neighborhood of an impinging shock (see Fig. 6) (first studied

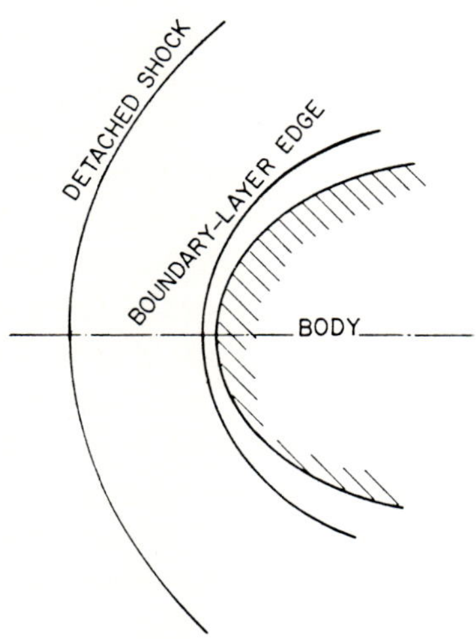

FIG. 6. Impinging shock.

experimentally by Ackeret et al. [22]; recently studied theoretically with many assumptions by Lees and Reeves [23]) and the viscous flow at the rear edge of a finite plate (see Fig. 7).

There is no doubt that the wake behind the infinitely thin plate must

FIG. 7. Finite plate with boundary layer.

influence the boundary-layer profiles close to the rear edge. They cannot be of the Blasius type.†

The moment we retain the derivative u_{xx} in the first momentum equation of a steady flow we have to consider this equation to have elliptic character. This led Plotkin and me to develop a difference scheme which replaced differential quotients by difference quotients which are a weighted composition of those used in parabolic and those used in elliptic equations. In the case described in Fig. 7, the larger weight is given to the difference type used in parabolic equations as long as one is in the plate region, and this should be gradually changed when one enters the region behind the plate. At present we are still having difficulties with the assumption that a Blasius-type profile (we are first studying the incompressible case) exists at some station $x = L - \Delta$, that means at some distance Δ from the erar edge. The difficulties show up in an unreasonable behavior of the normal velocity component, which to all appearance is caused by assumptions on $\partial p/\partial y$ (pressure gradient) which are no longer valid.

In conclusion, I would like to stress that whatever system of difference equations we choose to substitute for the differential equations, a careful consideration of the magnitude of the deviation between difference and differential quotients is necessary. I personally would like it if a computer-science group would take in hand the following basic problem:

A system of n coupled partial differential equations is given. Each of them should display one of the n unknowns prominently and the influence of the others in a form which would allow the attribution of a "type" to each equation based on the derivatives of the prominent unknown. If the types are different (some parabolic, some elliptic, some hyperbolic), one should choose various grid types, one after the other, and study error propagation in the difference scheme chosen for obtaining a solution of the system. Such an investigation would give an urgently needed guidance to engineers who face the task of solving viscous compressible flow problems.

† This assumption was made in papers cited in Ames, "Nonlinear Partial Differential Equations in Engineering," p. 365. Academic Press, New York, 1965.

References

1. Schlichting, H., "Boundary-Layer Theory." McGraw-Hill, New York, 1960.
2. Van Dyke, M., Second-Order Compressible Boundary-Layer Theory with Application to Blunt Bodies in Hypersonic Flow, in: "Hypersonic Flow Research" (F. R. Ridell, ed.), pp. 37-76. Academic Press, New York (1962).
3. Erdelyi, A., An Expansion Procedure for Singular Perturbation, *Atti Accad. Sci. Torino* **95** (1961).
   Van Dyke, M., "Perturbation Methods in Fluid Mechanics." Academic Press, New York (1964).
4. Crocco, L., Lo Strato Limite Laminare Nei Gas, Monografie Sci. Aeronaut. 3 (1946) (*Transl. in* North American Aviation Aerophysics Lab. Rep. AL-684 (July, 1948)). A brief description is contained in N. Curle, The Laminar Boundary Layer Equations, Oxford Math. Monographs, pp. 94–99. Clarendon Press, Oxford (1962).
5. Poots, G., Compressible Laminar Boundary Layer Flow at a Point of Attachment, *J. Fluid Mech.* **22**, Part 1, 197–208 (1965).
6. Baxter, D. C. and Flügge-Lotz, I., "The Solution of Compressible Laminar Boundary-Layer Problems by Finite-Difference Methods, Part II–Further Discussion of the Method and Computation of Examples," Div. Eng. Mech., Stanford University, Stanford, California, Tech. Rep. No. 110 (October, 1957).
7. Smith, A. M. O., Clutter, D. W., Machine Calculation of Compressible Laminar Boundary Layers, *AIAA J.* **3**, no. 4, 639–647 (1965).
8. Courant, R., Friedrichs, K. O., and Lewy, H., Ueber die partiellen Differenzengleichungen der mathematischen Physik, *Math. Ann. Berlin* **100**, 32 (1928).
9. Richtmyer, R. D., "Difference Methods for Initial-Value Problems." Wiley (Interscience), New York, 1957.
10. O'Brien, G. G., Hyman, M. A., and Kaplan, S., A Study of the Numerical Solution of Partial Differential Equations, *J. Math. Phys.* **29**, 223 (1950).
11. Petrovsky, I. G., Lecture on Partial Differential Equations." Wiley (Interscience), New York, 1954.
12. Davis, R. T., and Flügge-Lotz, I., Second-Order Boundary-Layer Effects in Hypersonic Flow past Axisymmetric Blunt Bodies, *J. Fluid Mech.* **20**, part. 4, 593–623 (1964); see in particular p. 598.
13. Rheinboldt, W., Über die äussere Randbedingung bei den Grenzschichtgleichungen, *in*: "50 Jahre Grenzschichtforschung." Vieweg, Braunschweig, Germany, 1955.
14. Blottner, F. G., and Flügge-Lotz, I., Finite-Difference Computation of the Boundary Layer with Displacement Thickness Interaction, *J. Mecanique* **2**, 397–423 (1963).
15. Davis, R. T., and Flügge-Lotz, I., Laminar Compressible Flow Past Axisymmetric Blunt Bodies (Results of a Second-Order Theory), Div. Eng. Mech., Stanford University, Stanford, California, Tech. Rep. 143 (1964) (abbreviated version of part of this report is published as Ref. 12).
16. Fannelöp, T. K., and Flügge-Lotz, I., Two-Dimensional Viscous Hypersonic

Flow over Simple Blunt Bodies Including Second-Order Effects, Div. Eng. Mech., Stanford University, Stanford, California, Tech. Rep. 144 (1964). Material taken from this report appears as: 1) Two-Dimensional Hypersonic Stagnation Flow at Low Reynolds Numbers, *Z. Flugwissenschaften* **13**, 282–296 (1965) and as 2) Viscous Hypersonic Flow over Simple Blunt Bodies; Comparison of a Second-Order Theory with Experimental Results. *J. Mecanique* **5**, No. 1 (1966).

17. Davis, R. T., and Flügge-Lotz, I., The Laminar Compressible Boundary-Layer in the Stagnation-Point Region of an Axisymmetric Blunt Body Including the Second-Order Effect of Vorticity Interaction. *Intern. J. Heat Mass Transfer* **7**, 341–370 (1964). (See also Ref. 15).
18. Flügge-Lotz, I., Computation of the Laminar Compressible Boundary Layer. Final Rept. AFOSR 64–1628 (June, 1964).
19. Inouye, M., and Lomax, H., Comparison of Experimental and Numerical Results for the Flow of a Perfect Gas about Blunt-Nosed Bodies, NASA Tech. Note D-1426 (1962).
20. Thommen, H. U., A Method for the Numerical Solution of the Complete Navier–Stokes Equations for Steady Flows, *Symp. Numerical Solutions of Partial Differential Equations, Univ. Maryland, College Park, Maryland* Academic Press, New York and London (1966).
21. Crocco, L., A Suggestion for the Numerical Solution of the Steady Navier–Stokes Equations, *AIAA J.* **3**, No. 10, 1824–1832 (1965).
22. Ackeret, J., Feldmann, F., and Rott, N., Untersuchungen an Verdichtungstössen und Grenzschichten in schnell bewegten Gasen, *Inst. Aero. ETH, Zürich*, Rept. 10 (1946); see also *NACA Tech. Mem.* 1113 (1947).
23. Lees, L., and Reeves, B. L., Supersonic Separated and Reattaching Laminar Flows: I. General Theory and Application to Adiabatic Boundary-Layer/Shock-Wave Interactions, *AIAA J.* **2**, No. 11, 107–120 (November, 1964).
24. Davis, R. T., and Chyu, W. J., Laminar Flow past a Sphere at High Mach Numbers, *J. Fluid Mech.* **24**, part 3, 481–495 (1966).

Integral Equations for Nonlinear Problems in Partial Differential Equations

DONALD H. HYERS

Department of Mathematics
University of Southern California
Los Angeles, California

Introduction

Integral equations have been used in the study of differential equations, both ordinary and partial, since time immemorial. In this article we will give some illustrative examples from the recent literature involving problems in partial differential equations where either the equation or the boundary conditions contain nonlinearities. In most cases, the nonlinear problem is attacked by first solving a related linear problem. The solution to the linear problem involves integral operators, and when the nonlinearities are introduced, the original problem is replaced by that of solving a nonlinear integral equation. I am going to present a few examples and ideas from the recent literature, with respect to analytic methods.

I. Boundary Value Problems for Elliptic Equations

(a) Consider the Dirichlet Problem:

$$\Delta u = \phi(u) \quad \text{in} \quad D, \qquad u = \psi(s) \quad \text{on} \quad C = \partial D, \tag{1.1}$$

where D is the region bounded by a simple closed curve C in the plane, such that Green's function $G(X, Y)$ is known for this domain, and ψ is continuous. The case when ϕ is analytic:

$$\phi(u) = a_0 + \sum_{p=1}^{\infty} a_p U^p, \qquad |U| < R \tag{1.2}$$

has been studied† by Gouyon [1] by Poincare's method, as follows. Introduc-

† He has also studied a more general quasilinear equation using the same method in [2].

ing a parameter λ, he considers:

$$\Delta u = \lambda\phi(u) \quad \text{in} \quad D, \qquad U = \lambda\psi(s) \quad \text{on} \quad C. \tag{1.3}$$

It is well known that if f is somewhat smooth (e.g., if it is of class C^1 in D and continuous in the closure of D) then the linear Dirichlet problem

$$\Delta u = f \quad \text{in} \quad D, \qquad u = \lambda\psi(s) \quad \text{on} \quad C \tag{1.4}$$

may be solved by the formula

$$u(X) = -\int_D G(X, Y)f(Y)\,d\sigma_y + \lambda V\psi(X). \tag{1.5}$$

Here $V\psi(X)$ is the harmonic function with the boundary values $\psi(s)$ on C. Replacing the function f by $\lambda\phi(u)$ he obtains the integral equation

$$u(X) = -\lambda\int_D G(X, Y)\,\phi[u(Y)]\,d\sigma + \lambda V\psi(X) \tag{1.6}$$

for the solution of the problem (1.3).

He looks for solutions which are analytic in λ. Since the solution corresponding to $\lambda = 0$ is zero, these analytic solution will be of the form $U = \sum_{n=1}^{\infty}\lambda^n U_n$. It follows that

$$U^p = \sum_{n=p}^{\infty} \lambda^n S_n^{\ p}, \tag{1.7}$$

where

$$S_n^{\ p} = \sum_{\substack{\alpha_i \geq 1 \\ \alpha_1 + \cdots + \alpha_p = n}} U_{\alpha_1} U_{\alpha_2} \ldots U_{\alpha_p}. \tag{1.8}$$

Thus

$$\phi(u) = a_0 + \sum_{p=1}^{\infty} a_p \sum_{n=p}^{\infty} \lambda^n S_n^{\ p} = a_0 + \sum_{n=1}^{\infty} \lambda^n \sum_{p=1}^{n} a_p S_n^{\ p}. \tag{1.9}$$

Substituting (1.9) into (1.6) we obtain

$$\sum_{n=1}^{\infty} \lambda^n U_n(X) = -\lambda\int_D G(X, Y)\left[a_0 + \sum_{n=1}^{\infty} \lambda^n \sum_{p=1}^{n} a_p S_n^{\ p}(Y)\right] d\sigma + \lambda V\psi(X).$$

Hence, by equating like powers of λ he finds that

$$U_1 = -a_0 \int_D G(X, P)\, d\sigma + V\psi(X)$$

$$U_{n+1} = -\int_D G(X, Y) \sum_{p=1}^{n} a_p S_n^{\,p}(Y)\, d\sigma. \tag{1.10}$$

The formal solution thus derived will be valid for $|\lambda| < \propto$ if

$$\sum \propto^n \sup |U_n| \tag{1.11}$$

converges, and

$$\sum \propto^n \sup |U_n| < R, \tag{1.12}$$

where R is the radius of convergence of the series (1.2).

By an induction argument Gouyon proves an inequality of the form

$$\sup |U_n| \le \frac{AK^n}{5n^2},$$

where A and K are positive constants. The "key inequality" is:

$$\sum_{q=1}^{n-1} \frac{n^2}{q^2(n-q)^2} < 5 \qquad \text{for all} \quad n > 1.$$

For the case $\lambda = 1$ and $R = \infty$ which corresponds to the original problem when $\phi\ (u)$ is an entire function, one can obtain the following sufficient condition for the convergence of the formal series:

$$4v \sum_{p=1}^{\infty} |a_p| h^{p-1} < 1,$$

where

$$v = \sup \left[\int_D G(X, Y)\, dY \right],$$

and

$$h = 5 \sup \left| -a_0 \int_D G(X, Y)\, dY + V\psi(X) \right|.$$

As an example, if the boundary values $\psi(s)$ are zero, D is a circle of radius r, and $\phi(u) = e^u$, we have $4v = r^2$, $4h = 5r^2$. The method will apply when $r < 0.8$.

In the case of zero boundary values, the method shows the existence of a solution when the diameter of the region D is sufficiently small.

Unless the size of the region is restricted, the solution may not exist, or when it does exist, it may not be unique. For example in the one-dimensional case, consider the b.v. problem: $u_{xx} = f(u)$, for $0 < x < \pi$ and $u(0) = 0 = u(\pi)$. When $f(u) = -u$, we have the one parameter family of solutions $u = \lambda \sin x$. In this example, f is decreasing in u. If f is an *increasing* function of u, it can be shown (e.g., by Hammerstein's uniqueness theorem [3]) that the integral equation (1.5) has at most one solution.

(b) The opposite case, when the right member of (1.1) is a decreasing function of u has been investigated, in a somewhat more general setting, by Krasnosel'skii and Stecenko [4]. They consider the integral equation of Hammerstein type:

$$\phi(t) = \int_\Omega K(t, s) f[s, \phi(s)]\, ds \equiv A\phi, \tag{1.13}$$

where $K(t, s)$ is a nonnegative function for t and s in Ω, with Ω denoting a bounded closed set in the space of N dimensions, and where $f(s, u)$ is monotonically increasing in u. Note that this corresponds to the case in which the right member of (1.1) is monotonically *decreasing*. Thus we may expect that there may be many solutions (or no solutions) to (1.13). The kernel $K(s, t)$ is assumed to be such that the operator

$$Kv = \int_\Omega K(t, s) v(s)\, ds \tag{1.14}$$

takes the space $L_p(\Omega)$ into the space $C(\Omega)$ and is completely continuous, where $p > N/2$. Moreover, it is assumed that there exists a positive number ε_0 and a measurable subset Ω_1 of Ω such that

$$\int_{\Omega_1} K(t, s)\, ds \geqq \varepsilon_0 \qquad \text{for} \quad t \quad \text{in} \quad \Omega_1. \tag{1.15}$$

It should be observed that all these properties of the kernel hold for the Green's function for the first boundary value problem. Denote the norm of the linear operator defined by (1.14) in the space $C(\Omega)$ by $\| K \|$. (Observe that K also maps $C(\Omega)$ onto itself and is completely continuous).

LEMMA. Let $f(s, 0)$ be *non-negative for* s *in* Ω *and let* u *and* v *be real numbers such that the following inequalities hold*: $0 \leq u \leq v$, $\beta u \leq f(s, u)$, $f(s, v) \leq \alpha v$ *for* s *in* Ω. *Furthermore, let the numbers* α *and* β *in the above inequalities satisfy*

$$\propto \| K \| \leq 1, \qquad \beta \varepsilon_0 \geq 1,$$

where ε_0 *is the number satisfying* (1.15).

Then the integral equation (1.13) *has at least one solution* $\phi(t)$ *such that*

$$A[uh(t)] \leq \phi(t) \leq v,$$

where $h(t)$ *is the characteristic function of the set* Ω_1.

Proof. Consider the functions $u(t) = uh(t)$ and $v(t) \equiv v$ for t in Ω. From the hypotheses of the lemma and the inequality (1.15) we have

$$\begin{aligned} Au(t) &= \int_\Omega K(t, s) f[s, uh(s)]\, ds \\ &= \int_\Omega K(t, s) f[s, u]\, ds \geqq \beta \int_\Omega K(t, s) uh(s)\, ds \geqq \beta \varepsilon_0 uh(t) \geqq u(t). \end{aligned}$$

Since f is increasing in its second argument and K is nonnegative, it follows that

$$A(Au(t)) \geqq Au(t). \tag{1.16}$$

Again, from the hypotheses of the lemma, we have

$$Av(t) = \int_\Omega K(t, s) f[s, v(s)]\, ds \leq \propto v \int_\Omega K(t, s)\, ds \leqq \propto \| K \| v \leq v(t). \tag{1.17}$$

Now consider in the space $C(\Omega)$ the set M of functions $x(t)$ satisfying

$$Au(t) \leq x(t) \leq v(t). \tag{1.18}$$

The set M is evidently a bounded, closed, convex subset of $C(\Omega)$. From the monotonicity property of the operator A and the inequalities (1.16) and (1.17) it follows that $Au(t) \leq A(Au(t)) \leq Ax(t) \leq Av(t) \leq v(t)$, i.e., the completely continuous operator A transforms the closed bounded convex set M into itself. Hence by the Schauder fixed point theorem, the operator has at least one fixed point ϕ in M. Thus $\phi(t)$ is a continuous solution of (1.13) which satisfies the inequality

$$Auh(t) \leq \phi(t) \leq v(t)$$

and the lemma is proved. From the lemma there follows the

THEOREM. *Let there be n pairs of numbers u_i, v_i $(i = 1, \ldots, n)$ such that*

$$\begin{aligned} 0 \leq u_i \leq v_i \qquad & i = 1, \ldots, n, \\ v_i < u_{i+1} \qquad & i = 1, \ldots, n-1, \\ \beta u_i \leq f(s, u_i), & \\ f(s, v_i) \leq \propto v_i \qquad & s \in \Omega, \quad i = 1, \ldots, n, \end{aligned} \tag{1.19}$$

where $\alpha \| K \| \leq 1$ and $\beta \varepsilon_0 \geq 1$. *Then the integral equation* (1.13) *has in the space* $C(\Omega)$ *at least n solutions* $\phi_j(t)$ *which satisfy*

$$Au_j h(t) \leq \phi_j(t) \leq v_j \qquad j = 1, \ldots, n.$$

The authors apply this theorem to the boundary value problem:

$$Lu = f(t, u) \quad \text{for} \quad t \in D \qquad \text{and} \qquad u = 0 \quad \text{for} \quad t \in \partial D, \tag{1.20}$$

where

$$Lu = -\sum_{i,j=1}^{N} a_{ij}(t_1, \ldots t_N) \frac{\partial^2 u}{\partial t_i \, \partial t_j} + \sum_{i=1}^{N} a_i(t_1, \ldots, t_N) \frac{\partial u}{\partial t_i} + a(t_1, \ldots, t_N)u,$$

is a uniformly elliptic operator, and where the derivatives of the coefficients a_{ij}, together with the functions a_i and a are Holder continuous in D, while a is nonnegative. Here D is an open bounded region in N-dimensional space whose boundary is smooth enough so that the Green's function for the first boundary value problem for L exists. This Green's function will have all the properties ascribed to the Kernel $K(t, s)$ of the integral equation (1.13), so that the above theorem will apply when the function $f(t, u)$ satisfies the hypotheses of this theorem. An illustration of the theorem is given in the Appendix.

Krasnosel'skii and Stecenko also prove similar existence theorems when the function $f(t, u)$ is not monotomic, but is bounded above and below by monotonically increasing functions, and also when f depends on the first derivatives of u. The same general idea has also been applied by Bakel'man and Krasnosel'skii [5] to be the Dirichlet problem for certain partial differential equations involving the Monge–Ampere operator.

In attempting to push the type of problem discussed in part (a) of this section further, beyond the local results given by Gouyon [1], many authors, beginning with Lyapunov [6], have studied the phenomenon of the branching (or bifurcation) of the local solution, as the controlling parameters (such as R, ν, h above) grow so large that the local theory fails. An account of the methods which have been developed by Lyapunov and his successors is given by Vainberg and Trenogin [7]. Applications to hydrodynamics and elasticity of these methods are discussed by Hyers in chapter 6 of Saaty's "Modern Nonlinear Equations," McGraw-Hill, New York, 1967. See also the article by Hyers entitled "Some Nonlinear Integral Equations of Hydrodynamics" in [8], and the paper [9] by Hyers and Ferling.

2. Upper and Lower Function for Volterra Equations with Monotonic Integrands

The idea of upper and lower functions which is classical (goes back to Perron's work) in the case of ordinary differential equations is easily extended to the case of a Volterra integral equation of the form:

$$u(t) = g(t) + \int_0^t f(t, s, u(s))\, ds. \tag{2.1}$$

For details see Chapter I of the book on differential and integral inequalities by Walter [10]. The upper and lower functions are, of course, easier to deal with in the case when f is monotonically increasing in u. Some of the authors who have used these methods are: L. Collatz, K. Nickel, J. Schroder and W. Walter.

When f is monotonically *decreasing*, the following results can be obtained on the basis of the first chapter of the book by Walter just cited.

LEMMA. *Let* $J = [0, T]$. *For the monotonically increasing case. Assume*

$$\lim_{t \to 0} \int_0^t f(t, s, v(s))\, ds = 0.$$

If

$$v(t) < g(t) + \int_0^t f(t, s, v(s))\, ds$$

and

$$w(t) > g(t) + \int_0^t f(t, s, w(s))\, ds$$

then $v(t) < w(t)$ *in* J.

Proof. $v(0) < g(0) < w(0)$. We proceed by *reductio ad absurdum*. If the lemma is not true, there will be a first point $t_0 > 0$ in J at which it fails. Then we will have $v(t_0) = w(t_0)$. But,

$$\begin{aligned} v(t_0) &< g(t_0) + \int_0^{t_0} f(t, s, v(s))\, ds \\ &\leq g(t_0) + \int_0^{t_0} f(t, s, w(s))\, ds \qquad \text{since } f\uparrow \\ &< w(t_0) \quad \text{which is a contradiction.} \end{aligned}$$

When f satisfies a Lipschitz condition (or even a generalized Lipschitz condition), it can be shown that *equality signs* may be permitted (i.e., the sign $\leq$) in the above. This leads to upper and lower functions.

THEOREM. *Let* $f(t, s, u)$ *be defined for* $0 \leq s < t < T$ *and all real* u, *and let* f *be monotonically decreasing in* u *for fixed* s *and* t. *Let* $f(t, s, \phi(s))$ *be absolutely integrable on the interval* $0 < s < t$ *for each positive* $t \leq T$ *and each continuous function* ϕ *on the closed interval* $0 \leq t \leq T$, *and let*

$$\lim_{t\to 0} \int_0^t f(t, s, \phi(s))\, ds = 0$$

for each such ϕ. *If* $u(t)$ *is any solution of the integral equation* (2.1) *and if* g, v *and* w *are continuous functions on the closed interval* $0 \leq t \leq T$ *which there satisfy the inequalities*:

$$v(t) < g(t) + \int_0^t f(t, s, w(s))\, ds, \tag{2.2}$$

$$w(t) > g(t) + \int_0^t f(t, s, v(s))\, ds, \tag{2.3}$$

then the relations

$$v(t) < u(t) < w(t) \tag{2.4}$$

hold on the given interval.

COROLLARY. *If the hypotheses of the theorem are satisfied and if in addition the function f satisfies a Lipschitz condition:*

$$|f(t, s, u_1) - f(t, s, u_2)| \leqq M(t, s)\,|u_1 - u_2|,$$

where $M(t, s)$ is integrable with respect to s when $0 < s < t \leq T$, then the equality signs may be permitted in the inequalities (2.2), (2.3), (2.4).

The theorem may also be proved by a simple "real variable" argument.

When the hypotheses of the theorem and corollary are satisfied one can obtain an interesting result concerning the method of successive approximations. Put

$$v_{n+1}(t) = g(t) + \int_0^t f(t, s, v_n(s))\, ds$$

for $n \leq 0$, and assume that $v_0(t)$ can be chosen so that:

$$v_0(t) \leqq v_1(t),\ v_0(t) \leqq v_2(t)$$

on the interval $0 < t \leq T$. By repeated use of the monotonicity properties of the integrand together with the corollary and the given "starting inequalities," it can be shown that on the same interval,

$$v_0(t) \leq v_2(t) \leq v_4(t) \leq \cdots \leq u(t) \leq \cdots \leq v_5(t) \leq v_3(t) \leq v_1(t).$$

That is, all the approximating functions with even subscripts are lower functions, and those with odd subscripts are upper functions. Moreover, we have an immediate estimate of the error at the nth approximation:

$$|v_n(t) - u(t)| \leqq |v_n(t) - v_{n+1}(t)|.$$

(The proof starts as follows. By hypothesis,

$$v_0 \leq v_2 = g + \int f v_1, \qquad \text{and} \qquad v_1 = g + \int f v_0;$$

apply the corollary with $v = v_0$ and $w = v_0$ to get:

$$v_0 \leq u \leq v_1.$$

Since f is decreasing, and $v_0 \leq v_2$, we have $v_2 = g + \int f v_1$ and

$$v_1 = g + \int f v_0 \geq g + \int f v_2 \,(= v_3)$$

so the corollary applies with $v = v_2$ and $w = v_1$ to give $v_2 \leq u \leq v_1$. So for this, we have

$$v_0 \leq v_2 \leq u \leq v_1, \qquad \text{and} \qquad v_3 \leq v_1, \quad \text{etc.}$$

The proof can be accomplished by induction.)

An Illustration

These ideas and results can be applied to the following heat flow problem (which was solved in 1951 by Mann and Wolf [11]) with a nonlinear boundary condition:

$$u_t = u_{xx}, \qquad x > 0, \quad t > 0 \tag{2.5}$$

$$u(x, 0) = 0, \qquad x > 0 \tag{2.6}$$

$$-u_x(0, t) = G[u(0, t)], \qquad t > 0. \tag{2.7}$$

The function $G[u]$ is assumed to be continuous, monotonically decreasing, and to have the value zero at $u = 1$, corresponding to the medium outside the body being kept at temperature "one."

First, we observe that the related linear problem consisting of (2.5), (2.6), and the boundary condition

$$-u_x(0, t) = h(t)$$

can be solved, for example, by Laplace transforms, and the result is

$$u(x, t) = \int_0^t \frac{h(s) \exp[x^2/4(t - s)]\, ds}{(\pi(t - s))^{1/2}}. \tag{2.8}$$

Thus, we are lead to the nonlinear integral equation

$$u(t) = \int_0^t \frac{G[u(s)]\, ds}{(\pi(t - s))^{1/2}} \tag{2.9}$$

where we have written $u(t) = u(0, t)$. When the function $G(u)$ satisfies a Lipschitz condition on the interval $0 \leq u \leq 1$ with the Lipschitz constant L, one can apply the above results on Volterra equations with monotonically decreasing integrands. This results in the error bound:

$$| v_{n+1}(t) - u(t) | \leq | v_{n+1}(t) - v_n(t) | \leq \frac{L^{n+1} t^{n+1/2}}{\Gamma\left(\frac{n+3}{2}\right)}$$

for the approximating iterates to the solution of (2.9), where we have taken $v_0 \equiv 0$. (Since the right member is positive on the interval $0 \leq u \leq 1$, it follows that the two starting inequalities are satisfied.)

Several other applications of the same general ideas are given in the book by Walter [10, Ch. I].

3. A Nonlinear Initial Value Problem

The initial value problem:

$$\begin{aligned} &u_{xx} - u_{tt} - 2ku_t - \propto u = b(x, t) + \varepsilon u^3, \qquad -\infty < x < \infty, \quad t > 0, \\ &u(x, 0) = f(x), \quad u_t(x, 0) = g(x), \qquad -\infty < x < \infty, \end{aligned} \tag{3.1}$$

where $\propto$, k, ε are constants, has been investigated by Ficken and Fleishman [12] and we shall indicate their results here. The problem may be interpreted physically as that of the forced vibrations of a string in a medium with damping and constrained by nonlinear springs, or alternatively, as a nonlinear transmission line problem.

In order to reduce the problem to that of solving an integral equation, one first solves the linear problem:

$$\begin{aligned} Lu \equiv u_{xx} - u_{tt} - 2ku_t - \propto u &= r(x, t) \\ u(x, 0) = f(x), \qquad u_t(x, 0) &= g(x). \end{aligned} \tag{3.2}$$

The system (3.2) may be solved, for example, by the method of Riemann, or by the method given in Courant–Hilbert [13]; cf. also Frank and von Mises [14]. The solution to the linear problem (3.2) can be conveniently described in terms of the linear operators H, M, G, and F which are defined as follows:

$$Hw(x, t, \tau) = \tfrac{1}{2} e^{-k(t-\tau)} \int_{x-t+\tau}^{x+t-\tau} w(\xi, \tau)\, J_0(\omega\zeta)\, d\xi,$$

$$\zeta = ((t - \tau)^2 - (x - \xi)^2)^{1/2}, \qquad \omega = (\propto - k^2)^{1/2}$$

$$Mw(x, t) = -\int_0^t Hw(x, t, \tau)\, d\tau \tag{3.3}$$

$$Gg(x, t) = Hw(x, t, 0), \qquad \text{where} \quad w(x, 0) = g(x)$$

$$Ff(x, t) = G_t f(x, t) + 2kGf(x, t).$$

More explicitly:

$$Gg(x, t) = \tfrac{1}{2} e^{-kt} \int_{x-t}^{x+t} g(\xi)\, J_0\, [\omega(t^2 - (x - \xi^2))^{1/2}]\, d\xi$$

$$Ff(x, t) = \tfrac{1}{2} e^{-kt} \Bigg[f(x + t) + f(x - t)$$

$$+ k \int_{x-t}^{x+t} f(\xi)\, J_0\, [\omega(t^2 - (x - \xi)^2)^{1/2}]\, d\xi$$

$$+ \int_{x-t}^{x+t} f(\xi)\, \frac{\partial J_0[\omega(t^2 - (x - \xi)^2)^{1/2}]}{\partial t}\, d\xi \Bigg].$$

When the initial data f, g satisfy: $f(x) \in C^1$ (class of bounded function with bounded continuous first derivatives) and $g(x) \in C^2$ (class of bounded functions with bounded continuous first and second derivatives) and $r(x, t)\, \varepsilon C^1 C$ (class of functions which are bounded and continuous together with their first derivatives with respect to x for all real x and all non-negative t), then the problem (3.2) is solved by:

$$u = Ff + Gg + Mr.$$

We observe that the sum of the first two terms is a solution of the homogenous equation $Lu = 0$, and also satisfies the intial conditions. The last term satisfies the inhomogeneous equation $Lu = r$, together with zero initial data.

Thus we are led to the integral equation

$$u = Ff + Gg + M(b + \varepsilon u^3) \equiv Uu \tag{3.4}$$

for the determination of the solution of (3.1).

In solving (3.4) one employs the space CC of functions of x and t which are bounded and continuous for all real x and all $t \geq 0$, with the norm $\| w \| = \sup | w(x, t) |$. It will always be assumed that $k > 0$ and $\propto > 0$.

One tries to solve (3.4) by successive approximations, using the contraction mapping theorem. Thus one seeks a closed ball

$$S_R = \{w : \| w \| \leq R\}$$

in the space CC such that the operator U maps S_R into itself, and such that U satisfies a Lipschitz condition

$$\| U(u - v) \| \leq \theta \| u - v \|$$

with $0 < \theta < 1$ when u and v belong to S_R.

In order to carry out this method one first finds the norms of the linear operators M, G, and F. This can be done quite easily from the formulas (3.3). In particular one has

$$\| Mw \| \leq m \| w \|, \quad \text{where} \quad m = k^{-2} \quad \text{when} \quad 0 < k^2 \leq \propto,$$
$$m = [k - (k^2 - \propto)^{1/2}]^{-2} \quad \text{when} \quad 0 < \propto < k^2.$$

Now, let a number θ be chosen between zero and one, and let R and ε satisfy the inequality $3 | \varepsilon | mR^2 \leq \theta$. Then if $\| u \| \leq R$ and $\| v \| \leq R$, we have

$$\| U(u - v) \| = \| \varepsilon M(u^3 - v^3) \| \leq | \varepsilon | m \| u^2 + uv + v^2 \| \| u - v \|$$
$$\leq 3 | \varepsilon | mR^2 \| u - v \| \leq \theta \| u - v \|,$$

so that the Lipschitz condition is satified in S_R,

With R and ε subject to the same inequality, and with $\| u \| \leq R$,

$$\| Uu \| \leq \| F \| \| f \| + \| G \| \| g \| + m \| b \| + | \varepsilon | mR^3$$
$$\leq \| F \| \| f \| + \| G \| \| g \| + m \| b \| + \frac{\theta R}{3}.$$

Thus the operator U will be a contraction mapping of S_R into itself providing that the inequalities

$$\| F \| \| f \| + \| G \| \| g \| + m \| b \| \leq \left(1 - \frac{\theta}{3}\right) R \tag{3.5}$$

$$3mR^2 | \varepsilon | \leq \theta < 1 \tag{3.6}$$

are both satisfied. Under these conditions the method of successive approximations, starting with a function $u_0 \in S_R$, will converge to a solution $u(x, t)$ of (3.4).

This method gives one a good grip on the solution, so that a number of its properties may be studied. After demonstrating a global uniqueness theorem, the authors define p-preferred solutions u as those whose data satisfy $f \in C^2$, $g \in C^1$ and such that $\| u \| \leq p$, and refer to the data of such a solution as p-preferred data. Here p is a positive number such that

$$p < \frac{1}{(3m \mid \varepsilon \mid)^{1/2}},$$

in accordance with the inequality (3.6). It is shown that $f = 0$ and $g = 0$ are p-preferred data if $\| b \|$ is suitably restricted in size:

$$3m\,(3m \mid \varepsilon \mid)^{1/2} \, \| b \| < 2. \tag{3.7}$$

One of the main results of their paper concerns the existence of periodic solutions of the partial differential equation $Lu = b + \varepsilon u^3$ when $b(x, t)$ is periodic in t of period λ. It is shown that this differential equation has a solution periodic in t with period λ if the periodic forcing function $b(x, t)$ belong to C^1C and if the inequality (3.7) is satisfied. This solution is p-preferred and is unique among all p-preferred solutions.

Appendix: An Example

To illustrate the theorem of Krasnosel'skii and Stecenko, for simplicity, we consider a one-dimensional case:

$$\phi(t) = \int_0^1 K(t, s) f[s, \phi(s)] \, ds$$

where

$$K(t, s) = \begin{matrix} t(1 - s), & 0 \leq t \leq s \\ s(1 - t), & s \leq t \leq 1. \end{matrix}$$

We have $\max K(t, s) = \frac{1}{4}$; thus, $\| K \| = \frac{1}{4}$. Choose Ω_1 to be the interval $(\frac{1}{4}, \frac{3}{4})$. Then if $t \in \Omega_1$,

$$\int_{\Omega_1} K(t, s) \, ds = \int_{1/4}^{3/4} K(t, s) \, ds = \int_{1/4}^{t} s(1 - t) \, ds + \int_t^{3/4} t(1 - s) \, ds$$

$$= \int_{1/4}^{t} s \, ds - \int_{1/4}^{t} st \, ds + t \int_t^{3/4} ds - \int_t^{3/4} st \, ds$$

$$= \frac{t^2}{2} - \frac{1}{32} + t\left(\frac{3}{4} - t\right) - t\int_{1/4}^{3/4} s\,ds$$

$$= -\frac{t^2}{2} + \frac{3t}{4} - \frac{1}{32} - t\left(\frac{9}{32} - \frac{1}{32}\right)$$

$$\int_{\Omega_1} K(t, s)\,ds = -\frac{t^2}{2} + \frac{t}{4} - \frac{1}{32} = \frac{1}{2}t(1 - t) - \frac{1}{32}.$$

Thus, if we take $t = 1/4$, we get $1/8(3/4) - 1/32 = (3 - 1)/32 = 1/16$. Hence we may take $\varepsilon_0 = 1/16$

We also choose $\beta = 17$, $\alpha = 4$. Then $\alpha \| K \| = 1$ and $\beta\varepsilon_0 = 17/16 > 1$. We have obtained: $\| K \| = 1/4$, $\varepsilon_0 = 1/16$, $\beta = 17$, $\propto = 4$. Let us define the function $f(s, u)$ as follows:

$$\begin{aligned} f(s, u) &= 1 \quad \text{for} \quad 0 \le u \le 1 \\ &= 1 + 36(u - 1) \quad \text{for} \quad 1 < u \le 2 \\ &= 37 \quad \text{for} \quad 2 < u \le 10 \\ &= 37 + 150(u - 10) \quad \text{for} \quad 10 < u \le 11 \\ &= 187 \quad \text{for} \quad u > 11. \end{aligned}$$

Put $u_1 = 0, v_1 = 1, u_2 = 2, v_2 = 10, u_3 = 11, v_3 = 47.$

Then we have:

$$\begin{aligned} 0 &= \beta u_1 < f(s, u_1) = 1, \\ 34 &= \beta u_2 < f(s, u_2) = 37, \\ 187 &= \beta u_3 \le f(s, u_3) = 187. \end{aligned}$$

Also:

$$\begin{aligned} 1 &= f(s, v_1) < \propto v_1 = 4 \\ 37 &= f(s, v_2) < \propto v_2 = 40 \\ 187 &= f(s, v_3) < \propto v_3 = 188, \end{aligned}$$

and clearly $0 \le u_j \le v_j$ $(j = 1, 2, 3)$ and $v_1 < u_2$, $v_2 < u_3$. Hence all the inequalities of the theorem are satisfied with $n = 3$.

References

1. Gouyon, R., Sur le problème de Dirichlet pour l'équation $\Delta v = \phi(v)$ *Compt. Rend.* **251**, 26–28 (1960).

2. GOUYON, R., Sur le problème de Dirichlet pour l'équation elliptique reduite la plus generale a second membre analytique, *Compt. Rend.* **251**, 2645–2647 (1960).
3. HAMMERSTEIN, A., Nichtlineare Integralgleichungen nebst Anwendungen, *Acta Math. (Stockholm)* **54**, 118–172 (1930).
4. KRASNOSEL'SKII, M. A., and STECENKO, V. YA., Some Nonlinear problems with Many Solutions, (in Russian), *Sibirski Mat.* **4**, 120–137 (1963).
5. BAKELMAN, I. YA., and KRASNOSEL'SKII, M. A., Nontrivial Solutions of Dirichlet's Problem for Equations with the Monge–Ampere Operator, *Soviet Math.* **2**, 369–372 (1961).
6. LYAPUNOV, A. M., Sur les figures d'equilibre peu différentes des ellipsoides. ... *Zap. Akad. Nauk (St. Petersburg)*, 1–225 (1906).
7. VAINBERG, M. M., and TRENOGIN, The Lyapunov and Schmidt Methods in the Theory of Nonlinear Equations and Their Subsequent Development, *Math. Surveys* **17**, No. 2, 1–60 (1963).
8. ANSELONE, P., (ed.), Nonlinear Integral Equations, Univ. Wisconsin Press, Madison, Wisconsin, 1964.
9. HYERS, D. H., and FERLING, J. A., On the Local Uniqueness Problem for Periodic Surface Waves of Permanent Type in a Channel of Infinite Depth, *Math. Mag.* **31**, 61–74 (1957).
10. WALTER, W., "Differential- und Integralgleichungen." Springer, Berlin, 1964.
11. MANN, W. R., and WOLF, F., Heat Transfer between Solids and Gases under Nonlinear Boundary Conditions, *Quart. Appl. Math.* **9** 163–184 (1951).
12. FICKEN, F. A., and FLEISHMAN, B., Initial Valve Problems and Time-Periodic Solutions for a Nonlinear Wave Equation, *Comm. Pure Appl. Math.* **10**, 331–356 (1957).
13. COURANT, R., and HILBERT, D., "Methods of Mathematical Physics," Vol. II, pp. 691–695. Wiley (Interscience), New York, 1961.
14. FRANK, P., and VON MISES, R., Differentialgleichungen der Physik, Vol. I, pp. 812–814. New York, 1943.

Electrical Problems Modeled by Nonlinear Partial Differential Equations

W. J. CUNNINGHAM

Department of Engineering and Applied Science
Yale University
New Haven, Connecticut

Nonlinearities in electrical problems requiring partial differential equations can be broadly classified into those where some physical property of the surrounding region is not constant, and those where interaction takes place between an electromagnetic field and some sort of charged particles. Several examples of these kinds of problems are given, and the effects of the nonlinearities are described.

Many physical problems in all areas of science and engineering lead to nonlinear partial differential equations. Those investigators who work with electrical phenomena may be particularly fortunate in that many of their problems can be described with fairly good accuracy by using only linear equations. On the other hand, there are electrical phenomena which are inherently nonlinear. Even those which appear to be linear where only small variations take place usually become nonlinear with large variations. Thus, the investigator of electrical problems ultimately is faced with nonlinear equations as he pushes his systems toward extremes of performance. If the system is distributed, it will be described by nonlinear partial differential equations. This is the sort of problem to be considered here.

In the following pages several different nonlinear electrical problems are discussed. The treatment of each is quite brief, and the reader should refer to the original papers for more information. An effort is made to explain the source of the nonlinearity in each system being considered, and the effect which it produces. Few details of a mathematical analysis are given. In most cases, the results quoted were obtained through the use of numerical methods.

An appropriate way to begin is to cite Maxwell's equations [1-3], which provide the basis for studying all sorts of electromagnetic phenomena. In the form of differential equations, these may be written as follows

$$\nabla \times \mathbf{E} = -\dot{\mathbf{B}} \tag{1}$$

$$\nabla \times \mathbf{H} = \mathbf{J} + \dot{\mathbf{D}} \tag{2}$$

$$\nabla \cdot \mathbf{D} = \rho \tag{3}$$

$$\nabla \cdot \mathbf{B} = 0 \tag{4}$$

These equations relate five vector quantities

E, the electric intensity
H, the magnetic intensity
D, the electric flux density
B, the magnetic flux density
J, the electric current density

and one scalar quantity, ρ, the electric charge density.

The three vector densities, **D**, **B**, and **J**, are all per unit area, while the scalar density ρ is per unit volume. On the left sides of the equations are the vector operators, $\nabla\times$ and $\nabla\cdot$, which of course require partial differentiation with respect to the space coordinates. On the right sides of the equations the dot indicates differentiation with respect to time. In general, therefore, there are three independent space coordinates and one time coordinate, so that these are partial differential equations with four independent variables. These equations must, of course, hold at every point in the electromagnetic field.

In addition to Eqs. (1)–(4), there are three relations taking into account physical properties of the materials in the region where fields exist. The electric permittivity ε, the magnetic permeability μ, and the electric conductivity σ, are defined by the relations

$$\mathbf{D} = \varepsilon \mathbf{E} \tag{5}$$

$$\mathbf{B} = \mu \mathbf{H} \tag{6}$$

$$\mathbf{J} = \sigma \mathbf{E} \tag{7}$$

In simple cases, ε, μ, and σ are all constant scalar quantities. There are, however, physical materials for which these quantities are not constants, but change with one or more of the variables. When the quantities are not constants, Eqs. (1)–(4) typically become nonlinear. Thus, one type of problem which characteristically leads to a nonlinear partial differential equation is that in which one or more of the quantities ε, μ, or σ are not constants.

A second type of problem with nonlinear equations is that in which some sort of interaction takes place between an electromagnetic field and particles having both mass and electric charge, as for example, electrons or ions. The equations for such interactions are usually nonlinear, and thus the partial differential equations for the entire system are nonlinear.

Examples of problems of the foregoing types are given in the discussion which follows. Before proceeding, however, certain combinations of the Maxwell equations should be pointed out.

If $\nabla \times \mathbf{E} = 0$, which requires $\dot{\mathbf{B}} = 0$ from Eq. (1), a scalar electric potential V can be defined so that

$$\mathbf{E} = -\nabla V = \mathbf{D}/\varepsilon \tag{8}$$

If the value of $\mathbf{D}$ from Eq. (8) is used in Eq. (3), the result can be written

$$\nabla \cdot \mathbf{D} = -\nabla \cdot (\varepsilon \nabla V) = -\varepsilon \nabla^2 V - \nabla V \cdot \nabla \varepsilon = \rho$$

If ε is a constant, so that $\nabla \varepsilon = 0$, this becomes Poisson's equation

$$\nabla^2 V = -\rho/\varepsilon \tag{9}$$

Furthermore, if the charge density ρ is zero, it is Laplace's equation

$$\nabla^2 V = 0 \tag{10}$$

In a similar manner, if $\nabla \times \mathbf{H} = 0$, which requires $\mathbf{J} + \dot{\mathbf{D}} = 0$ from Eq. (2), a scalar magnetic potential ψ can be defined so that

$$\mathbf{H} = -\nabla\psi = \mathbf{B}/\mu \tag{11}$$

If the value of $\mathbf{B}$ from Eq. (11) is used in Eq. (4), the result is

$$\nabla \cdot \mathbf{B} = -\nabla \cdot (\mu \nabla \psi) = 0 \tag{12}$$

Again, if μ is a constant, Laplace's equation in ψ can be identified.

Finally, if $\rho = 0$, $\mathbf{J} = 0$, and ε and μ are constants, then Eqs. (1)–(3) can be manipulated as follows

$$\nabla \times \nabla \times \mathbf{E} = \nabla(\nabla \cdot \mathbf{E}) - \nabla^2 \mathbf{E} = -\nabla^2 \mathbf{E} = -\mu(\nabla \times \dot{\mathbf{H}}) = -\mu\varepsilon\ddot{\mathbf{E}}$$

and the wave equation can be identified as

$$\nabla^2\mathbf{E} = (1/c^2)\,\ddot{\mathbf{E}} \tag{13}$$

with $c = (\mu\varepsilon)^{-1/2}$ being the velocity of the wave.

Example I. Magnetic Saturation

A type of nonlinear problem that was recognized very early in the study of electrical phenomena is that of magnetic saturation. Virtually all electrical machinery, such as motors, generators, and transformers, make use of iron structures to produce intense and controlled magnetic fields. Coils of wire carrying electric current are wound around the iron structures. Current in a coil produces a magnetic intensity **H**, which in turn leads to a magnetic flux density **B** in the iron. It was soon recognized that as **H** is increased from an initial small value, at first **B** increases almost proportional to **H**. As **H** increases further, **B** increases more slowly, and ultimately becomes essentially constant, as shown in Fig. 1. Since

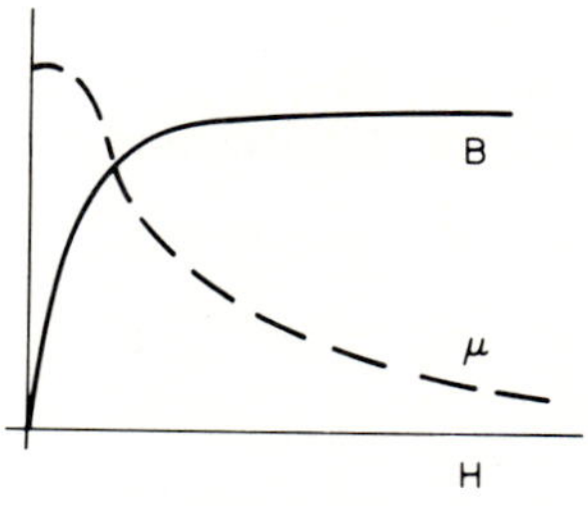

FIG. 1. Magnetic flux density B and permeability μ as a function of magnetic intensity H, showing saturation in a ferromagnetic material.

the permeability μ is the ratio of **B** to **H**, it is at first large, but then becomes small. This effect is known as saturation, and the iron is said to saturate magnetically. The equation describing the magnetic field in the iron is thus Eq. (12), which may be written for a two-dimensional system as

$$\nabla \cdot (\mu\,\nabla\psi) = \frac{\partial}{\partial x}\left(\mu\frac{\partial\psi}{\partial x}\right) + \frac{\partial}{\partial y}\left(\mu\frac{\partial\psi}{\partial y}\right) = 0 \tag{14}$$

where

$$H_x = -\frac{\partial \psi}{\partial x} \qquad H_y = -\frac{\partial \psi}{\partial y}$$

$$H = (H_x^2 + H_y^2)^{1/2} \qquad \mu = \mu(H)$$

If μ were constant, Eq. (14) would become Laplace's equation, $\nabla^2 \psi = 0$. If saturation effects occur, μ is not constant and Eq. (14) is nonlinear. In general, only numerical solutions are possible [4,5].

A two-dimensional example is shown in Fig. 2, which may represent a portion of a transformer lamination having holes in it. Because of symmetry, only that portion bounded by the dotted lines need be considered.

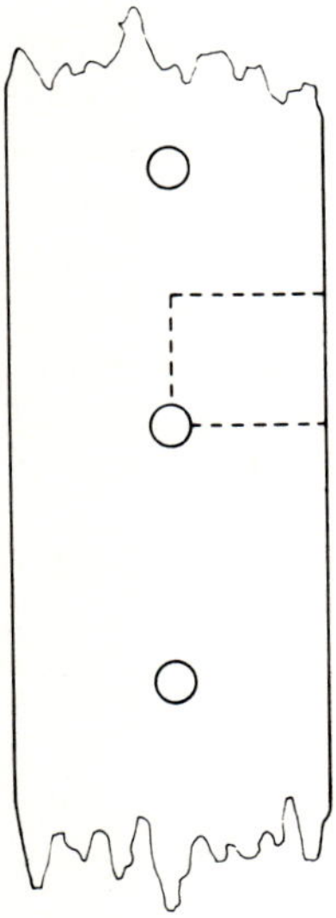

FIG. 2. Iron transformer lamination, with holes.

This portion is shown enlarged in Figs. 3 and 4.

In Fig. 3, the permeability is assumed to be constant. There is no saturation, Laplace's equation applies, and the magnetic field can be plotted as a set of curvilinear squares. The generally vertical lines represent lines of magnetic flux, with the total flux in Fig. 3 being 10 units. The generally horizontal lines represent lines of scalar magnetic potential. A numerical

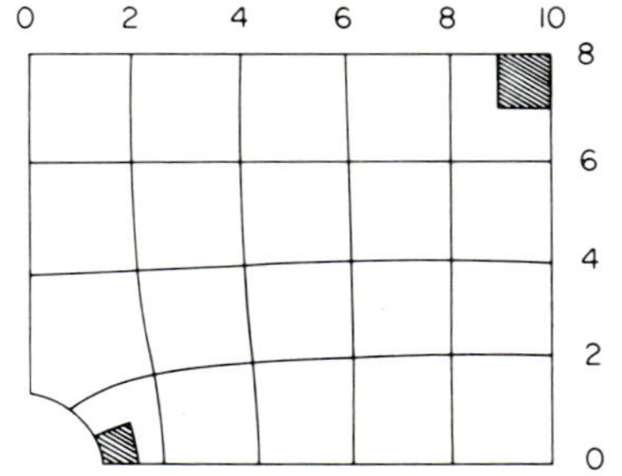

FIG. 3. Map of magnetic field in lamination of Fig. 2, with constant permeability (from Poritsky [4]).

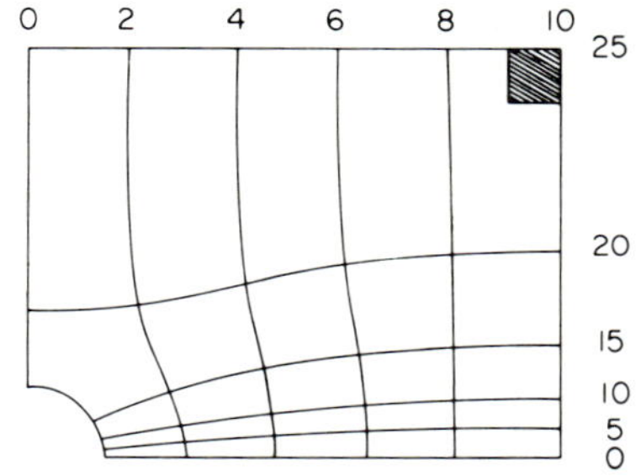

FIG. 4. Map of magnetic field in lamination of Fig. 2, with magnetic saturation, as near the knee of Fig. 1. (from Poritsky [4]).

value of the permeability has been chosen so that a total difference of potential of eight units is required. Two shaded areas, one unit on a side, are shown in diagonally opposite corners of Fig. 3. These areas take the form of curvilinear squares. Since the lines of magnetic flux are more closely spaced at the bottom of Fig. 3, the flux density **B** is greater there than at the top of the figure.

In Fig. 4 the permeability is allowed to vary, and operation is assumed to be near the knee of the curve for **B** in Fig. 1. At the top of Fig. 4, flux density is small enough that operation is below the knee, and the permeability is large. At the bottom of Fig. 4 the increased flux density causes operation to be beyond the knee, so that the permeability is small. The most apparent difference between Figs. 3 and 4 is that the field map of Fig. 4 is not composed of curvilinear squares. The contours of magnetic potential are closely spaced where the permeability is small, and more widely spaced where it is larger. A total potential difference of 25 units is required to give the same total flux as produced by only eight units in Fig. 3. The shaded area, one unit on a side, is a square in the upper right hand corner, but is a very thin rectangle in the lower left hand corner.

Example 2. Parametric Amplifier

The same kind of saturation effect, which occurs in certain magnetic materials as in Example 1, also occurs in some dielectric materials. Such a saturating dielectric can be used to form a capacitor. A voltage difference v across the capacitor leads to a charge q on the electrodes of the capacitor. The voltage is proportional to the electric intensity **E**, and the charge is proportional to the electric flux density **D**. For a nonlinear dielectric, q and v are related as shown in Fig. 5. The variational capacitance of the

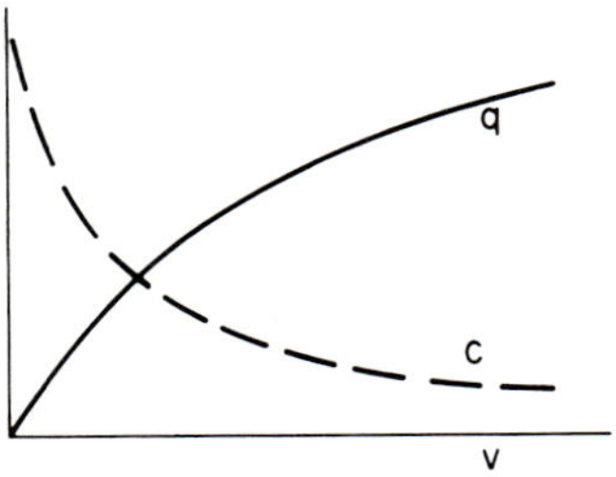

FIG. 5. Charge q and variational capacitance c as a function of voltage v across a saturating capacitor.

capacitor is defined as $c = dq/dv = c(v)$ and clearly is dependent upon the voltage v. This statement is equivalent to saying that the permittivity ε varies with **E**.

One way of amplifying an alternating signal voltage is to use what is known as a parametric amplifier. A simple parametric amplifier consists of a series circuit of inductance and capacitance to which the signal voltage is applied. The elements are selected to be in resonance at the frequency of the input signal. The output signal is taken across the capacitor. The output power for such a circuit cannot exceed the input power. However if the capacitor is caused to vary in time at a frequency twice that of the signal frequency, and with the proper phase relation with respect to the signal, the output power will exceed the input power. Such a device is then a parametric amplifier. The operation of varying the capacitance is called pumping, and the phase of the pumping must be such that the capacitance is reduced whenever the magnitude of voltage across it is maximum. Power is required to pump the capacitance, and this power is converted from the pump frequency to the signal frequency [6].

Parametric amplifiers find their use at frequencies far too high to allow mechanical variation of the capacitance. Instead, a capacitor with nonlinear dielectric, as illustrated in Fig. 5, can be used. The capacitor can be pumped by applying to it a relatively large pumping voltage of proper frequency, amplitude, and phase. Suitable filters will be needed, of course, to separate the signal voltage from the pumping voltage.

A logical extension from a single resonant circuit used in the manner described is to use several such circuits in tandem, as shown in Fig. 6.

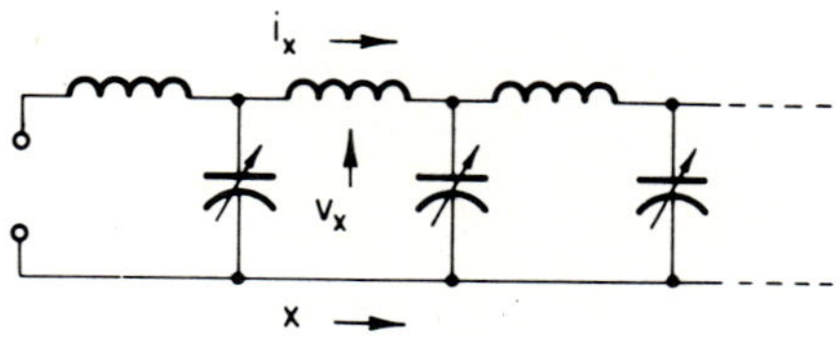

FIG. 6. Multisection parametric amplifier with varying capacitances. As more sections of smaller dimensions are used, a distributed amplifier results.

Each capacitor in such a system must be pumped. Since the phase of the signal voltage will be different as it is developed across each capacitor, the pumping phase must be correspondingly different at each capacitor.

A further extension would be to replace the discrete elements of Fig. 6 by distributed elements. In particular, the capacitance might be provided by a single continuous slab of the proper dielectric material. Both the signal voltage and the pumping voltage could be applied at one end of the slab. It would appear that they should travel along the slab together, maintaining proper phase relations. The power in the signal should increase as it travels along, and increased power should be delivered at the other end of the slab [7].

Current i and voltage v in the system of Fig. 6 are governed by the equations

$$\partial v/\partial x = -l\, \partial i/\partial t \qquad \partial i/\partial x = -c(v)\, \partial v/\partial t \tag{15}$$

which can be combined to give

$$\partial^2 v/\partial x^2 = l \frac{\partial}{\partial t}[c(v)\, \partial v/\partial t] \tag{16}$$

This is a kind of wave equation with the propagation velocity

$$dx/dt = [lc(v)]^{-1/2} \tag{17}$$

Since the capacitance $c(v)$ varies with voltage v, so also the velocity varies. In Fig. 7 is shown a plot of the pump voltage at a given instant in time

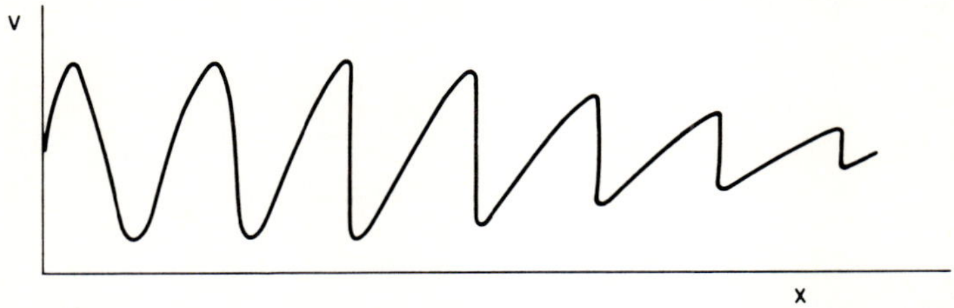

FIG. 7. Instantaneous pump voltage along a distributed parametric amplifier showing shock wave (from Landauer [9]).

as it is distributed along the dielectric slab. A sinusoidal pump voltage was originally applied. The velocity of propagation at the peaks of this wave is greater than that at the troughs, as is evident from Eq. (17). Thus, as the pump wave moves along, the peaks overtake the troughs, ultimately a shock wave results, and then finally the wave disappears. A pump wave of larger amplitude would lead to more amplification of the signal, but would tend to form a shock wave and disappear sooner. As a result, only a limited amount of amplification can be achieved by a system of this sort [8, 9]. It will not produce indefinitely large amplification as might have been hoped from a simple consideration.

This discussion should not be construed as asserting that a distributed parametric amplifier is not possible. The simple form described here does not, however, produce unlimited amplification.

Example 3. Electron Tubes

Many high-vacuum electron tubes [10] consist of a heated cathode, plus one or more additional electrodes, all contained within an evacuated envelope. The cathode produces free electrons, having definite negative electric charge and definite mass, in the space around it. The simplest tube is a diode, with an anode, or plate, located near the cathode. If the plate is made positive with respect to the cathode, electrons may flow between the electrodes and through an external circuit. If the plate is negative, no current flows. Among other functions, therefore, the diode is useful as a rectifier. A section through a diode with plane parallel electrodes is shown in Fig. 8. It is desirable to find a relation among the current in the diode, the voltage across it, and its geometry.

The diode relation can be found by combining several equations. The

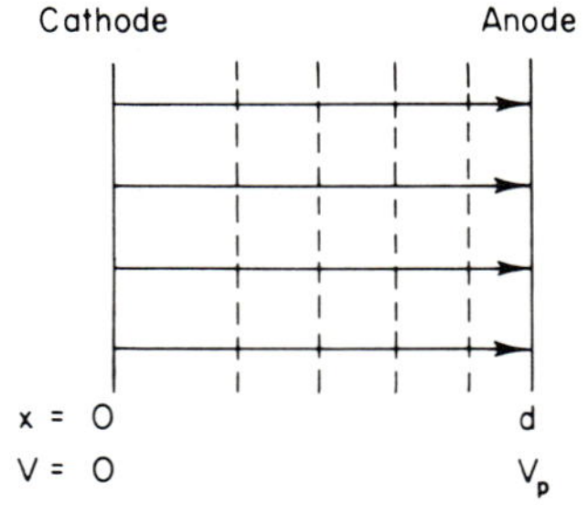

FIG. 8. High vacuum, plane parallel diode showing electron paths and potential contours.

first of these describes how the electric field is influenced by the presence of the charged electrons, and is Poisson's equation,

$$\nabla^2 V = -\rho/\varepsilon \tag{18}$$

Here, V is the electric potential at any point, ρ is the charge density due to the electron cloud, and ε is the permittivity. The second relation describes how the motion of the charged particles is influenced by the electric field, and may be written

$$Vq = mu^2/2 \tag{19}$$

where q is the charge and m the mass of the electron, and u is the velocity of the electron where the potential has the value V. In Eq. (19) it is assumed that $u = 0$ where $V = 0$. This is the equating of the potential energy change as the electron moves through the field, to the kinetic energy it acquires. The velocity is assumed small enough so that relativity effects can be neglected. Finally, the third relation is a kind of continuity equation

$$J = -\rho u \tag{20}$$

where J is current density at the point where ρ and u are evaluated. Three relations of the sort typified by Eqs. (18)–(20) are characteristic of problems where fields and charged particles interact. Furthermore, Eq. (19) is clearly nonlinear.

If ρ and u are eliminated among these equations, the result can be written

$$\nabla^2 V = \left[\frac{m}{2q\varepsilon^2}\right]^{1/2} \frac{J}{V^{1/2}} \tag{21}$$

This is a nonlinear equation in V. Only if the geometry is simple enough that a single space coordinate appears in $\nabla^2 V$ has this equation been solved.

The simplest case is that of the plane parallel diode, of Fig. 8, with edge effects being neglected. Here, $\nabla^2 V = \partial^2 V/\partial x^2$, and with $V = 0$ at $x = 0$ and $V = V_p$ at $x = d$, the resulting solution for J becomes

$$J = \frac{4}{9}\varepsilon\left[\frac{2q}{m}\right]^{1/2} \frac{V_p^{3/2}}{d^2} \tag{22}$$

This is known as Child's law for the diode. Somewhat similar forms for this relation have been derived also for concentric cylindrical and concentric spherical electrodes. In each case, J varies as $V_P^{3/2}$. The solid lines of Fig. 8 represent paths of electrons as they move between cathode and plate. The dotted lines represent contours of constant potential.

A somewhat more complicated diode structure is shown in cross section in Fig. 9. Here the cathode is a cylinder while the plate is a rectangle. It

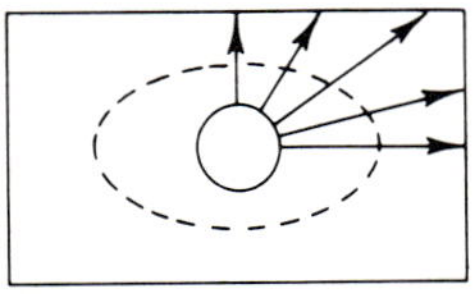

FIG. 9. Diode with cylindrical cathode and rectangular plate.

appears reasonable that the general shapes of electron paths and potential contours would be as shown. This case cannot be analyzed exactly, however.

A plane parallel triode structure is shown in cross section in Fig. 10.

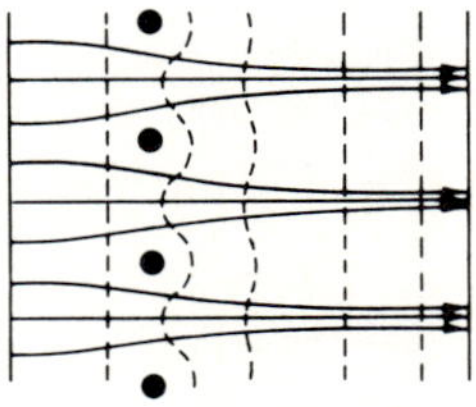

FIG. 10. Plane parallel triode, showing electron paths and potential contours.

In addition to the cathode and plate, the triode has a third electrode known as the grid. This consists of an array of parallel wires in a picket-fence arrangement, located between cathode and plate. In operation the plate is made positive with respect to the cathode, and the grid is usually made negative. The grid provides a sensitive control of electron current between cathode and plate. Reasonable shapes for electron paths and potential contours are shown in Fig. 10. Again, exact solutions are not possible.

Actually, at the time basic studies of tubes were first being made, ingenious methods of approximate analysis were devised [11, 12]. If no electrons are present, the space charge density ρ is zero, and Eq. (18) becomes merely Laplace's equation $\nabla^2 V = 0$. While the triode geometry is relatively complicated, nevertheless the electric field described by Laplace's equation can be found by processes of conformal mapping. Thus, the electric field with no electrons can be determined.

It was recognized that with electrons present the space charge density is most important between cathode and grid, where the electron velocity is low. Between grid and plate the electron velocity is high, and the space charge density is relatively small. The effect of the high density near the cathode can be accounted for by replacing the actual cathode with an equivalent cathode suitably located somewhat closer to the grid. The location is selected to make the potential gradient at the grid with the equivalent cathode and no electrons to be the same as with the actual cathode and electrons. Furthermore, the actual potential in the plane of the grid wires is of one value at the wires themselves, and a different value midway between the wires. A single equivalent potential for the grid plane can be determined, which leads to the proper potentials well away from the grid plane. By following these lines of reasoning it is possible to determine the operation of the triode to good accuracy. A great many successful tubes were designed on the basis of such approximate methods of analysis.

Example 4. Traveling Wave Amplifier

A type of electronic amplifier useful at high frequencies is known as a traveling wave amplifier [13]. It has the desirable property that it can amplify signals over a frequency band that is very broad in comparison with the frequency at the center of the band. This property results because the traveling wave amplifier does not employ resonant circuits, which are important components in most other high frequency amplifiers. The parametric amplifier of Example 2 uses resonant circuits, and it is inherently limited to rather narrow frequency bands.

The traveling wave amplifier is illustrated in very schematic form in Fig. 11. It consists of a structure which often takes the form of a long slender helix of wire, located in an evacuated space. The signal to be

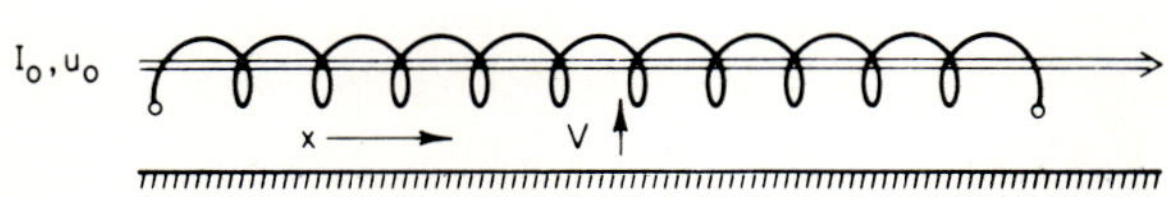

FIG. 11. Traveling wave amplifier, showing helix with electron beam moving along its axis. The signal to be amplified travels along the helix.

amplified is applied to one end of the helix. It travels as a wave along the helix and is removed at the other end. A beam of electrons is projected along the axis of the helix, traveling in the same direction as the signal. The electrons are collected and removed after they have passed along the length of the helix. The electric field produced by the signal in the helix interacts with the electrons, changing their velocity. The intention is that on the average the velocity of the electrons will be reduced. This represents a reduction in power in the electron beam, and this power must be transferred to the wave on the helix. The increased power in the helix appears as an amplification of the original signal. So long as operation takes place as intended, the signal power increases as an exponential function of distance along the helix.

In order to analyze the operation of the traveling wave amplifier [14, 15], the three basic sorts of equations are needed as appeared in the study of the electron tube. First, the electric potential V associated with the wave on the helix is modified by electrons in the beam according to the equation

$$\frac{\partial^2 V}{\partial t^2} - {c_0}^2 \frac{\partial^2 V}{\partial x^2} = c_0 Z_0 \frac{\partial^2 \rho}{\partial t^2} \tag{23}$$

The terms on the left of this equation are readily identified as those of the usual wave equation, with velocity of propagation c_0. The term on the right includes the charge density per unit length ρ in the electron beam, and a kind of characteristic impedance Z_0 for the device. Second, the motion of the electrons is modified by the field of the helix as described by the equation

$$m \frac{d^2 x}{dt^2} = \left[\frac{\partial V}{\partial x} - E_s \right] q \tag{24}$$

where $\partial V / \partial x$ is the gradient of the potential along the helix. In addition, the term E_s is the electric intensity due to the electrons within the beam itself. The beam is often sufficiently intense that this term is important. Finally, a kind of continuity equation can be written as

$$\rho(x, t)\, dx = \rho(x_0, 0)\, dx_0 = (I_0/u_0)\, dx_0 \tag{25}$$

where dx is an increment of length containing a given group of electrons for which the charge density is $\rho(x, t)$. At $t = 0$ and $x = x_0$, this same group of electrons was contained in length dx_0, and $\rho(x_0, 0) = I_0/u_0$ where I_0 is the current and u_0 is the velocity of the beam at $x = x_0$. Because of the strong interactions which occur, there may be electrons of widely differing velocities located at any one given point.

The details of the operation of the traveling wave amplifier may differ considerably, depending upon just what conditions exist. A typical mode of operation leads to the results shown in Figs. 12 and 13. The amplitude A of the signal as it progresses along the helix in the x direction is shown in Fig. 12. The amplitude does, indeed, increase exponentially during the first part of the travel of the signal along the helix. Soon, however, a maximum is reached and afterward the amplitude fluctuates around some mean value.

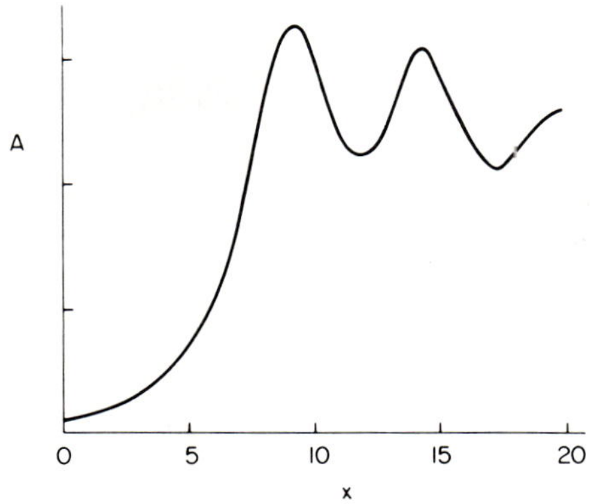

FIG. 12. Signal amplitude A as a function of position x along the helix for a particular mode of operation (from Tien *et al.* [15]).

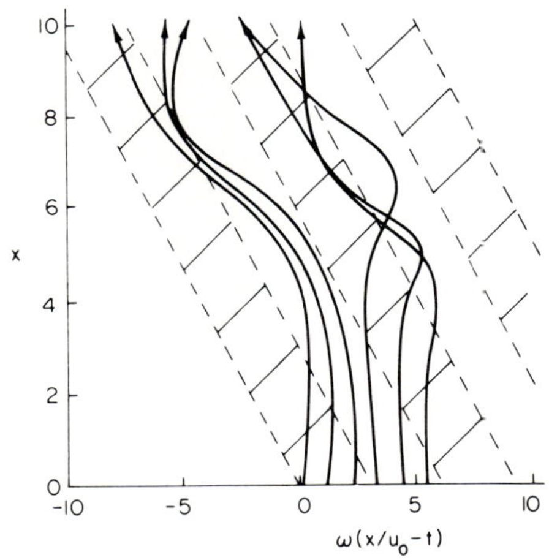

FIG. 13. Electron trajectories for Fig. 12 (from Tien *et al.* [15]).

The diagram in Fig. 13 is a way of representing the trajectories of several different typical electrons. The vertical coordinate is position x. The horizontal coordinate is a dimensionless number formed as the product of the signal frequency ω and the quantity $(x/u_0 - t)$. This latter quantity is the typical argument of the expression for a traveling wave moving in the plus x direction. Electrons injected into the helix at different phases of the signal wave will appear at different points along the horizontal axis. If they travel with constant velocity u_0, their trajectory is represented by a vertical straight line in Fig. 13. If their velocity decreases, the curve tends to the left; if their velocity increases, the curve tends to the right. The shaded areas in the figure are regions in which the field of the helix is such as to accelerate the electrons. The unshaded areas are regions where deceleration occurs. It is evident from a study of the trajectories shown in Fig. 13 that generally deceleration has occurred for x less than about seven units which corresponds to the point of maximum amplitude in Fig. 12. Beyond this point, acceleration of some of the electrons begins again, and the amplitude in Fig. 12 must decrease.

Example 5. Semiconductor Diode

A diode, having rectifying properties similar to the high vacuum diode of Example 3, can be constructed in a quite different way using semiconducting material [16]. In a semiconducting material, electric current may flow because of motion of electrons with negative charge, as is usual in such conducting materials as metals. In addition, there may be conduction by holes, the absence of electrons, which are equivalent to positive charges. Furthermore, electrons and holes may move because of the presence of an electric field, with its potential gradient, in which case a drift current is said to ensue. Electrons and holes may also move because their density varies from one point in the material to another. The resulting density gradient leads to a diffusion current. Thus, the current density $\mathbf{J}_p$ due to holes with positive charge, and $\mathbf{J}_n$ due to electrons with negative charge, can be written

$$\mathbf{J}_p = q(\mu_p p\,\mathbf{E} - D_p \nabla p) \tag{26}$$

$$\mathbf{J}_n = q(\mu_n n\mathbf{E} + D_n \nabla n) \tag{27}$$

where q is the electronic charge, $\mathbf{E}$ is the electric intensity, p is the hole density, n is the electron density, μ_p and μ_n are mobility coefficients, and D_p and D_n are diffusion coefficients. Since $\mathbf{E} = -\nabla V$, the first terms in Eqs. (26) and (27) represent drift currents, while the second terms represent diffusion currents. An important property is that

$$\mu_n/\mu_p = D_n/D_p = b \tag{28}$$

The total current density is, of course,

$$\mathbf{J}_T = \mathbf{J}_p + \mathbf{J}_n \tag{29}$$

Considerable theoretical and experimental physics is buried in Eqs. (26) and (27).

A diode can be created by adding a small amount of impurity to two regions of a slab of semiconducting material. One region, the *P*-region, has impurity added which greatly increases the density of holes present. The other region, the *N*-region, has impurity added which greatly increases the density of electrons. There is a sharp boundary, the junction, between the two regions, as illustrated in Fig. 14. Electrical connections are made

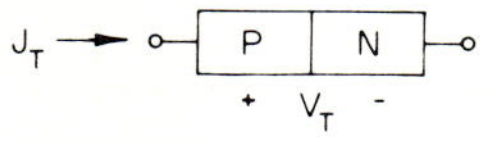

FIG. 14. Semiconductor P-N junction diode.

to the material well away from the junction. The resulting diode will conduct current most easily in one direction, from the *P*-region to the *N*-region. It has many properties analogous to those of the high vacuum diode, without requiring a hot cathode or an evacuated envelope.

Under the assumption that enough impurity is added to the *P*-region so that its conductivity is much higher than that of the *N*-region, it is sufficient to investigate only the density of holes in the *N*-region [17]. A continuity relation is

$$\partial p/\partial t = -(1/\tau)(p - p_0) - (1/q)\,\nabla \cdot \mathbf{J}_p \tag{30}$$

and an assumption of charge neutrality requires that

$$p - p_0 = n - n_0 \tag{31}$$

$$\nabla p = \nabla n \tag{32}$$

In these equations, τ is the lifetime of holes, and p_0 and n_0 are equilibrium densities of holes and electrons, respectively. In the *N*-region, $p_0 << n_0$. For this example, Eqs. (26), (27), and (30) serve the purposes of the three basic equations used in Examples 3 and 4.

A combination of Eqs. (26)–(29), (31), and (32) gives for the electric intensity

$$\mathbf{E} = \frac{\mathbf{J}_T - qD_p(b - 1)\,\nabla p}{q\mu_p[(b + 1)p + b(n_0 - p_0)]} \tag{33}$$

The numerator here represents the total drift current density, and the denominator is the total conductivity, both measured in the N-region.

Finally, by combining Eqs. (26), (30), and (33), the relation for the hole density in the N-region becomes

$$\partial p/\partial t = -\frac{(p - p_0)}{\tau} - \frac{b(n_0 - p_0)[\mathbf{J}_T \cdot \nabla p - qD_p(b-1)(\nabla p \cdot \nabla p)]}{q[(b+1)p + b(n_0 - p_0)]^2} + bD_p \frac{(2p + n_0 - p_0)\nabla^2 p}{(b+1)p + b(n_0 - p_0)} \tag{34}$$

Both the second and third terms of this equation are highly nonlinear.

The simple theory for semiconductor diodes is based on the assumption that $(p - p_0) << n_0$ and that $\mathbf{E} = 0$, in which case Eq. (34) is replaced by the much simpler linear equation

$$\partial p/\partial t = -(1/\tau)(p - p_0) + D_p \nabla^2 p \tag{35}$$

In particular, if $\mathbf{E} = 0$, the entire second term of Eq. (34) is missing.

A study of Eq. (34) leads to information about the total current density J_T in the diode as it is related to the voltage V_T appearing across the diode. The relation between steady state values is shown in Fig. 15. Simple

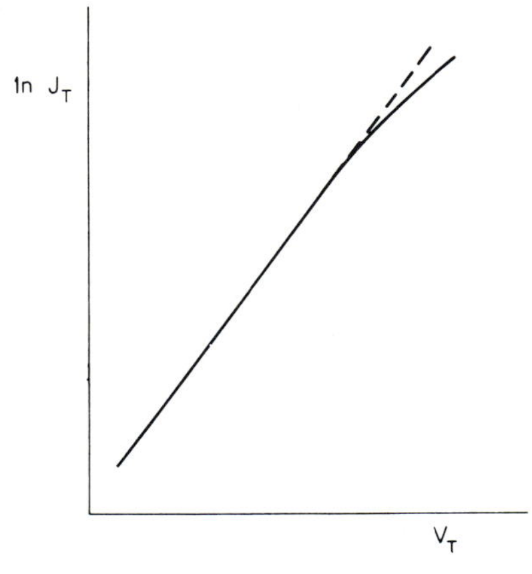

FIG. 15. Steady state voltage V_T and steady state current J_T for junction diode. Dotted curve is result of linear theory. Solid curve is result of more exact theory (from Kano and Reich [17]).

linear theory predicts an exponential relation between J_T and V_T, which would lead to a straight line on the logarithmic plot of Fig. 15. The more complete analysis shows that V_T increases faster than exponential as J_T increases. The increase in V_T results because **E** is not zero in the body of the semiconducting material.

The effect of suddenly applying a steady current of value J_T at zero time is shown in Fig. 16, If J_T is small, V_T rises from zero monotonically approaching a maximum final value. If J_T is large, however, V_T jumps

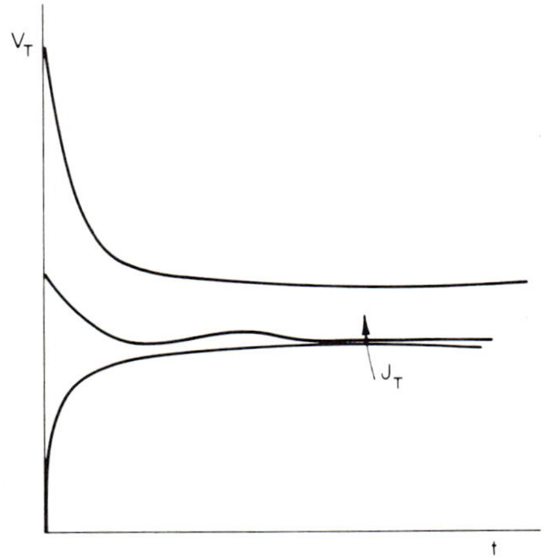

FIG. 16. Transient voltage V_T following the sudden application of a steady current J_T at $t = 0$. Larger values of J_T correspond to the upper curves (from Kano and Reich [17]).

to a large value immediately after the current is applied, and then falls to a final value. If J_T is intermediate in value, V_T initially jumps, falls, and finally rises to the steady state value. These initial changes in V_T again come about because of the fact that **E** is not zero. Instead **E** varies considerably as the current carriers adjust themselves from one condition to another.

These several examples illustrate nonlinear effects produced by the properties of certain materials and by the interaction between electric fields and charged particles. In each case, the operation of the nonlinear system is significantly different from that of a simpler system in which nonlinearity is assumed to be absent.

References

1. Maxwell, J. C., "Electricity and Magnetism." Oxford, 1904.
2. Harrington, R. F., "Electromagnetic Engineering." McGraw-Hill, New York, 1958.
3. Javid, M., and Brown, P. M., "Field Analysis and Electromagnetics." McGraw-Hill, New York, 1963.
4. Poritsky, H., Calculation of Flux Distributions with Saturation, *Trans. AIEE* **70**(I), 309–319 (1951).
5. Trutt, F. C., Erdelyi, E. A., and Jackson, R. F., The Nonlinear Potential Equation and Its Numerical Solution for Highly Saturated Electrical Machines, *IEEE Trans.* AS-1, 430–440 (1963).
6. Manley, J. M., and Rowe, H. E., Some General Properties of Nonlinear Elements, *Proc. IRE* **44**, 904–913 (1956).
7. Tien, P. K., and Suhl, H., A Traveling Wave Ferromagnetic Amplifier, *Proc. IRE* **46**, 700–706 (1958).
8. Roe, G. M., and Boyd, M. R., Parametric Energy Conversion in Distributed Systems, *Proc. IRE* **47**, 1213–1218 (1959).
9. Landauer, R., Shock Waves in Nonlinear Transmission Lines and Their Effect on Parametric Amplification, *IBM J. Res. Develop.*, **4**, 391–401 (1960).
10. Dow, W. G., "Fundamentals of Engineering Electronics." Wiley, New York, 1952.
11. Dow, W. G., Equivalent Electrostatic Circuits for Vacuum Tubes, *Proc. IRE* **28**, 548–556 (1940).
12. Thompson, B. J., Space Current Flow in Vacuum Tube Structures, *Proc. IRE* **31**, 485–491 (1943).
13. Pierce, J. R., "Traveling Wave Tubes." Van Nostrand, Princeton, New Jersey, 1960.
14. Nordsieck, A., Theory of the Large Signal Behavior of Traveling Wave Amplifiers, *Proc. IRE* **41**, 630–637 (1953).
15. Tien, P. K., Walker, L. R., and Wolontis, V. M., A Large Signal Theory of Traveling Wave Amplifiers, *Proc. IRE* **43**, 260–277 (1955).
16. Gaertner, W. W., "Transistors: Principles, Design, and Applications." Van Nostrand, Princeton, New Jersey, 1960.
17. Kano, K., and Reich, H. J., Forward Transient Behavior of P-N Junction Diodes at High Injection Levels, *IEEE Trans.* **ED11**, 515–523 (1964).

Difference Methods and Soft Solutions †

M. H. PROTTER

Department of Mathematics
University of California
Berkeley, California

I. Soft Solutions

We shall be concerned with methods for obtaining generalized solutions of certain classes of partial differential equations. As an example, we consider the equation

$$u_t + uu_x = 0 \tag{1}$$

for $t > 0$ subject to the initial condition

$$u(x, 0) = f(x), \qquad -\infty < x < \infty. \tag{2}$$

It is well known that any solution u of this problem must satisfy the functional relation

$$u(x, t) = f[x - u(x, t)t]. \tag{3}$$

If f is differentiable it is a simple matter to verify that relation (3) implies (1). However, if f is not differentiable on some subset of its domain of definition then u, as given by (3), will not be a strict solution of the differential equation (1) in a corresponding region of the upper half-plane.

Suppose that u is solution of (1), (2) and that $\bar{u}$ is a difference approximation to u defined on some rectangular network of points R: $\{x_n, t_n\}$. Since the domain of $\bar{u}$ is a discrete set, we may attempt to extend the definition of $\bar{u}$ to the entire upper half-plane so as to satisfy Eq. (1). Generally, such an extension is impossible. On the other hand, since (3) is a condition which may or may not hold at individual points of R, we can test whether or not $\bar{u}$ satisfies (3) without extending the definition of $\bar{u}$ to all points of the upper half-plane. Furthermore, this verification

† This work was supported by the Air Force Office of Scientific Research, Contract Number AFOSR 231-63.

can be made whether or not f is smooth. In difference methods for initial value problems it frequently happens that the initial function is known only at a discrete set of points. Under such circumstances it is still possible to verify relation (3) by extending the definition of the initial data to all points on the initial line. Note that this extension does not have to be smooth.

Definition

A function $u(x, t)$ is a *soft solution* of the problem (1), (2) if u satisfies relation (3).

Equation (1) is rather special, and we wish to explore classes of partial differential equations which have solutions satsifying a functional relation. As Noh and the author showed in [2], an equation or a system of equations which satisfies one or more functional relations leads in a natural way to a distinguished set of difference methods.

A simple computation shows that any function u which satisfies the relation

$$u(x, t) = f[x - t\varphi(u)] \tag{4}$$

is a solution of the equation

$$u_t + \varphi(u)u_x = 0 \tag{5}$$

subject to the initial condition (2,) provided that f is differentiable. For an arbitrary function f (not necessarily smooth) the expression (4), when it can be solved for u, is called a *soft solution* of Eq. (5).

It is natural to ask: how wide is the class of first order linear and nonlinear partial differential equations in two independent variables which have associated soft solutions? While this interesting question remains unsolved, we can give examples to show that the class is quite extensive. First, we observe that any equation of the form

$$u_t + [A(u)]_x = 0$$

is expressible in the form (5) whenever A is differentiable, and so must have an associated soft solution. Additionally, it is not difficult to verify that the equation

$$u_t + uu_x = u^n \tag{6}$$

subject to the initial condition (2) has the solution

$$u^{1-n} = (1-n)t + f\left\{x + \frac{1}{n-2}u^{2-n} + \frac{1}{2-n}\left[u^{1-n} + (n-1)t\right]^{(n-2)/(n-1)}\right\}, \qquad n \neq 1, 2. \tag{7}$$

Relation (7) defines a soft solution of (6) whether or not f is differentiable. For $n = 1{,}2$ we find

$$u(x, t) = e^t f[x - u + ue^{-t}] \qquad (n = 1),$$

$$u(x, t) = \frac{f[x + \log(1 - tu)]}{1 + tf[x + \log(1 - tu)]} \qquad (n = 2).$$

Again it is not difficult to get soft solutions for equations of the form

$$u_t + (A(u))_x = u^n$$

provided that A is differentiable. More generally, if $F(u)$ satisfies a simple quadrature condition, then soft solutions exist for the first order equation

$$u_t + (A(u))_x = F(u).$$

In treating first order systems, we use the equations of gas dynamics as a guide for obtaining soft solutions. The one-dimensional compressible flow of a fluid with velocity u, density ρ, and pressure p is governed by the first order system

$$u_t + uu_x + \frac{1}{\rho}p_x = 0 \tag{8a}$$

$$\rho_t + (\rho u)_x = 0 \tag{8b}$$

with p and ρ connected by an equation of state. In the simple (and non-physical) case of constant pressure, Eq. (8a) reduces to Eq. (1). If u and ρ satisfy the initial conditions

$$u(x, 0) = f(x) \qquad \rho(x, 0) = g(x) \tag{9}$$

and if f and g are differentiable, then the solution for ρ with pressure constant) is given by

$$\rho = \frac{g(x - tu)}{1 + tf'(x - tu)}. \tag{10}$$

Whether or not g is differentiable, the expression (10) represents a soft solution of Eq. (8b). More generally, a system of the form

$$u_t + \varphi(u)u_x = 0, \qquad \rho_t + (\rho\varphi)_x = 0$$

with initial conditions (9) has the soft solutions

$$u = f[x - t\varphi(u)], \qquad \rho = \frac{g[x - t\varphi(u)]}{1 + t\varphi'(u)f'[x - t\varphi(u))]}.$$

Further extensions to equations of the type (6) are possible.

We can obtain soft solutions for systems of first order equations in three or more independent variables. For example, the system for the functions $u(x, y, t)$, $v(x, y, t)$ given by

$$\begin{aligned} u_t + uu_x + vu_y &= 0 \\ v_t + uv_x + vv_y &= 0 \end{aligned} \tag{11}$$

subject to the initial conditions

$$u(x, y, 0) = f(x, y), \qquad v(x, y, 0) = g(x, y)$$

has the soft solution

$$u = f(x - ut, y - vt), \qquad v = g(x - ut, y - vt).$$

If we add to (11) an equation for the density ρ:

$$\rho_t + (\rho u)_x + (\rho v)_y = 0$$

so that we again have a simplified gas dynamical system, then it is possible to obtain a soft solution for ρ. If the initial density is given by

$$\rho(x, y, 0) = h(x, y),$$

we find that

$$\rho(x, y, t) = \frac{h(x - ut, y - vt)}{\Delta}$$

where

$$\Delta = (1 + tf_{,1})(1 + tg_{,2}) - t^2 f_{,2} g_{,1}.$$

Extensions of the above analysis to systems of the form

$$u_t + \varphi(u)u_x + \psi(v)u_y = 0 \qquad v_t + \varphi(u)v_x + \psi(v)v_y = 0$$

and the determination of the corresponding soft solutions are completely straightforward.

2. Weak Solutions

A *weak solution* of the problem (1), (2) is any function u such that

$$\iint_D u(v_t + \tfrac{1}{2} uv_x)\, dx\, dt + \int_{-\infty}^{\infty} v(x, 0) f(x)\, dx = 0$$

for every function v which is defined and twice differentiable in the upper half-plane and which vanishes for sufficiently large $t + |x|$. The domain D is the half plane $t > 0$.

Since the definition of a weak solution implies an integrability condition while that of a soft solution does not, it is clear that there is no simple equivalence between weak and soft solutions of a given equation. Furthermore, for nonsmooth initial data a soft solution is not necessarily unique. We observe that a functional relation such as (3) determines u explicitly (at least locally) only when an implicit function theorem is applicable to f. Of course in those regions where weak and soft solutions are smooth they necessarily coincide because of the uniqueness of strict solutions to the initial value problem for first order equations and systems.

A weak solution may be smooth in various regions separated by curves along which the solution is not differentiable or even continuous. Whenever the differential equation is in conservation form, weak solutions necessarily satisfy certain jump conditions across curves of discontinuity. There is no such requirement for soft solutions. However, it is not difficult to show that those particular soft solutions which are smooth in the regions separated by curves of discontinuity and which satisfy appropriate jump conditions across these curves are in fact weak solutions.

3. Exact Difference Methods

Soft solutions of differential equations lead in a natural way to difference methods. To illustrate the process we show how to obtain various difference methods from the soft solution

$$u = f(x - ut) \tag{12}$$

of the equation

$$u_t + uu_x = 0. \tag{13}$$

Of course, the initial values are

$$u(x, 0) = f(x), \qquad -\infty < x < \infty. \tag{14}$$

Suppose we introduce a fixed rectangular grid in the upper half plane with mesh sizes Δx and Δt in the x and t directions, respectively. Let $t_n = n\,\Delta t$ and denote by

$$u^n(x) \equiv u(x, t_n)$$

the solution of (13), (14) at time $t = t_n$. Because of (12), we have

$$u^n(x) = f[x - t_n u(x, t_n)] \equiv f[x - t_n u^n(x)]. \tag{15}$$

Hence, at the time $t_{n+1} = (n + 1)\,\Delta t$, we have from (15)

$$u^{n+1}(x) = f\{x - t_n u^n[x - u^{n+1}(x)\,\Delta t] - (\Delta t)u^{n+1}(x)\}. \tag{16}$$

On the other hand, using (12) directly, we find

$$u^{n+1}(x) = f[x - u^{n+1}(x)t_{n+1}] = f[x - u^{n+1}(x)t_n - u^{n+1}(x)(\Delta t)]. \tag{17}$$

Since (16) and (17) must coincide, we obtain

$$u^{n+1}(x) = u^n[x - u^{n+1}(x)(\Delta t)]. \tag{18}$$

Suppose we have a difference method which determines a solution of a difference equation corresponding to (13). The values of this solution at the mesh points $\{k\,\Delta x, n\,\Delta t\}$ $k = 0, \pm 1, \pm 2, \ldots, n = 0, 1, 2, \ldots$ are denoted by $\{u_k{}^n\}$. Then if this solution satisfies

$$u_k^{n+1} = u^n(x_k - u_k^{n+1}\,\Delta t) \tag{19}$$

where $x_k = k\,\Delta x$ and $u^n = u(x_k, n\,\Delta t)$, then $\{u_k^{n+1}\}$ will coincide precisely with the soft solution as given by (18). A difference method which, at the mesh points where it is defined, coincides with the corresponding soft solution is called an *exact difference method*. We now show how such exact methods can be obtained.

A soft solution is determined when the initial values (14) are prescribed. Suppose that $u_k{}^0$, $k = 0, \pm 1, \pm 2, \ldots$ are given and that f is determined by linear interpolation between mesh points on the initial line. That is, we have

$$u^0(x) = u(x,0) = f(x) = u_k{}^0 + (x - x_k)\frac{u_k{}^0 - u_{k+1}^0}{\Delta x}, \qquad x_{k-1} \le x \le x_k . \quad (20)$$

If we wish to satisfy (19), a difference method is obtained at once from (20). We find (if $u_k{}^1 \ge 0$)

$$u_k{}^1 = u_k{}^0 - u_k{}^1 \,\Delta t \,\frac{u_k{}^0 - u_{k-1}^0}{\Delta x} \qquad \text{for} \quad |u_k{}^1|\,\Delta t \le \Delta x$$

and a corresponding expression replacing $u_k{}^0 - u_{k-1}^0$ by $u_{k+1}^0 - u_k{}^0$ if $u_k{}^1 < 0$. Solving for $u_k{}^1$ we get

$$\begin{aligned} u_k{}^1 &= \frac{u_k{}^0}{1 + \dfrac{\Delta t}{\Delta x}(u_k{}^0 - u_{k-1}^0)} \qquad \text{if} \quad u_k{}^1 \ge 0, \\ &= \frac{u_k{}^0}{1 + \dfrac{\Delta t}{\Delta x}(u_{k+1}^0 - u_k{}^0)} \qquad \text{if} \quad u_k{}^1 < 0. \end{aligned} \tag{21}$$

The difference method (21) is exact in the sense described above. More generally, if $u^n(x)$ is a piecewise linear function at the nth time step and is taken as initial values for Eq. (13), then the difference method described by (20) with $n + 1$ in place of 1 and n in place of 0 gives an exact difference method at time step $n + 1$.

Each method of interpolation between successive initial values $\{u_k{}^0\}$, $k = 0, \pm 1, \pm 2, \ldots$ leads to a corresponding exact difference method. For example if quadratic interpolation is used at the nth step, then the corresponding exact method for (13), (14) at time step $n + 1$ is given by

$$u_k^{n+1} = \frac{1 + \frac{1}{2}\lambda(\Delta u_k{}^n) - [(1 + \frac{1}{2}\lambda\,\Delta u_k{}^n)^2 - 2u_k{}^n \lambda^2 \Delta^2 u_k{}^n]^{1/2}}{\lambda^2 \Delta^2 u_k{}^n}$$

where

$$\lambda = \Delta t/\Delta x, \qquad \Delta u_k{}^n = u_{k+1}^n - u_{k-1}^n, \qquad \Delta^2 u_k{}^n = u_{k+1}^n - 2u_k{}^n + u_{k-1}^n.$$

Since soft solutions have been obtained for a variety of simplified versions of the equations of gas dynamics, exact difference methods have been derived for these and related systems. The reader is referred to [2] for specific descriptions of these methods.

4. Second Order Equations

We consider the one-dimensional wave equation

$$u_{tt} = c^2 u_{xx} \tag{22}$$

subject to the initial conditions

$$u(x, 0) = f(x) \qquad u_t(x, 0) = g(x). \tag{23}$$

For convenience we choose units so that $c = 1$, in which case the general solution of (22), (23) is given by

$$u(x, t) = \frac{f(x + t) + f(x - t)}{2} + \frac{1}{2}\int_{x-t}^{x+t} g(\xi)\, d\xi. \tag{24}$$

For arbitrary f and merely integrable g, Eq. (24) defines a soft solution of the problem (22), (23). As in Section 3, these soft solutions lead to difference methods. Suppose that we are given the initial values $\{u_k{}^0\}$, $\{u_k{}^1\}$, $k = 0, \pm 1, \pm 2, \ldots$ for the problem (22), (23). If we use quadratic interpolation for f and assume that g is a step function, we obtain the exact difference method

$$u_k{}^2 = 2u_k{}^1 - u_k{}^0 + \left(\frac{\Delta t}{\Delta x}\right)^2 (u_{k+1}^1 - 2u_k{}^1 + u_{k-1}^1).$$

We recognize this difference method as the standard symmetric difference approximation to Eq. (22), known to be stable for $\Delta t \le \Delta x$. Each specific choice of interpolation functions for f and g leads to an exact difference method.

Equations of the form

$$u_{tt} = c^2 u_{xx} + \alpha(x, t) u_x + \beta(x, t) u_t$$

have soft solutions of the form

$$u(x, t) = f(ax + bt)$$

where a and b are functions of x and t determined by α and β.

In certain cases it is possible to obtain soft solutions of second order hyperbolic equations in more than two independent variables. Let $u(x, y, t)$ be a solution of the wave equation

$$u_{tt} = u_{xx} + u_{yy}. \tag{25}$$

Sobolev [5] has shown that for arbitrary f and g, the relation

$$x + y f(u) + (1 + f^2(u))^{1/2} t + h(u) = 0 \tag{26}$$

holds. If f and h are such that we can solve (26) for u in terms of the remaining quantities, then we have a soft solution of (25). Of course, such a formula does not solve the initial value problem and so is not well adapted for obtaining exact difference methods. A generalization of (26) to solutions of the wave equation in any number of variables was obtained by Erugin [1].

We consider the hyperbolic equation

$$a_{11}u_{xx} + 2a_{12}u_{xy} + a_{22}u_{yy} - u_{tt} + b_1 u_x + b_2 u_y - \beta u_t = 0 \tag{27}$$

with a_{11} and a_{22} positive. In the special case

$$a_{11}\, a_{22} - a_{12}^2 = 0, \tag{28}$$

it was shown by Smirnov [3, 4] that there exist functionally invariant solutions. In particular, if $b_1 = b_2 = \beta = 0$ and if the a_{ij} are constant, we find that

$$u(x, y, t) = F(x + \sqrt{a_{11}}\, t, y + \sqrt{a_{22}}\, t) + G(x - \sqrt{a_{11}}\, t, y - \sqrt{a_{22}}\, t)$$

is a general solution of (27). We now may define a soft solution, solve the initial value problem, and obtain exact difference methods. Of course, condition (28) is quite restrictive since it implies that (27) is hyperbolic-parabolic, and it would be desirable to obtain soft solutions for the wave Eq. (25).

If F, G, H, K, L, M are arbitrary functions, then

$$\begin{aligned} u(x, y, t) = {} & F(x + y + \sqrt{2}\, t) + G(x + y - \sqrt{2}\, t) + H(x + t) \\ & + K(x - t) + L(y + t) + M(y - t) \end{aligned} \tag{29}$$

is a solution of (25) whenever the above six functions are twice differentiable. More generally, (29) defines a soft solution of (25). If u satisfies the initial conditions

$$u(x, y, 0) = f(x, y) \qquad u_t(x, y, 0) = g(x, y), \tag{30}$$

we see that (29), being a combination of functions of a single variable, does not provide a general solution of the problem (25), (30). However, suppose that f and g are polynomials of degree no higher than three. Then it is easy to see that six functions in (29) can be chosen as polynomials so as to satisfy (30) at $t = 0$. From the general uniqueness theorem for solutions of the initial value problem (25), (30), we conclude that (29) provides the precise solution for f and g third degree polynomials. Since most interpolation functions used in difference methods are linear or quadratic, we find exact difference methods from the soft solutions (29).

References

1. Erugin, N. P., Functionally Invariant Solutions of Second Order Equations, *LGU, Math. Ser. NAUK*, **16** (1949).
2. Noh, W. F., and Protter, M. H., Difference Methods and the Equations of Hydrodynamics *J. Math. Mech.* **12** (1963).
3. Smirnov, M. M., Functionally Invariant Solutions of the Wave Equation, *Doklady Akad. Nauk SSSR*, **67**, No. 6 (1949).
4. Smirnov, M. M., Functionally Invariant Solutions of Equations of Hyperbolic-Parabolic Typet *Priklad. Mat. Mek.* **17** (1953).
5. Sobolev, S. L., *Matemat. Sbornik* **40**, No. 2 (1953).

Numerical Solution of the Nonlinear Equations for Two-Phase Flow through Porous Media

D. W. PEACEMAN

Esso Production Research Company
Houston, Texas

Introduction

Oil and gas are found in underground porous formations, usually in the company of water. The reservoir engineer needs to predict the movement of these fluids through a reservoir under the influence of various boundary conditions, in order to plan for recovery of the most hydrocarbon at the least cost. While the differential equations which govern multiphase flow through porous media have been known for many years, until recently only very simplified mathematical models could be used for predicting reservoir performance. With the advent of modern high speed computers and the development of suitable numerical methods, solutions of two-phase flow problems in two space dimensions and time with quite arbitrary geometry, reservoir properties, and fluid properties have become fairly routine.

The differential equations which describe multiphase flow are simultaneous, highly nonlinear partial differential equations. The numerical procedures for solving these equations, described briefly below, have evolved from methods developed for linear parabolic and elliptic differential equations. In many cases of practical interest, these procedures have proven quite satisfactory; in other cases, difficulties have arisen which sharply increase the amount of computing necessary to obtain sufficiently accurate solutions. Since these difficulties stem from the nonlinear nature of the differential equations, it seems appropriate for this symposium that we examine the numerical methods, with emphasis on their nonlinear aspects.

The Differential Equations

A brief summary of the mechanics of fluid flow through porous media will be given here that will be sufficient for derivation of the differential equations. More thorough discussions can be found elsewhere [1].

For a single phase, the mass rate of transport of fluid per unit area is given by Darcy's law:

$$\mathbf{q} = -\frac{\rho k}{\mu} \nabla p,$$

where ρ is the density of fluid, μ the viscosity of fluid, p the pressure, and k the permeability.

This equation serves as the definition of permeability, which is a property of the porous medium and is independent of the nature of the fluid. In the presence of gravity, the equation is modified by the addition of a gravity term:

$$\mathbf{q} = -\frac{\rho k}{\mu}(\nabla p + \rho g \nabla h), \tag{1}$$

where g is the acceleration due to gravity and h the height above an arbitrary datum level. Continuity (i.e., conservation of mass) requires that

$$-\nabla \cdot \mathbf{q} = \phi \frac{\partial \rho}{\partial t}, \tag{2}$$

where ϕ is the porosity (fraction of the bulk volume of the material occupied by voids) and t the time.

Combining (1) and (2) gives the differential equation for single-phase flow:

$$\nabla \cdot \left[\frac{\rho k}{\mu}(\nabla p + \rho g \nabla h)\right] = \phi \frac{\partial \rho}{\partial t}. \tag{3}$$

When two immiscible fluids (i.e., oil and water, or oil and gas) flow simultaneously through the pores in two-phase flow, the situation is more complicated. One of the fluids wets the rock more than the other; we refer to this as the wetting phase fluid (and use the subscript w), and we refer to the other as the nonwetting phase fluid (and use the subscript n). Because of surface tension and the curvature of the interfaces between the two fluids, the pressure in the nonwetting fluid is higher than in the wetting fluid. The difference between these two pressures is known as the capillary pressure, p_c:

$$p_c = p_n - p_w. \tag{4}$$

Saturation is defined as the fraction of the void volume of the porous medium filled by the fluid in question. Since the two fluids jointly fill the void space, we have

$$S_n + S_w = 1. \tag{5}$$

Empirically, it has been found that capillary pressure can be considered a unique function of saturation. Thus

$$p_n - p_w = p_c(S), \tag{6}$$

where

$$S = S_w = 1 - S_n. \tag{7}$$

A typical capillary pressure curve is shown in Fig. 1.

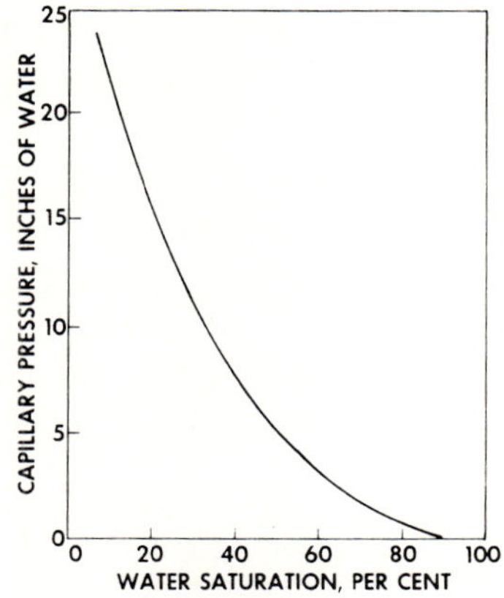

FIG. 1. Typical capillary pressure function.

Finally, Darcy's law must be modified for two-phase flow. Thus,

$$\mathbf{q}_n = -\frac{\rho_n k_n}{\mu_n}(\nabla p_n + \rho_n g \nabla h), \tag{8a}$$

$$\mathbf{q}_w = -\frac{\rho_w k_w}{\mu_w}(\nabla p_w + \rho_w g \nabla h). \tag{8b}$$

Because the simultaneous flow of the two fluids causes each to interfere with the flow of the other, we do not expect, nor do we find, that k_n or k_w is equal to the single-phase permeability, k. On the other hand, experiment shows that they are independent of flow rate and fluid properties and depend only on saturation, S. Relative permeabilities are defined by

$$k_{rn} = k_n/k, \tag{9a}$$

$$k_{rw} = k_w/k. \tag{9b}$$

A typical plot of these functions is given in Fig. 2.

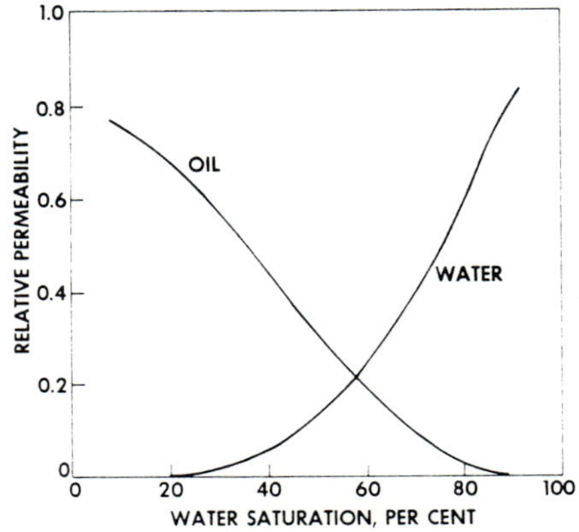

FIG. 2. Typical relative permeability functions.

With the assumption that the two fluids are immiscible and that there is no mass transfer between them, continuity for each fluid gives

$$-\nabla \cdot \mathbf{q}_n = \phi \frac{\partial}{\partial t}(\rho_n S_n), \tag{10a}$$

$$-\nabla \cdot \mathbf{q}_w = \phi \frac{\partial}{\partial t}(\rho_w S_w). \tag{10b}$$

Combining equations (7)–(10) gives

$$\nabla \cdot \left[\frac{\rho_n k\, k_{rn}}{\mu_n}(\nabla p_n + \rho_n g\, \nabla h)\right] = \phi \frac{\partial}{\partial t}[\rho_n(1 - S)], \tag{11a}$$

$$\nabla \cdot \left[\frac{\rho_w k\, k_{rw}}{\mu_w}(\nabla p_w + \rho_w g\, \nabla h)\right] = \phi \frac{\partial}{\partial t}[\rho_w S]. \tag{11b}$$

Associated with these differential equations are the following functional relations:

Functions of position: $k(x, y)$, $h(x, y)$, $\phi(x, y)$.
Functions of pressure: $\rho_n(p_n)$, $\rho_w(p_w)$, $\mu_n(p_n)$. $\mu_w(p_w)$.
Functions of saturation: $k_{rn}(S)$, $k_{rw}(S)$, $p_n - p_w = p_c(S)$.

Our programs are designed to solve Eqs. (11a) and (11b) in their full generality, subject to various boundary conditions which correspond to inputs (injection) and outputs (production) at wells placed at arbitrary locations. The numerical procedures for the general case are given in Douglas *et al.* [2] and Blair and Peaceman [3].

However, for our present discussion of solution of nonlinear equations, it suffices to consider a simplified form of Eqs. (11). We ignore the influence of gravity and the effect of pressure on density and viscosity. Further, let us write

$$KN = \frac{k\,k_{rn}}{\mu_n}, \qquad KW = \frac{k\,k_{rw}}{\mu_w}, \tag{12a}$$

$$PN = p_n, \qquad PW = p_w. \tag{12b}$$

Then Eqs. (11) become

$$\nabla \cdot KN\, \nabla PN = -\phi \frac{\partial S}{\partial t} = \phi S' \left(\frac{\partial PW}{\partial t} - \frac{\partial PN}{\partial t} \right), \tag{13a}$$

$$\nabla \cdot KW\, \nabla PW = \phi \frac{\partial S}{\partial t} = \phi S' \left(\frac{\partial PN}{\partial t} - \frac{\partial PW}{\partial t} \right), \tag{13b}$$

where

$$S' = \frac{dS}{dp_c}. \tag{14}$$

Note that S' is always negative.

Thus, we have a set of simultaneous nonlinear partial differential equations in the variables PN and PW, with nonlinear coefficients KN, KW, and S' which are functions of the difference PN-PW.

An alternate form of Eqs. (13) will be of interest. Let us define

$$M = KN + KW, \tag{15a}$$

$$N = KN - KW, \tag{15b}$$

$$P = (PN + PW)/2, \tag{15c}$$

$$R = (PN - PW)/2. \tag{15d}$$

Note that R is half the capillary pressure, and is therefore a unique function of S. By adding and subtracting Eqs. (13a) and (13b), we obtain

$$\nabla \cdot M \nabla P + \nabla \cdot N \nabla R = 0, \tag{16a}$$

$$\nabla \cdot M \nabla R + \nabla \cdot N \nabla P = -2\phi \frac{\partial S}{\partial t} = -4\phi S' \frac{\partial R}{\partial t}. \tag{16b}$$

Here we have a set of simultaneous nonlinear partial differential equations in the variables P and R, with nonlinear coefficients M, N, and S' which are functions of R.

Solution by Finite Difference Equations

First we consider finite difference analogues of the system of equations (13). We define the following symbols:

$$x_i = i\,\Delta x, \qquad y_j = j\,\Delta y, \qquad t_n = n\,\Delta t,$$

$$PN_{i,j,n} = PN(x_i, y_j, t_n),$$

$$KN_{i+\frac{1}{2},j,n} = k(x_{i+\frac{1}{2}}, y_j, t_n)\,(k_{rn_{ijn}} + k_{rn_{i+1,j,n}})/2\mu_n,$$

$$\Delta_x(KN_n\,\Delta_x PN_{n'})_{ij} = \frac{1}{\Delta x^2}\,[KN_{i+\frac{1}{2},j,n}(PN_{i+1,j,n'} - PN_{i,j,n'}) - KN_{i-\frac{1}{2},j,n}(PN_{i,j,n'} - PN_{i-1,j,n'})],$$

$$\Delta(KN_n\,\Delta PN_{n'})_{ij} = \Delta_x(KN_n\,\Delta_x PN_{n'})_{ij} + \Delta_y(KN_n\,\Delta_y PN_{n'})_{ij}.$$

Since Eqs. (13) resemble parabolic differential equations, it is natural to consider using techniques well known for solving such equations. See, for example, Richtmyer [4]. The first procedure that comes to mind is an explicit one:

$$\Delta(KN_n\,\Delta PN_n)_{ij} = (\phi S'_{ij}/\Delta t)(PW_{i,j,n+1} - PW_{ijn} - PN_{i,j,n+1} + PN_{ijn}), \tag{17a}$$

$$\Delta(KW_n\,\Delta PW_n)_{ij} = (\phi S'_{ij}/\Delta t)(PN_{i,j,n+1} - PN_{ijn} - PW_{i,j,n+1} + PW_{ijn}). \tag{17b}$$

However, a linearized von Neumann stability analysis [4] of this difference system, assuming that KN, KW, and S' are constant, shows that this procedure is unstable. Another approach would be to use the Peaceman–Rachfold alternating-direction procedure [5] modified for use with two dependent variables. Again, a linearized von Neumann stability analysis shows that this would not be stable, either. In spite of the similarity of Eqs. (13) to the heat flow equation, there is something about the way they are coupled together through their right-hand sides that causes this system to behave in a nonparabolic way. This can best be seen, perhaps, from Eqs. (16), which are equivalent to Eqs. (13). While Eq. (16b) appears parabolic in nature, the absence of an accumulation term in Eq. (16a) shows it to be elliptic in nature.

Thus, a fully implicit procedure is indicated, as given by the following:

$$\Delta(KN_n \, \Delta PN_{n+1})_{ij} = (\phi S'_{ij}/\Delta t)\,(PW_{i,j,n+1} - PW_{ijn} - PN_{i,j,n+1} + PN_{ijn}), \tag{18a}$$

$$\Delta(KW_n \, \Delta PW_{n+1})_{ij} = (\phi S'_{ij}/\Delta t)\,(PN_{i,j,n+1} - PN_{ijn} - PW_{i,j,n+1} + PW_{ijn}). \tag{18b}$$

The linearized von Neumann stability analysis shows this to be stable. Douglas [6] demonstrated the convergence of (18) to (13), but had to ignore the lower order terms in the differential operator. (For example, $(\partial K/\partial x)\,(\partial P/\partial x)$ is ignored in $(\partial/\partial x)\,(K\,(\partial P/\partial x)) = K(\partial^2 P/\partial x^2) + (\partial K/\partial x)\,(\partial P/\partial x)$). He points out that it is a commonly accepted, though heuristic, principle that lower order terms in a differential operator do not affect the convergence or divergence of a difference analogue of the differential operator.

Use of Eqs. (18) leads to a system of linear algebraic equations which are solved by an extension of the Peaceman–Rachford alternating-direction iteration procedure for elliptic equations [5]. This extension to two dependent variables at each grid point is presented fully in the appendix of Douglas *et al.* [2].

Another procedure is given in [2] which can reduce the computing labor required for solution. This involves use of the following difference analogues for Eqs. (16):

$$\Delta(M_n \, \Delta P_{n+1})_{ij} + \Delta(N_n \, \Delta R_n)_{ij} = 0, \tag{19a}$$

$$\Delta(M_n \, \Delta R_{n+1})_{ij} + \Delta(N_n \, \Delta P_{n+1})_{ij} = -4(\phi S'_{ij}/\Delta t)\,(R_{i,j,n+1} - R_{ijn}). \tag{19b}$$

If the second term of Eq. (19a) had been $\Delta(N_n \, \Delta R_{n+1})_{ij}$, then these equations would be exactly equivalent to Eqs. (18). However, the use of the old value of R at this point means that P is the only unknown in Eq. (19a), and R is the only unknown in Eq. (19b). Thus, the less laborious alternating-direction iteration procedure for one dependent variable can be used to solve, first Eq. (19a), and then Eq. (19b). In many problems, we have found the use of (19) satisfactory. However, it has a larger time truncation error than (18), and in some problems we have found that comparable, or better, accuracy can be obtained with the same computer labor using (18), because larger time steps and/or fewer iterations could be used.

From the point of view of nonlinear behavior, both difference systems (18) and (19) are similar, and we shall confine our attention to (18). We shall examine several features of this finite difference method which arise from the nonlinear nature of the differential equations.

Evaluation of the Nonlinear Coefficient S′

While the coefficients KN and KW have been assigned subscripts of n in Eqs. (18) to indicate that they are evaluated at t_n, no such assignment was made for S'_{ij} in order to leave that question open. Since the time derivatives of (13) have been approximated by backward differences, Eqs. (18) are only first-order correct in time. Consequently, our first approach was to evaluate S'_{ij} at t_n. Results of such calculations indicated rather large time truncation errors which could be sufficiently reduced only by excessively small time steps. This time truncation error manifested itself particularly in the calculated material balance, which will be discussed further below. Thus, even though the use of S'_{ijn} brings about an error of the same order, $O(\Delta t)$, as that caused by use of backward difference equations, it was clear that a better procedure for evaluating S' was needed.

Douglas [7, 8], Douglas and Jones [15], and Lees, (see Chapter 11 of this volume) have introduced several predictor-corrector methods for quasilinear equations. These methods could have been applied here to estimate S'_{ij} at $t_{n+\frac{1}{2}}$ and most probably would have strongly reduced the time truncation error. However, we can take advantage of the fact that, due to the multidimensional nature of the problem, we need to iterate for $PN_{i,j,n+1}$ and $PW_{i,j,n+1}$. Within this iteration, we can also include an iteration for S'_{ij}. This has been done in two ways.

First Method

After k iterations, iterates $p_{c_{ij}}^{(k)} = PN_{i,j,n+1}^{(k)} - PW_{i,j,n+1}^{(k)}$ are available. Then $S_{ij}^{(k)}$ is evaluated from $p_{c_{ij}}^{(k)}$ using the known functional relation between S and p_c. Finally,

$$S_{ij}'^{(k)} = \frac{S_{ij}^{(k)} - S_{ij,n}}{p_{c_{ij}}^{(k)} - p_{c_{ijn}}}. \tag{20}$$

This value of $S_{ij}'^{(k)}$ is used in the solution for the next set of iterates, $PN_{i,j,n+1}^{(k+1)}$ and $PW_{i,j,n+1}^{(k+1)}$. To start the process off, we take

$$S_{ij}'^{(0)} = S'_{ijn}.$$

For the most part, this method has worked well and does not seem to slow down the iteration for the pressures. Occasionally, when curves of p_c vs. S with large second derivatives are used, and large time steps are taken, this procedure fails to converge. Consequently another method has been used which has been found to work over a wider range of conditions.

Second Method

After the kth iteration, which used an $S_{ij}'^{(k-1)}$, we have obtained iterates $P_{c_{ij}}^{(k)} = PN_{i,j,n+1}^{(k)} - PW_{i,j,n+1}^{(k)}$. We obtain $S_{ij}^{(k)}$ from

$$S_{ij}^{(k)} - S_{ijn} = S_{ij}'^{(k-1)}(p_{c_{ij}}^{(k)} - p_{c_{ijn}}). \tag{21}$$

Then $p_{c_{ij}}^{*(k)}$ is evaluated from $S_{ij}^{(k)}$ using the known functional relation between S and p_c. Finally,

$$S_{ij}'^{(k)} = \frac{S_{ij}^{(k)} - S_{ijn}}{p_{c_{ij}}^{*(k)} - p_{c_{ijn}}}. \tag{22}$$

This is used to obtain the next set of iterates. Again, $S_{ij}'^{(0)} = S'_{ijn}$ is used to start the process off.

After convergence of the iteration, both of these methods yield the result that

$$S'_{ij} = \frac{S_{i,j,n+1} - S_{ijn}}{p_{c_{i,j,n+1}} - p_{c_{ijn}}}.$$

Substituting into Eqs. (18) gives

$$\Delta(KN_n\,\Delta PN_{n+1})_{ij} = -(\phi/\Delta t)\,(S_{i,j,n+1} - S_{ijn}), \tag{23a}$$

$$\Delta(KW_n\,\Delta PW_{n+1})_{ij} = (\phi/\Delta t)\,(S_{i,j,n+1} - S_{ijn}). \tag{23b}$$

Thus, to the extent that the iteration converges, we have eliminated completely the error introduced by any approximation for the nonlinear coefficient S'.

Finally, by this device, we have reduced our material balance equation to an identity. To see this, we need to take into account the inputs and outputs, which we have ignored up to now. These are accounted for by including input-output terms QN_{ij} and QW_{ij} at each point at which there exists a well. QN represents a volume rate of input of nonwetting phase (negative, if output); QW, input of wetting phase. Taking these into account modifies Eqs. (23) to appear thus:

$$\Delta(KN_n\,\Delta PN_{n+1})_{ij} + \frac{QN_{ij}}{\Delta x\,\Delta y} = -\frac{\phi}{\Delta t}(S_{i,j,n+1} - S_{ijn}), \tag{24a}$$

$$\Delta(KW_n\,\Delta PW_{n+1})_{ij} + \frac{QW_{ij}}{\Delta x\,\Delta y} = \frac{\phi}{\Delta t}(S_{i,j,n+1} - S_{ijn}). \tag{24b}$$

The material balance equation states that total rate of input of wetting phase should equal

$$\phi\frac{\partial}{\partial t}\iint S\,dx\,dy.$$

In finite difference form, the material balance equation is

$$\sum_i\sum_j QW_{ij} = \frac{\phi}{\Delta t}\sum_i\sum_j (S_{i,j,n+1} - S_{ijn})\,\Delta x\,\Delta y. \tag{25}$$

Now, if we add up Eq. (24b) for every point, noting that

$$\sum_i\sum_j \Delta(KW_n\,\Delta PW_{n+1})_{ij} = 0,$$

then clearly we will get Eq. (25). This identity would not have been true had we not iterated on S'.

Because the material balance equation is an identity, it is used as a check on the calculations, as well as a check on the convergence of the iteration.

Limiting Form of Equations at Zero Capillary Pressure

While the above implicit difference methods, together with iteration on S', have been applied successfully to many reservoir studies, there have been some situations where application has been difficult, if not impossible. The difficult situations seem to separate into two areas: one is characterized by high fluid velocities and unstable or oscillatory behavior of the solution which requires very small time steps to eliminate; the other area of difficulty can occur at low fluid velocities and is characterized by apparently large time truncation error and poor convergence to certain known analytical solutions.

In these situations, it has been found that increasing the capillary pressure function (in a multiplicative way, so that the magnitude of the slope of p_c vs. S increases) will reduce the difficulty. This suggests that the difference equations should be studied in the limiting case where capillary pressure goes to zero in order to expose the difficulty more clearly. Since the differential equations (13) and (16) and the difference equations (18) and (19) involve the pressures as unknowns, they cannot be used in their present form in order to study this limiting behavior. For this purpose, another form of these equations is necessary.

Let us define a "total flow vector" $\mathbf{Q}$ such that

$$\mathbf{Q} = -(M\,\nabla P + N\,\nabla R) = -(KN\,\nabla PN + KW\nabla PW). \tag{26}$$

Solving for ∇P gives

$$\nabla P = -\frac{1}{M}\mathbf{Q} - \frac{N}{M}\nabla R.$$

Substituting into Eq. (16b) gives

$$\nabla \cdot \left(M - \frac{N^2}{M}\right)\nabla R - \nabla \cdot \left(\frac{N}{M}\right)\mathbf{Q} = -2\phi\,\frac{\partial S}{\partial t}. \tag{27}$$

But

$$\nabla \cdot \left(\frac{N}{M}\right) \mathbf{Q} = \frac{N}{M} \nabla \cdot \mathbf{Q} + \nabla \left(\frac{N}{M}\right) \cdot \mathbf{Q}.$$

Now, by Eq. (16a), $\nabla \cdot \mathbf{Q} = 0$. Also

$$\frac{N}{M} = \frac{KN - KW}{KN + KW} = 1 - 2f, \tag{28}$$

where

$$f = \frac{KW}{KN + KW}. \tag{29}$$

Note that f is a unique function of S. (It is equal to the fraction flowing of the wetting phase.) Further, let

$$2g = M - \frac{N^2}{M} = \frac{4\,KN\,KW}{KN + KW}, \tag{30}$$

also a unique function of saturation. Then Eq. (27) can be written as

$$\nabla g\, \nabla R + \nabla f \cdot \mathbf{Q} = -\phi \frac{\partial S}{\partial t}. \tag{31}$$

As we let $p_c \to 0$, we have, from (16a), (26), and (31),

$$\nabla \cdot \mathbf{Q} = \nabla \cdot (M\, \nabla P) = 0, \tag{32a}$$

$$\mathbf{Q} \cdot \nabla f = -\phi \frac{\partial S}{\partial t}. \tag{32b}$$

Now we see that our system of equations consists of an elliptic equation and a first order, or hyperbolic, equation. It should be clear that any analysis of the above difference equations based upon the resemblance of the differential equations to parabolic equations can no longer be expected to hold. More specifically, Douglas' analysis [6] of convergence of difference equations (18) to (13), which ignored the low order terms in the differential operator, can not be considered applicable, in view of the fact that the low order terms are all that remain in Eq. (32b).

At this point, it may be appropriate to make a general remark concerning nonlinear systems. This situation, where the character of the system changes (in this case, from parabolic-elliptic to hyperbolic-elliptic), depending on the size of one of the nonlinear coefficients (S'), probably occurs in many other problems. Since methods for studying stability and convergence for the most part depend on linearizing assumptions, we have to be extremely careful in using them, even when numerical experiments appear to substantiate their conclusion. For it may well be the case, as occurs here, that such analyses are valid only for some range of the coefficients and are not valid over other ranges.

Let us consider, now, the limiting behavior of the difference system (23), which is equivalent to (18), as the capillary pressure approaches zero. Taking sums and differences, respectively, of Eqs. (23a) and (23b), and combining with the definitions (15), gives,

$$\Delta(M_n\,\Delta P_{n+1})_{ij} + \Delta(N_n\,\Delta R_{n+1})_{ij} = 0, \tag{33a}$$

$$\Delta(M_n\,\Delta R_{n+1})_{ij} + \Delta(N_n\,\Delta P_{n+1})_{ij} = -\frac{2\phi}{\Delta t}(S_{i,j,n+1} - S_{ijn}). \tag{33b}$$

Let

$$QX_{i+\frac{1}{2},j} = \frac{1}{\Delta x}\big[M_{i+\frac{1}{2},j,n}(P_{i,j,n+1} - P_{i+1,j,n+1}) + N_{i+\frac{1}{2},j,n}(R_{i,j,n+1} - R_{i+1,j,n+1})\big], \tag{34a}$$

$$QY_{i,j+\frac{1}{2}} = \frac{1}{\Delta y}\big[M_{i,j+\frac{1}{2},n}(P_{i,j,n+1} - P_{i,j+1,n+1}) + N_{i,j+\frac{1}{2},n}(R_{i,j,n+1} - R_{i,j+1,n+1})\big]. \tag{34b}$$

Then

$$P_{i+1,j,n+1} - P_{i,j,n+1} = -\frac{QX_{i+\frac{1}{2},j}\,\Delta x}{M_{i+\frac{1}{2},j,n}} - \frac{N_{i+\frac{1}{2},j,n}}{M_{i+\frac{1}{2},j,n}}(R_{i+1,j,n+1} - R_{i,j,n+1}).$$

Similarly, for $P_{i,j+1,n+1} - P_{i,j,n+1}$. Substituting into Eqs. (33), and making use of (28), (29), and (30), gives

$$\frac{1}{\Delta x}(QX_{i+\frac{1}{2},j} - QX_{i-\frac{1}{2},j}) + \frac{1}{\Delta y}(QY_{i,j+\frac{1}{2}} - QY_{i,j-\frac{1}{2}}) = 0, \tag{35a}$$

$$\Delta(g_n\,\Delta R_{n+1})_{ij} + \frac{1}{\Delta x}(QX_{i+\frac{1}{2},j}f_{i+\frac{1}{2},j,n} - QX_{i-\frac{1}{2},j}f_{i-\frac{1}{2},j,n}) + \frac{1}{\Delta y}(QY_{i,j+\frac{1}{2}}f_{i,j+\frac{1}{2},n} - QY_{i,j-\frac{1}{2}}f_{i,j-\frac{1}{2},n}) = -\frac{\phi}{\Delta t}(S_{i,j,n+1} - S_{ijn}). \tag{35b}$$

As we let $p_c \to 0$, then the Eqs. (35) become

$$\Delta(M_n\,\Delta P_{n+1}) = 0, \tag{36a}$$

$$\frac{1}{\Delta y}(QX_{i+\frac{1}{2},j}f_{i+\frac{1}{2},j,n} - QX_{i-\frac{1}{2},j}f_{i-\frac{1}{2},j,n})$$
$$+ \frac{1}{\Delta y}(QY_{i,j+\frac{1}{2}}f_{i,j+\frac{1}{2},n} - QY_{i,j-\frac{1}{2}}f_{i,j-\frac{1}{2},n}) = -\frac{\phi}{\Delta t}(S_{i,j,n+1} - S_{ijn}). \tag{36b}$$

Remembering that $KN_{i+\frac{1}{2},j,n}$, $KW_{i+\frac{1}{2},j,n}$, etc., and therefore $f_{i+\frac{1}{2},j,n}$, etc., are evaluated at the midpoints of the various intervals, we see that Eq. (36b) is an explicit, centered-in-distance difference analogue of Eq. (32b). What is the stability of this difference equation?

Consider the case of one-dimensional flow. In that case, $Q = QX$ is a constant, QY is zero, and we have

$$\frac{Q}{\Delta x}(f_{i+\frac{1}{2},n} - f_{i-\frac{1}{2},n}) = -\frac{\phi}{\Delta t}(S_{i,n+1} - S_{in}). \tag{37}$$

Assume

$$f_{i+\frac{1}{2}} = \frac{KW_i + KW_{i+1}}{KN_i + KN_{i+1} + KW_i + KW_{i+1}} \approx \frac{f_i + f_{i+1}}{2}. \tag{38}$$

Then

$$\frac{Q}{\Delta x}\frac{f_{i+1,n} - f_{i-1,n}}{2} = -\frac{\phi}{\Delta t}(S_{i,n+1} - S_{in}). \tag{39}$$

For the stability analysis, we linearize Eq. (39) by assuming

$$f' = \frac{df}{dS} \tag{40}$$

is a constant. Then

$$\frac{Qf'\,\Delta t}{2\phi\,\Delta x}(S_{i+1,n} - S_{i-1,n}) + S_{i,n+1} - S_{in} = 0. \tag{41}$$

We can estimate the eigenvalues of the amplification matrix of the difference operator (41) by assuming exponential solutions of the form

$$S_{in} = A_n \exp[i\alpha\, x_i] \tag{42}$$

and substituting into Eq. (41). This yields

$$\begin{aligned} A_{n+1} &= A_n\left(1 - \frac{Qf'\,\Delta t}{2\phi\,\Delta x}[\exp[i\alpha\,\Delta x] - \exp[-i\alpha\,\Delta x]]\right), \\ \frac{A_{n+1}}{A_n} &= 1 - \frac{Qf'\,\Delta t}{\phi\,\Delta x}(i \sin \alpha\,\Delta x). \end{aligned} \tag{43}$$

The amplification ratio, A_{n+1}/A_n, is of modulus greater than unity; hence, Eq. (41) is unstable. Heuristically, we can extend this result to the nonlinear case and expect (36b) to be unstable. Finally, for small enough capillary pressure, we can expect (35b) and, equivalently, (18) to be unstable.

We can see now, more explicitly, the danger of accepting without reservation a stability analysis for nonlinear equations. Whereas one approach, that of Douglas [6], of linearizing the system of Eqs. (18) showed it to be unconditionally stable, another approach shows that over some range of the coefficients it can be unstable.

Use of "Upstream" Values of Coefficients KN and KW

It has been found empirically that sometimes better solutions can be obtained by evaluating KN and KW for each interval at the upstream end of that interval. That is, if flow is from point (i, j) to point $(i + 1, j)$ then, for example,

$$KN_{i+\frac{1}{2},j,n} = k(x_i, y_j, t_n)\, k_{rn_{ijn}}/\mu_n .$$

Similarly for KW. Then

$$f_{i+\frac{1}{2},j,n} = f_{ijn} .$$

Substituting into (37) gives

$$\frac{Q}{\Delta x}(f_{in} - f_{i-1,n}) = -\frac{\phi}{\Delta t}(S_{i,n+1} - S_{in}). \tag{44}$$

This is also an explicit equation, but not centered-in-distance. Again linearizing for a stability analysis, we have

$$\frac{Qf'\,\Delta t}{\phi\,\Delta x}(S_{in} - S_{i-1,n}) + S_{i,n+1} - S_{in} = 0, \tag{45}$$

which is identical in form to the Courant *et al.* equation [9]. Proceeding as before, we substitute (42) into (45) to obtain

$$A_{n+1} = A_n(1 - \lambda[1 - \exp[-i\alpha\,\Delta x]]),$$

where

$$\lambda = \frac{Qf'\,\Delta t}{\phi\,\Delta x}.$$

$$\frac{A_{n+1}}{A_n} = 1 + \lambda(\cos\alpha\,\Delta x - 1 - i\sin\alpha\,\Delta x).$$

$$\left|\frac{A_{n+1}}{A_n}\right|^2 = 1 - 2\lambda(1-\lambda)(1 - \cos\alpha\,\Delta x).$$

Thus, the amplification ratio, A_{n+1}/A_n, is always of modulus less than unity only if $\lambda < 1$. Hence stability requires that

$$\Delta t < \frac{\phi\,\Delta x}{Qf'}, \tag{46}$$

which result is in agreement with the conclusion of Courant *et al.*

Use of the upstream coefficients, then, can insure stability, provided Δt is sufficiently small. However, when Δt is appreciably less than $\phi\,\Delta x/Qf'$, it can be shown, as well as observed from computational results, that the solution, which should have a discontinuity (as discussed below), can be quite badly smeared. Stone and Brian [14] explain the smearing from the fact that Eq. (44) is only first-order correct in distance and so implicitly includes a dispersion term

$$\frac{Q\,\Delta x}{2\phi}\frac{\partial^2 f}{\partial x^2}.$$

We can also see from the fact that Q appears in the denominator of (46) why even the use of upstream coefficients does not bring about stability in the situation where the fluid velocity is high.

Existence of Discontinuity

In one dimension, Eq. (32b) can be solved quite easily by the method of characteristics. An equivalent method of solution (based on using S

and t as independent variables and x for the dependent variable) was first proposed for the two-phase flow problem in the absence of capillary pressure by Buckley and Leverett [10]. For one-dimensional flow, **Q** can be considered a constant scalar, so that (32b) becomes

$$Q\frac{\partial f}{\partial x} = -\phi\frac{\partial S}{\partial t}, \tag{47}$$

or

$$\frac{Qf'}{\phi}\frac{\partial S}{\partial x} + \frac{\partial S}{\partial t} = 0. \tag{48}$$

Applying the method of characteristics gives

$$\frac{dx}{dt} = \frac{Qf'}{\phi}, \tag{49}$$

with S constant along each characteristic path. A typical plot of f vs. S is shown in Fig. 3; the derivative of that function, f', is shown in Fig. 4.

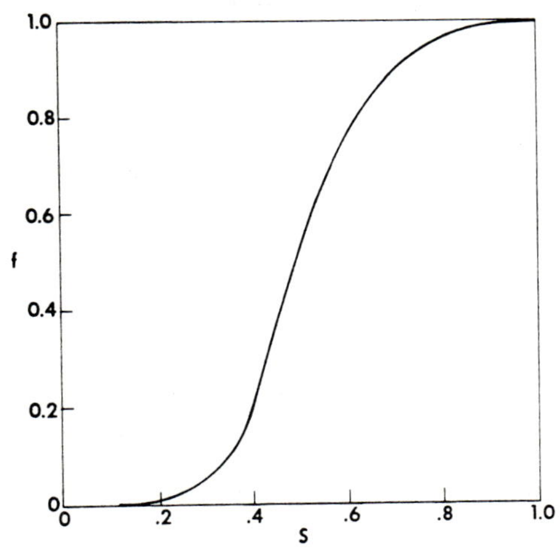

FIG. 3. Typical curve of f vs. S.

Because of the shape of the f' curve, an uncritical application of the method of characteristics yields triple values for the saturation at some values of x; an example of such a solution is shown in Fig. 5. An original

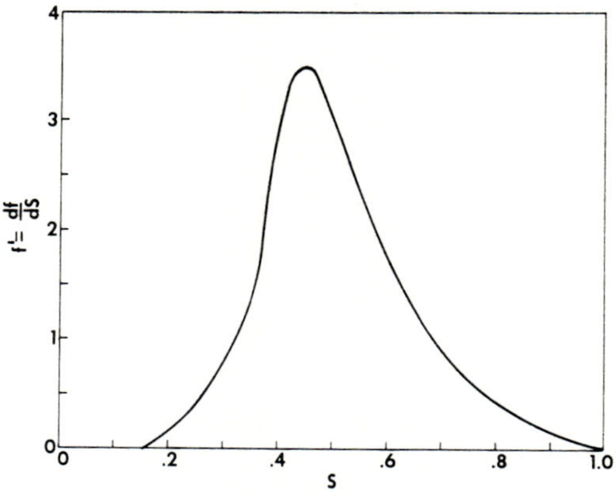

FIG. 4. Curve of $f' = df/dS$ vs. S.

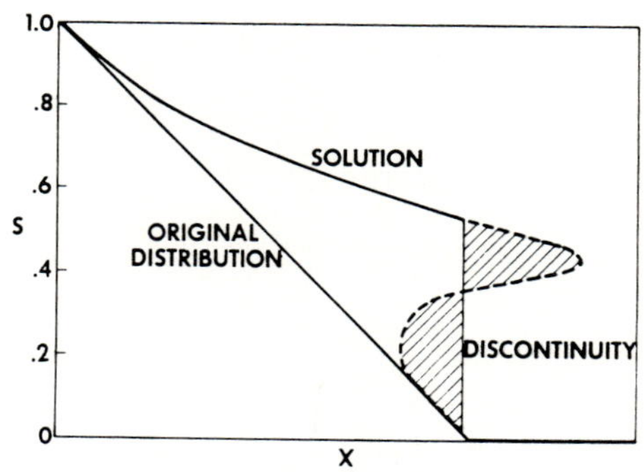

FIG. 5. Solution of S vs. x.

linear distribution of S vs. x starts out and continues for a short while as a single-valued function of x but, after some time, triple values are developed. Such a situation is, of course, physically impossible. Buckley and Leverett got around this by postulating a discontinuity whose location satisfies material balance; that is, the two areas between the curve and the line of discontinuity (the cross-hatched areas in Fig. 5) are equal.

A plot of the characteristic paths, as shown in Fig. 6, shows more clearly why a discontinuity must develop. Equation (49) shows the slope of each characteristic is constant, but a function of saturation. Because of the shape of f' vs. S, some of these characteristics must intersect and

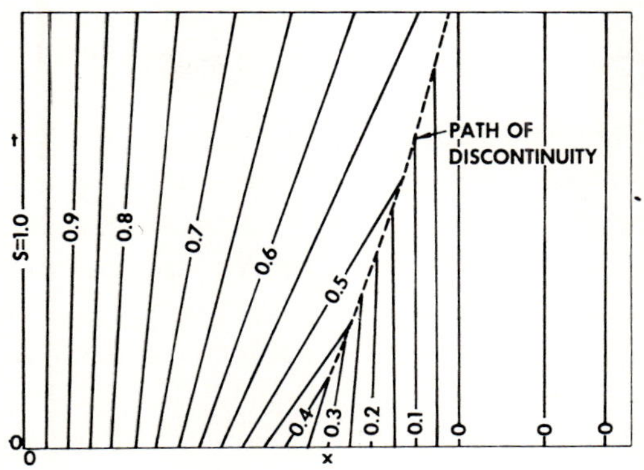

FIG. 6. Characteristic paths.

cross. Since crossings of characteristics cannot be permitted, a discontinuity must form.

The Russians have done much work on the existence and properties of solutions of nonlinear first-order equations. A good summary of this work is given by Gel'fand [11]. He points out that discontinuities are very common in the solution of quasilinear equations as, for example, the existence of shock waves in hydrodynamics. If any diffusional, or second-order terms, appear in addition to the first-order terms, the solutions are continuous, but they should approach the discontinuous solutions as the second-order terms are reduced to zero.

A similar situation exists here. The presence of capillary pressure causes the solution to be continuous, but as capillary pressure is decreased, the solution should tend to the discontinuous solution of Eq. (47). With relatively small values of capillary pressure, we expect steep fronts in the solution. We can expect, and indeed we find, great difficulty in matching such solutions by the use of finite difference methods without exceedingly fine definition in the x (and y) grid.

Possible Improvements

There are several avenues being investigated for obtaining improved solutions to the two-phase flow problem. One might suppose that some variant of the method of characteristics would be useful, but there is no direct way of extending it to two dimensions, nor of including the capillary pressure. Rachford [12] has proposed a "moving reference point" method which has some resemblance to the method of characteristics but takes capillary pressure into account. While the method has been proposed for two or more dimensions, it has so far been tested only for one dimension in the absence of gravity. Further work needs to be done to include the effect of gravity and to test it under multidimensional conditions. While it is a somewhat cumbersome method, it does overcome many of the disadvantages of the finite difference methods that have been discussed above.

Another approach is to consider variations in the difference equations themselves. One such variation is equivalent to the Lax–Wendroff method [13], in that an additional second-order term is added to a first-order difference equation to improve the accuracy of the solution. Other second-order methods for solving nonlinear first-order problems have been proposed in Chapter 15 of this volume.

Another variation of the difference equation under consideration is to avoid the explicit first-order difference equation by evaluating the nonlinear coefficients KN and KW in Eqs. (18) at time $n + 1$; that is, to make them implicit. Such an approach promises to improve the stability of the resulting difference equations, but of course raises the problems of solving nonlinear sets of simultaneous equations. These problems are currently being investigated.

References

1. Collins, R. E., "Flow of Fluids through Porous Media." Reinhold, New York, 1961.
2. Douglas, J., Jr., Peaceman, D. W., and Rachford, H. H., Jr., A Method for Calculating Multi-Dimensional Immiscible Displacement, *Trans. AIME* **216**, 297–308 (1959).
3. Blair, P. M., and Peaceman, D. W., An Experimental Verification of a Two-Dimensional Technique for Computing Performance of Gas-Drive Reservoirs, *Soc. Petrol. Eng. J.*, **3**, 19–27 (1963).
4. Richtmyer, R. D., "Difference Methods for Initial Value Problems," Wiley (Interscience), New York, 1957.
5. Peaceman, D. W., and Rachford, H. H., Jr., "The Numerical Solution of Parabolic and Elliptic Differential Equations, *J. Soc. Indust. Appl. Math.* **3**, 28–41 (1955).

6. Douglas, J., Jr., A Numerical Method for the Solution of a Parabolic System, *Numer. Math.* **2**, 91–98 (1960).

7. Douglas, J., Jr., The Application of Stability Analysis in the Numerical Solution of Quasi-Linear Parabolic Differential Equations, *Trans. Am. Math. Soc-* **89**, 484–518 (1958).

8. Douglas, J., Jr., Alternating Direction Methods for Three Space Variables, *Numer. Math.* **4**, 41–63 (1962).

9. Courant, R., Isaacson, E., and Rees, M., On the Solution of Nonlinear Hyper. bolic Differential Equations by Finite Difference Methods, *Comm. Pure Appl. Math.* **5**, 243–255 (1952).

10. Buckley, S. E., and Leverett, M. C., Mechanism of Fluid Displacements in Sands, *Trans. AIME* **146**, 107–116 (1942).

11. Gel'fand, I. M., Some Problems in the Theory of Quasilinear Equations, Uspehi Mat. Nauk. (N.S.) **14**, No. 2 (86), 87–158 (1959). (English transl. in: *Am. Math. Soc. Translations, Ser. 2* **29**, 295–381 (1963)).

12. Rachford, H. H., Jr., Numerical Calculation of Immiscible Displacement by a Moving Reference Point Method, *Soc. Petrol. Engrs. J.* **6**, 87–101 (1966).

13. Lax, P. D., and Wendroff, B., Difference Schemes for Hyperbolic Equations with High Order of Accuracy, *Comm. Pure Appl. Math.* **17**, 381–398 (1964).

14. Stone, H. L., and Brian, P. L. T., Numerical Solution of Convective Transport Problems, *AIChE J.* **9**, 681–688 (1963).

15. Douglas, J., Jr., and Jones, B. F., Jr., On Predictor-Corrector Methods for Nonlinear Parabolic Differential Equations, *J. Soc. Indust. Appl. Math.* **11**, 195–204 (1963).

An Extrapolated Crank-Nicolson Difference Scheme for Quasilinear Parabolic Equations †

MILTON LEES

Department of Mathematics
Case Institute of Technology
Cleveland, Ohio

Let us $w(x, t)$ be a smooth solution of the quasilinear partial differential equation

$$K\left(y, t, \text{w}, \frac{\partial w}{\partial x}\right)\frac{\partial w}{\partial t} = \frac{\partial^2 w}{\partial x^2} + f\left(x, t, w, \frac{\partial w}{\partial x}\right) \tag{1a}$$

in the rectangular region $\Omega = \{(x, t) \mid 0 \leq x \leq 1, 0 \leq t \leq T\}$ such that

$$\begin{aligned} w(x, 0) &= \varphi(x), & 0 &\leq x \leq 1, \\ w(0, t) &= \varphi^-(t), & 0 &\leq t \leq T, \\ w(1, t) &= \varphi^+(t), & 0 &\leq t \leq T, \end{aligned} \tag{1b}$$

where φ, φ^-, and φ^+ are specified functions. We shall assume, first, that the functions $K(x, t, p, q)$ and $f(x, t, p, q)$ are defined and smooth for $(x, t) \in \Omega$ and $p^2 + q^2 < +\infty$ and that there is a positive constant μ such that, in this region,

$$K(x, t, p, q) \geq \mu > 0. \tag{2}$$

It follows from (2) that the differential equation (la) is uniformly of parabolic type and, consequently, $w(x, t)$ is the only solution of the boundary value problem (1). The differential equation (la) is called *semilinear* if the functions K and f are independent of q.

If N is a positive integer and $h > 0$ is defined by the relation $(N + 1)h = 1$, then we put

$$\bar{G}(h) = \{ih \mid i = 0, 1, \ldots, N + 1\}.$$

If $0 < 2k < T$ then we let J be the largest integer such that $kJ \leq T$.

† This research was supported in part by NSF Grant GP-5553.

We denote by $\mathscr{C}(h)$ the (real) linear space of all functions $u(x)$ defined on the finite set $\bar{G}(h)$. Obviously, $\mathscr{C}(h)$ has dimension $N + 2$. In this paper we are concerned with the construction of a difference scheme for the boundary value problem (1) which, for all sufficiently small h and k, determines uniquely a sequence $u_0, u_1, \ldots, u_J$ of functions from $\mathscr{C}(h)$ such that

(I) $u_0 = \varphi$ on $\bar{G}(h)$ and, for each $j \geq 1$, the N numbers $u_j(h)$, $u_j(2h)$, $\ldots$, $u_j(Nh)$ are determined *directly* as the solution of a *linear* system of equations.

(II) There is a $\delta > 0$ and a constant $A > 0$, independent of h and k, such that

$$\max_{\substack{0 \leq j \leq J \\ x \in \bar{G}(h)}} | u_j(x) - w(x, jk) | \leq A(h^2 + k^2),$$

whenever $(h^2 + k^2)^{1/2} \leq \delta$.

It is implied in (I) that the numbers $u_j(h)$, $u_j(2h)$, $\ldots$, $u_j(Nh)$ need not be determined by an iterative process applied to nonlinear system of equations. Condition (II) states that the sequence $u_0, u_1, \ldots, u_j$ furnishes a second order approximation to the solution of the boundary value problem (1).

It is known [2] that the standard Crank–Nicolson difference scheme for (1) satisfies condition (II). It is also known that it fails to satisfy condition (I), except in the special case where (1a) is a linear differential equation, a case which we exclude from consideration. In a 1963 paper [1], Douglas and Jones formulated two modifications of the standard Crank–Nicolson difference scheme for (1), which satisfy condition (I). These difference schemes are based on well-known predictor-corrector difference schemes for ordinary differential equations. When (1a) is a semilinear equation, they proved that their predictor-corrector Crank–Nicolson difference schemes for (1) also satisfy condition (II). In the general quasilinear case, however, they obtained only the weaker error estimate

$$| u_j(x) - w(x, jk) | \leq A(h^2 + k^{3/2}).$$

The purpose of this paper is to formulate and analyze another modification of the standard Crank–Nicolson difference scheme for (1) which, in addition to satisfying the conditions (I) and (II) in the general quasilinear case, requires approximately half the number of evaluations of the functions K and f as the predictor-corrector methods in [1]. However, in the general quasilinear case, we find it necessary to add to condition (II) the mild restriction $k \leq \lambda h$, for some fixed positive constant λ.

In the interests of notational simplicity, we make two completely inessential simplifications; namely, we assume that the functions K and f are independent of x and t that φ^- and φ^+ vanish identically. Thus, we suppose that $w(x, t)$ satisfies

$$K\left(w, \frac{\partial w}{\partial x}\right)\frac{\partial w}{\partial t} = \frac{\partial^2 w}{\partial x^2} + f\left(w, \frac{\partial w}{\partial x}\right) \tag{3a}$$

$$\begin{aligned} w(x, 0) &= \varphi(x), & 0 \leq x \leq 1, \\ w(0, t) &= w(1, t) = 0, & 0 \leq t \leq T. \end{aligned} \tag{3b}$$

We define the standard difference operators D_+, D_- and D_0 in the usual way, i.e.,

$$\begin{aligned} h\, D_+\, u(x) &= u(x + h) - u(x), \\ h\, D_-\, u(x) &= u(x) - u(x - h), \\ 2h\, D_0\, u(x) &= u(x + h) - u(x - h). \end{aligned}$$

Let $\mathscr{C}_0(h)$ be the N-dimensional subspace of $\mathscr{C}(h)$ consisting of those functions $u \in \mathscr{C}(h)$ for which $u(0) = u(1) = 0$. Let $G(h)$ denote the intersection of $\bar{G}(h)$ with the open interval $0 < x < 1$, and let $\tau = (h^2 + k^2)^{1/2}$.

We shall construct two sequences $u_0, u_1, \ldots, u_j$ and $u_{\frac{1}{2}}, u_{\frac{3}{2}}, \ldots, u_{J-\frac{1}{2}}$ in $\mathscr{C}_0(h)$ by induction. First, we put $u_0 = \varphi$ and

$$u_{1/2} = \varphi + \frac{k}{2K(\varphi, D_0\varphi)}\left[D_+\, D_-\, \varphi + f(\varphi, D_0\, \varphi)\right]. \tag{4a}$$

Now, assume that $u_0, u_1, \ldots, u_j$ and $u_{\frac{1}{2}}, u_{\frac{3}{2}}, \ldots, u_{j+\frac{1}{2}}$ have already been defined. Then u_{j+1} is defined to be the unique solution of the linearized Crank–Nicolson difference equation

$$\begin{aligned} K(u_{j+1/2}, D_0 u_{j+1/2})\,(u_{j+1} - u_j) &= \frac{k}{2} D_+ D_-(u_{j+1} + u_j) \\ &\quad + kf(u_{j+1/2}, D_0 u_{j+1/2}) \end{aligned} \tag{4b}$$

for $x \in G(h)$, and $u_{j+\frac{3}{2}}$ is defined directly by the linear extrapolation formula

$$u_{j+3/2} = \tfrac{3}{2}u_{j+1} - \tfrac{1}{2}u_j. \tag{4c}$$

It is clear that the two sequences $\langle u_j \rangle$ and $\langle u_{j+\frac{1}{2}} \rangle$ are well defined, and the sequence $\langle u_j \rangle$ satisfies condition (I). Because of (4c) we call the foregoing difference scheme an extrapolated Crank–Nicolson difference scheme.

It turns out that if f is a linear function of p, it is permissible to replace $f(u_{j+\frac{1}{2}}, D_0 u_{j+\frac{1}{2}})$ by $f((u_{j+1} + u_j)/2, D_0 u_{j+\frac{1}{2}})$, and if f is a linear function of q, it is permissible to replace $f(u_{j+\frac{1}{2}}, D_0 u_{j+\frac{1}{2}})$ by $f(u_{j+\frac{1}{2}}, \frac{1}{2} D_0(j_{j+1} + u_j))$. We shall not concern ourselves with these special cases.

Now, we record some properties of the linear space $\mathscr{C}_0(h)$, which will be useful in proving that the difference scheme (4) satisfies condition (II). If u and v belong to $\mathscr{C}_0(h)$, then their l_2 inner product is given by

$$(u, v) = h \sum_{i=1}^{N} u(ih)\, v(ih).$$

The l_2 norm of $u \in \mathscr{C}_0(h)$ induced by this inner product is

$$\| u \| = (u, u)^{1/2} = \left(h \sum_{i=1}^{N} u^2(ih) \right)^{1/2}.$$

We define the l_∞ norm of u $\in \mathscr{C}_0(h)$ by

$$\| u \|_\infty = \max_{1 \le i \le N} | u(ih) |.$$

We remark that $\| u \|$ and $\| u \|_\infty$ are only seminorms on the full space $\mathscr{C}(h)$. Finally, if u and v belong to $\mathscr{C}_0(h)$, then it is easy to see that

$$(u, v)_D = h \sum_{i=1}^{N+1} D_- u(ih)\, D_- v(ih)$$

defines another inner product on $\mathscr{C}_0(h)$; the induced norm is $\| u \|_D = (u, u)_D^{1/2}$.

For $u \in \mathscr{C}_0(h)$ and any function v defined on $G(h) \cup \{1\}$, the bilinear form $(u, D_+ v)$ is well defined. Summation by parts yields the formula

$$(u, D_+ v) = -h \sum_{i=1}^{N+1} v(ih)\, D_- u(ih).$$

In particular, for $u, v \in \mathscr{C}_0(h)$ we have the identity

$$(u, D_+ D_- v) = -h \sum_{i=1}^{N+1} D_- u(ih)\, D_- v(ih) = -(u, v)_D. \tag{5}$$

Also, it is readily verified [2] that the three norms just defined satisfy the relations

$$\| u \| \leq \| u \|_\infty \leq \| u \|_D . \tag{6}$$

For the solution w of (3) we put $w_\alpha(x) = w(x, \alpha k)$, where $0 \leq \alpha \leq$ J. A straightforward but very lengthy computation, based on Taylor's formula, yields the following relations satisfied by w:

$$w_{\frac{1}{2}} = u_{\frac{1}{2}} + R_{\frac{1}{2}}, \tag{7a}$$

$$K(w_{j+\frac{1}{2}}, D_0 w_{j+\frac{1}{2}})(w_{j+1} - w_j) = \frac{k}{2} D_+ D_-(w_{j+1} + w_j)$$

$$+ kf(w_{j+\frac{1}{2}}, D_0 w_{j+\frac{1}{2}}) + kR_{j+1}, \tag{7b}$$

$$w_{j+\frac{3}{2}} = \tfrac{3}{2} w_{j+1} - \tfrac{1}{2} w_j + R_{j+\frac{1}{2}}, \tag{7c}$$

$$D_0 w_{j+\frac{3}{2}} = \tfrac{3}{2} D_0 w_{j+1} - \tfrac{1}{2} D_0 w_j + R^0_{j+\frac{1}{2}}, \tag{7d}$$

$$D_0 w_{\frac{1}{2}} = D_0 u_{\frac{1}{2}} + R_{\frac{1}{2}}{}^0, \tag{7e}$$

where, for $0 \leq j < J$,

$$\| R_{j+1} \|_\infty < \| R_{j+\frac{1}{2}} \|_\infty + \| R^0_{j+\frac{1}{2}} \|_\infty \leq A_1 \tau^2. \tag{8}$$

Here A_1 is a constant independent of h, k, and j. Without loss of generality, we also assume that

$$\| w_{j+1} - w_j \|_\infty \leq A_1 k. \tag{9}$$

We introduce the error functions $z_j = w_j - u_j$ and $z_{j+\frac{1}{2}} = w_{j+\frac{1}{2}} - u_{j+\frac{1}{2}}$. Then z_j and $z_{j+\frac{1}{2}}$ belong to $\mathscr{C}_0(h)$. It follows easily from (4) and (7) that

$$z_{\frac{1}{2}} = R_{\frac{1}{2}}, \qquad D_0 z_{\frac{1}{2}} = R_{\frac{1}{2}}{}^0, \tag{10a}$$

$$z_{j+\frac{3}{2}} = \tfrac{3}{2} z_{j+1} - \tfrac{1}{2} z_j + R_{j+\frac{3}{2}}, \tag{10b}$$

$$D_0 z_{j+\frac{3}{2}} = \tfrac{3}{2} D_0 z_{j+1} - \tfrac{1}{2} D_0 z_j + R^0_{j+\frac{3}{2}}, \tag{10c}$$

$$K(w_{j+\frac{1}{2}}, D_0 w_{j+\frac{1}{2}})(z_{j+1} - z_j) = \frac{k}{2} D_+ D_-(z_{j+1} + z_j) + k\left(\frac{\partial f}{\partial p}\right) z_{j+\frac{1}{2}}$$

$$+ k\left(\frac{\partial f}{\partial q}\right) D_0 z_{j+\frac{1}{2}} + \left(\frac{\partial k}{\partial p}\right) z_{j+\frac{1}{2}} [w_{j+1} - w_j - (z_{j+1} - z_j)]$$

$$+ \left(\frac{\partial k}{\partial q}\right) D_0 z_{j+\frac{1}{2}} [w_{j+1} - w_j - (z_{j+1} - z_j)] + kR_{j+1}, \tag{10d}$$

where the partial derivatives appearing in (10d) are evaluated at points called for by the mean value theorem of the calculus.

For the solution w of (3) we put

$$m = \max_{\Omega}\left[\,|\,w\,| + \left|\frac{\partial w}{\partial x}\right|\,\right].$$

Then, we denote by A_2 the maximum of

$$\left|\frac{\partial k}{\partial p}\right|, \left|\frac{\partial k}{\partial q}\right|, \left|\frac{\partial f}{\partial p}\right|, \left|\frac{\partial f}{\partial q}\right|$$

over the closed bounded region determined by $|\,p\,| + |\,q\,| \leq 2m$.

If $k \leq \lambda h$, then there is a $\delta > 0$ such that $\tau \leq \delta$ implies that

$$2A_1\sigma\,\tau \leq 1, \tag{11a}$$

$$2\sigma\,\tau \leq m, \tag{11b}$$

$$2A_1\,\tau \leq \mu, \tag{11c}$$

$$2(A + A_1)\,\tau \leq 1, \tag{11d}$$

where $\sigma = 5 + 4\lambda^2$ and

$$A = \frac{(A_1 + A_2)}{\sqrt{\mu}}\,[\sigma + 2T(\mu^2 + 1)]^{1/2} \exp\left[\frac{16(A_1 + A_2)^2 T}{\mu}\right]. \tag{12}$$

Now, we shall prove by induction that, for $\tau \leq \delta$,

$$\|\,z_{j+\frac{1}{2}}\,\|_\infty + \|\,D_0 z_{j+\frac{1}{2}}\,\|_\infty \leq \sigma\tau \tag{13a}$$

$$\|\,z_j\,\|_\infty \leq A\tau^2, \qquad 0 \leq j \leq J. \tag{13b}$$

In view of (8) and (11a) it follows that the inequalities in (12) hold for $j = 0$. Assuming that (12) holds for all integers $\leq j$, we shall prove that (12) also holds with j replaced by $j + 1$. Our inductive assumption and (11b) imply that the modulus of each partial derivative in (10d) is not greater than A_2. Hence, forming the l_2 inner product of (10d) with $2(z_{j+1} - z_j)$ and making use of (2), we find that

$$2\mu\|\,z_{j+1} - z_j\,\|^2 \leq k(z_{j+1} - z_j, D_+D_-(z_{j+1} + z_j))$$
$$+ 2k\left(z_{j+1} - z_j, \left(\frac{\partial f}{\partial p}\right) z_{j+\frac{1}{2}}\right)$$

$$+\ 2k\left(z_{j+1} - z_j, \left(\frac{\partial f}{\partial q}\right) D_0 z_{j+\frac{1}{2}}\right)$$

$$+\ 2\left(z_{j+1} - z_j, \left(\frac{\partial k}{\partial p}\right) z_{j+\frac{1}{2}}[w_{j+1} - w_j - (z_{j+1} - z_j)]\right)$$

$$+\ 2\left(z_{j+1} - z_j, \left(\frac{\partial k}{\partial q}\right) D_0 z_{j+\frac{1}{2}}[w_{j+1} - w_j - (z_{j+1} - z_j)]\right)$$

$$+\ 2k(z_{j+1} - z_j, R_{j+1}). \tag{13}$$

Since

$$(z_{j+1} - z_j, D_+D_-(z_{j+1} + z_j)) = -\| z_{j+1} \|_D^2 + \| z_j \|_D^2,$$

in view of (5), we deduce from (13) the estimate

$$\begin{aligned} 2\mu\| z_{j+1} - z_j \|^2 + \| z_{j+1} \|_D{}^2 \le \| z_j \|_D{}^2 &+ 2k(A_1 + A_2)\| z_{j+j} - z_j \| \\ &+ [\| z_{j+\frac{1}{2}} \| + \| D_0 z_{j+\frac{1}{2}} \|] \\ &+ 2A_1 \| z_{j+1} - z_j \|^2 [\| z_{j+\frac{1}{2}} \| + \| D_0 z_{j+\frac{1}{2}} \|] \\ &+ k \| z_{j+1} - z_j \| \, \| R_{j+1} \|, \end{aligned} \tag{14}$$

after a liberal use of Schwarz's inequality together with (9). From the general inequality

$$2ab \le \varepsilon a^2 + \frac{1}{\varepsilon} b^2, \qquad \varepsilon > 0,$$

we obtain from (14) and our induction assumption the estimate

$$\begin{aligned} [2\mu - 2A_1\sigma\tau - 2\varepsilon] \| z_{j+1} - z_j \|^2 + \| z_{j+1} \|_D{}^2 \le \| z_j \|_D{}^2 \\ + k^2(A_1 + A_2)^2 \varepsilon^{-1} [\| z_{j+\frac{1}{2}} \|^2 + \| D_0 z_{j+\frac{1}{2}} \|^2] + k^2 \varepsilon^{-1} \| R_{j+1} \|^2. \end{aligned}$$

Now, choose $2\varepsilon = \mu$. Then, in view of (11c) and the inequality $\| R_{j+1} \| \le A_1\tau^2$, which follows from (6) and (8), it follows that

$$\| z_{j+1} \|_D{}^2 \le \| z_j \|_D{}^2 + kA_3 [\| z_{j+\frac{1}{2}} \| + \| D_0 z_{j+\frac{1}{2}} \|]^2 + \frac{2kA_1{}^2}{\mu} \tau^4, \tag{15}$$

where

$$A_3 = 2\mu^{-1} (A_1 + A_2)^2. \tag{16}$$

Since $z_0 = 0$, we may iterate the inequality (15) to get

$$\| z_{j+1} \|_D^2 \leq k \sum_{n=0}^{j} \left\{ A_3[\| z_{n+\frac{1}{2}} \| + \| D_0 z_{n+\frac{1}{2}} \|]^2 + \frac{2A_1^2}{\mu} \tau^4 \right\}$$

$$\leq (kA_3A_1^2\sigma^2 + 2\mu^{-1} A_1^2 T)\tau^4 + k \sum_{n=1}^{j} A_3[\| z_{n+\frac{1}{2}} \|$$

$$+ \| D_0 z_{n+\frac{1}{2}} \|]^2, \tag{17}$$

where we have made use of (8) and the relation $jk \leq Jk \leq T$. As

$$\| z_{n+\frac{1}{2}} \| + \| D_0 z_{n+\frac{1}{2}} \| \leq \tfrac{3}{2} \| z_n \| + \tfrac{1}{2} \| z_{n-1} \| + \tfrac{3}{2} \| D_0 z_n \| + \tfrac{1}{2} \| D_0 z_{n-1} \| + 2A_1 \tau^4,$$

in view of (10b) and (10c), it follows from (17) and (6) that

$$\| z_{j+1} \|_D^2 \leq A_4 \tau^4 + 16\, A_3 k \sum_{n=0}^{j} \| z_n \|_D^2,$$

where

$$A_4 = kA_3A_1^2\sigma^2 + 2\mu^{-1}A_1^2 T + 4\, A_1^2 A_3 \tau^2 T. \tag{18}$$

Using the discrete version of Gronwall's lemma [2], we find that

$$\| z_{j+1} \|_D^2 \leq A_4 \tau^4 \exp(16\, A_3 T)$$

hence,

$$\| z_{j+1} \|_\infty \leq \| z_{j+1} \|_D \leq \sqrt{A_4} \tau^2 \exp(8\, A_3 T).$$

It follows from (11a), (11c), (16) and (18) that

$$\sqrt{A_4} \leq \frac{(A_1 + A_2)}{\sqrt{\mu}} [\sigma + 2T(\mu^2 + 1)]^{1/2},$$

which implies that $\| z_{j+1} \|_\infty \leq A\tau^2$; hence, (13b) holds, with j replaced by $j + 1$.

It follows, now, from (10b) and (10c) that

$$\| z_{j+\frac{3}{2}} \|_\infty + \| D_0 z_{j+\frac{3}{2}} \|_\infty \leq 2A\tau^2 + 2A_1\tau^2 + \tfrac{3}{2} \| D_0 z_{j+1} \|_\infty + \tfrac{1}{2} \| D_0 z_j \|_\infty$$
$$\leq 2(A + A_1)\tau^2 + 4h^{-1}\tau^2$$

since $\| D_0 z_j \|_\infty \leq 4h^{-1} \| z_j \|_\infty$; hence, by (11d), we have

$$\| z_{j+\frac{3}{2}} \|_\infty + \| D_0 z_{j+\frac{3}{2}} \|_\infty \leq 2(A + A_1)\tau^2 + 4(1 + \lambda^2)h$$
$$\leq \tau + 4(1 + \lambda^2)\tau = (5 + 4\lambda^2)\tau = \sigma\tau,$$

which verifies (13a), with j replaced by $j + 1$. Thus, we have proved that *the extrapolated Crank–Nicolson difference scheme* (4) *satisfies conditions* (I) *and* (II).

We mention that, in the semilinear case, we may put $\sigma = 1$ and there is no restriction on h and k other than $(h^2 + k^2)^{1/2} \le \delta$.

Given a positive number η, we denote by S_η the set of all points (x, t, p, q) for which $(x, t) \in \Omega$, $|p - w(x, t)| \le \eta$ for some $(x, t) \in \Omega$, and $|q - \delta w(x_1 t)/\delta x| \le \eta$ for some $(x, t) \in \Omega$. An examination of the foregoing proof will show that it remains valid (possibly with a smaller δ) under the weaker assumption that there is an $\eta > 0$ such that the functions K and f are defined and smooth in S_η and that (2) holds in S_η. This generalization shows that the extrapolated Crank–Nicolson difference scheme is second order accurate when applied, for example, to approximate positive smooth solutions of the equation

$$\sqrt{w}\,\frac{\partial w}{\partial t} = \frac{\partial^2 w}{\partial x^2}.$$

In practice, one frequently meets parabolic equations of the form

$$K\left(w, \frac{\partial w}{\partial x}\right)\frac{\partial w}{\partial t} = \frac{\partial}{\partial x}\left(P\left(w, \frac{\partial w}{\partial x}\right)\frac{\partial w}{\partial x}\right) + f\left(w, \frac{\partial w}{\partial x}\right) \tag{19}$$

where K and f are bounded away from zero in a neighborhood of a solution $w(x, t)$. Although we give no details, it can be shown that our extrapolated Crank–Nicolson difference scheme, as applied directly to (19), satisfies conditions (I) and (II), provided $k \le \lambda h$. Even though (19) can be transformed into an equation of the form (3), it is often desirable to difference it as it stands.

References

1. Douglas, J., and Jones, B. F., On Predictor-Corrector Methods for Nonlinear Parabolic Equations, *J. Soc. Indust. Appl. Math.* **11**, 195–204 (1963).
2. Lees, M., Approximate Solution of Parabolic Equations, *J. Soc. Indust. Appl. Math.* **7**, 167–183 (1959).

Heat Transfer to the Endwall of a Shocktube
A Variational Analysis

JOHN R. FERRON and
KAZUTOSHI FUJIMURA†

Department of Chemical Engineering
University of Delaware
Newark, Delaware

Heat is conducted to the endwall of a shocktube, following reflection of a shock wave from the wall, through a steep temperature gradient. If temperature of the interface can be measured, one can deduce thermal conductivity of the gas as a function of temperature. The problem of analyzing data from such experiments is solved here by an implicit variational method, and the numerical solution is illustrated. Extension to other transport experiments is discussed.

Introduction

In the experimental determination of transport properties of fluids and of solids, one ordinarily establishes suitable boundary conditions, measures the local strength of the resulting field or flux and deduces numerical values of the appropriate transport coefficients. Thus, one might set temperatures (perhaps steady, perhaps varying in time) on two faces of a cubical solid, insulate the other faces and measure the flux of thermal energy through the solid. From the measurements and the energy equation for the process, thermal conductivity of the solid would be obtained. Similarly, one might release a pulse or a flow of tracer fluid into another fluid and from measurements of concentrations at points some distance from the tracer source, deduce appropriate diffusion coefficients.

It is necessary in such experiments that gradients of temperature, concentration, etc., be imposed. For the usual, nonideal material the transport coefficients are functions of temperature and concentration and hence vary

† *Present address*: Sumitomo Chemicals Company Ltd., New York.

from point to point, and frequently with time as well, in the type of experiment referred to above. It is not uncommon to choose experimental conditions so that variations may be neglected, allowing one to characterize an experimental transport coefficient by means of a single average temperature or concentration. Evidently such an assumption leads to inexact results. It is also wasteful of experimental effort, for one might wish to obtain the transport coefficient as a function of temperature or of concentration for the full ranges of these variables employed in the experiment.

The purpose of this work is to demonstrate, by example, that the functional dependence of a transport coefficient can be conveniently deduced from experiment. The method employed is an adaptation of a classical variational procedure [1] to problems constrained by partial differential equations. The method has been described in detail [2, 3], and this paper is devoted to illustration of the application to a specific problem.

The experiment referred to is that in which one follows the temperature of a solid surface during a brief interval of time immediately following normal reflection of a shock wave from the surface. The time-temperature relationship can be used to deduce the thermal conductivity of the gas adjacent to the solid and through which the shock wave passes [4, 8]. The experiment leads naturally to a large gradient of temperature in the gas next to the solid surface, and one is obliged to seek thermal conductivity as a function of temperature rather than as a single value associated with a single temperature. Moreover, experimental uncertainties do occur; and it is useful to repeat the laboratory work several times and to seek the thermal conductivity–temperature function which, according to a least-squares criterion, best fits all of the data.

The mathematical problem of data analysis is defined in general terms in the next section, and the solution procedure is outlined. Details of the experiment are next described, following which the solution is carried out in detail. Some numerical results are presented, and extensions to other experiments are discussed.

A Least–Error Problem for Transport Experiments

Let $u(x, t)$ represent the field strength (temperature or concentration, for example) in some transport experiment. Ranges of the independent variables are:

$$x_0 \leq x \leq x_1$$

and

$$t_0 \leq t \leq t_1.$$

Suppose that boundary measurements at x_1 are available for each of N experiments. Call these $u_i^*(x_1, t)$, $i = 1, \ldots, N$.

It is supposed that the field is described by the partial differential equation:

$$\phi(u, u_x, u_{xx}, u_t, x, t; \propto(u)) = 0 \tag{1}$$

where $\propto(u)$ is the transport coefficient for which we seek a functional form. Additional conditions are available in the form:

$$f_i(u(x_0, t), u_x(x_0, t), \propto_{0i}) = 0 \tag{2a}$$

$$g_i(u(x_0, t)) = 0 \tag{2b}$$

$$h_i(u(x, t_0)) = 0 \tag{2c}$$

where $\propto_{0i}$ represents the transport coefficient evaluated at the point x_0 in experiment i.

Denote by $u_i(x_1, t)$ the field strength at point x_1 calculated for experiment i from Eqs. (1) and (2). We seek the function $\propto(u)$ for which:

$$V(u_1, u_2, \ldots, u_N) = \sum_{i=1}^{N} [u_i^*(x_1, t) - u_i(x_1, t)]^2 \tag{3}$$

is minimum.

For a particular value of i, that is, for any particular experiment, Eqs. (1) and (2) define a well-set boundary value problem. This could be solved so that the difference:

$$[u_i^*(x_1, t) - u_i(x_1, t)]$$

would be identically zero. The function $\propto_i(u)$ thus obtained would differ, in general, from that obtained for some other experiment. Minimizing the expression (3) by choice of a single function $\propto(u)$ provides a suitable compromise among all of the functions $\propto_i(u)$ for which we might have solved.

The problem defined by (1), (2), and (3) is of variational type. One can define a variation, $\delta\alpha$, which yields, through Eqs. (1) and (2), corresponding variations, δu_i, for each experiment. These in turn provide the variation δV from (3). The requirement that $\delta V = 0$ leads finally to a set of necessary conditions, analogous to Euler–Lagrange equations [2, 3]. These are partial differential equations which are conveniently solved by finite-difference methods or by reduction to ordinary differential equations followed by numerical integration. Alternatively one may first transform Eqs. (1) and (2) to a suitable ordinary differential equation form and then apply the variational procedure. This latter scheme, which is followed here, has the advantage that the form obtained is like that of optimal control problems which have lately been studied thoroughly [9, 10]. Computational methods arising from these studies can be applied.

The procedure here, thus, is first to reduce the problem to one-dimensional form by suitable transformations. The result is a set of equations:

$$dy_j/dz = f_j(y_1, y_2, \ldots, y_m; q(z)) \tag{4}$$

where $j = 1, 2, \ldots, m > N$. The range of the new independent variable is $z_0 \leq z \leq z_1$. The initial conditions are obtained from (2a) to (2c) in the form:

$$y_j(z_0) = c_j, \qquad j = 1, \ldots, m. \tag{5}$$

The function $q(z)$ plays the role of a control variable. We will seek $q(z)$, $z_0 \leq z \leq z_1$, such that V is minimum.

The variational problem is conveniently defined in terms of the Hamiltonian:

$$H(y_1, \ldots, y_m; \beta_1, \ldots, \beta_m; q) = \sum_{j=1}^{m} \beta_j f_j(y_1, \ldots, y_m, q(z)). \tag{6}$$

Canonical equations of the type:

$$dy_j/dz = \partial H/\partial \beta_j$$

are given by (4), for which the initial conditions are (5). The canonical equations for the adjoint variables, $\beta_j(z)$, are:

$$d\beta_j/dz = -\partial H/\partial y_j, \qquad j = 1, 2, \ldots, m. \tag{7}$$

These have final conditions given by:

$$\beta_j(z_1) = \partial V/\partial y_j(z_1). \tag{8}$$

Following a suitable maximum principle [3], we seek $q(z)$ so that the Hamiltonian is stationary with respect to choice of $q(z)$. That is,

$$\partial H/\partial q = \sum_{j=1}^{m} \beta_j \, \partial f_j/\partial q = 0, \qquad z_0 \leq z \leq z_1 . \tag{9}$$

This condition holds if q is unconstrained or if one can establish that the solution is to be found inside the region to which q is constrained rather than on the boundary. Solution of (4), (5), (7), (8), and (9) leads to the optimal function, $q(z)$, which, when one inverts the transformations previously applied, yields the "best" transport coefficient, $\propto (u)$. The computational procedure we employ is an iterative one. Before dealing with this, we describe the experiment and data analysis in more detail.

The Shocktube Experiment

Temperatures in the vicinity of a gas-solid interface immediately following reflection of a shock wave from the solid are suggested in Fig. 1.

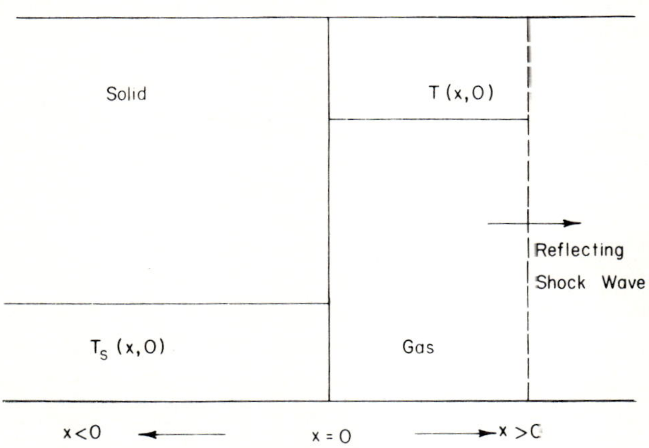

FIG. 1. Initial temperatures near the gas-solid interface.

The initial solid temperature, $T_s(x, 0)$, is uniform and of the order of room temperature. Depending upon the strength of the shock wave employed, the initial gas temperature, $T(x, 0)$, may be near room temperature or it may be very high, perhaps several thousand degrees. This temperature is known, either from direct measurement or, more commonly, from calculations. The latter are based on the assumption that the shock wave propogates isentropically. From measurements of the shock speed and of pressures on either side of the shock wave, one can compute the temperature rise in the compressed gas. Provided nonidealities of the gas are adequately accounted for, computed and directly measured temperatures are in good agreement [11].

Following reflection of the shock wave, thermal energy transfers from the hot gas into the cooler solid. The transfer mechanism in the rigid solid is entirely conductive; but both conduction and convection (as well as radiation, for conditions more extreme than those considered here) take part in heat transfer through the gas, as indicated in Fig. 2.

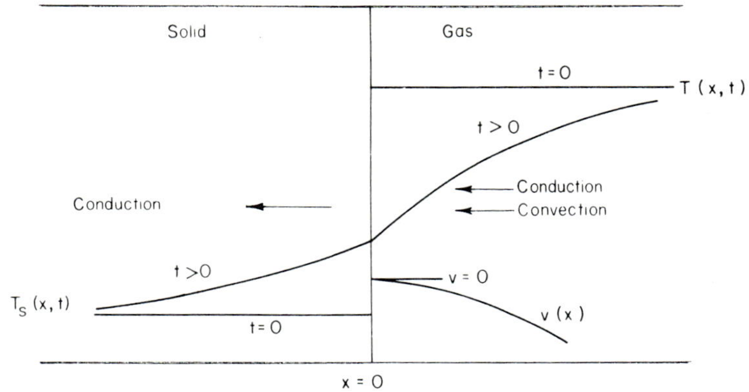

FIG. 2. Temperature and velocity profiles near the interface. (Schematic).

The convective flux arises because the gas pressure near the gas-solid interface is essentially constant for the brief experimental period following reflection. This constant pressure is maintained, despite cooling and increasing gas density, by convective flow into the region near the interface.

The temperature of the interface, $T(0, t)$, can be measured conveniently by a thin-film, platinum resistance thermometer of a type which has been employed frequently for shocktube experiments, [4–8, 12–14]. The observed response of the thermometer is a simple transient (caused by electronic delays) lasting for a few microseconds, followed by a constant value for a period of 100 μ sec or more. It is assumed here that this constant value represents the interface temperature, which is thus independent of time.

Constancy of the interface temperature (following an instantaneous jump from the initial value, $T_s(x, 0)$) is consistent with the assumption that gas pressure is constant and is to be expected. This can be seen as follows. Consider, first, conditions in the solid phase, that is, for $x \leq 0$. If the solid is homogeneous and is warmed solely by conduction, we expect its temperature to rise continuously with time at every point. One can thus conclude that the time derivative of temperature within the solid is nonnegative; and, furthermore, for the interface:

$$\lim_{x \to 0^-} (\partial T/\partial t) \geq 0. \tag{10}$$

Consider, now, conditions in the gas phase, that is, for $x \geq 0$. The continuity requirement yields:

$$\rho_t + (\rho\, v)_x = 0 \tag{11}$$

where ρ is gas density and $v(x, t)$ is gas velocity. Let us assume that the gas is ideal with density given by:

$$\rho = P/RT \tag{12}$$

where R is the gas constant. If pressure, P, is constant, one can rewrite (11) in the form:

$$T_t = -vT_x + Tv_x. \tag{13}$$

At the interface $v = 0$; and the requirement of momentum conservation suggests a velocity profile like that indicated in Fig. 2, where, at the interface, $v_x < 0$. One is thus led from (13) to the conclusion that the time derivative of temperature at the interface is nonpositive. That is, for the interface:

$$\lim_{x \to 0^+} (\partial T/\partial t) = \lim_{x \to 0^+} (Tv_x) \leqslant 0. \tag{14}$$

Conditions (10) and (14) are both satisfied if the interface temperature is constant.

The Thermal Conduction Model

It is now feasible to formulate the mathematical problem which defines the shocktube experiment, taking into account assumptions which have already been enumerated as well as additional ones to be pointed out.

For the solid phase we use the heat-conduction equation:

$$\partial T_s/\partial t = \propto_s \partial^2 T_s/\partial x^2. \tag{15}$$

The thermal diffusivity of the solid, $\propto_s$, is presumed to be a known constant. (Its temperature dependence may, however, be taken into account for very strong shocks [14].) Initial and boundary temperatures are:

$$T_s(x, 0) = T_s(-\infty, t) = T_{s\infty} \tag{16}$$

$$T_s(0, t) = T(0, t) = T_0 \tag{17}$$

where $T_{s\infty}$ and T_0 are measured experimentally.

In addition there is the requirement that thermal flux across the interface be equal for solid and gas; thus:

$$k_s(\partial/\partial x)\,(T_s\,(0, t)) = k(T_0)\,(\partial/\partial x)\,(T(0, t)) \tag{18}$$

where k_s is the (known) thermal conductivity of the solid; and $k(T_0)$ is the (as yet unknown) thermal conductivity of the gas at the interface temperature.

For the gas phase the energy equation is:

$$\rho c(\partial T/\partial t) + \rho c v(\partial T/\partial x) = (\partial/\partial x)\,[k(T)\,(\partial T/\partial x)] \tag{19}$$

$c(T, P)$ is the heat capacity per unit mass at constant pressure. The continuity equation (11) and the equation of state (12) also hold. Besides (17) and (18), one has the auxiliary conditions:

$$T(x, 0) = T_\infty \tag{20a}$$

$$T(\infty, t) = T_\infty \tag{20b}$$

where T_∞ is the temperature of the reflected shock.

Equations (15)–(17) can now be solved explicitly for $T_s(x, t)$. This can be differentiated and the derivative substituted in (18) to give:

$$T_x(0, t) = F(T_0, \propto_s, k_s). \tag{21}$$

The function on the right-hand side is:

$$F = 2\pi^{-1/2} (T_0 - T_{s\infty}) [k_s\rho_s c_s/k(T_0)\rho(T_0)c(T_0)]^{1/2}. \tag{22}$$

The gas-phase problem may now be defined by (19), (11), (12), (17), (20a), and (21). To solve it for a single experiment one searches for $k(T)$, the thermal conductivity-temperature function, such that (20b) is satisfied, T_∞ having been measured in the experiment.

Normally one will carry out several experiments, say N of them. The variables describing the gas-phase problem will behave differently for the various experiments; and one will have N sets of equations of the form of (19), (11), (12), (17), (21), and (20). In this case we seek $k(T)$ such that:

$$V(T^1, \ldots, T^N) = \sum_{i=1}^{N} [T_\infty{}^i - T^i(\infty, t)]^2$$

is minimum. Here the superscript, $i = 1, 2, \ldots, N$, denotes a value characteristic of the ith experiment.

Energy Equation Transformations

It is convenient to carry out a series of transformations of the energy equation (19) for the gas phase. The first of these is a Lagrangian transformation, in which $T(x, t)$ of (19) is transformed to $T(\gamma, t)$, where:

$$\gamma(x, t) = \int_0^x \rho(x, t)\, dx \tag{23}$$

and we assume that $\gamma \to \infty$ as $x \to \infty$. The energy equation takes the form:

$$c(T)\,(\partial T/\partial t) = (\partial/\partial\gamma)\,(k\rho\;\partial T/\partial\gamma) \tag{24}$$

with:

$$T(0, t) = T_0 \tag{25a}$$

$$T(\gamma, 0) = T_\infty \tag{25b}$$

$$T_\gamma(0, t) = F/\rho(T_0) = G(T_0, \propto_s, k_s). \tag{25c}$$

Next a similarity transformation:

$$\mu(\gamma, t) = \gamma/2(\propto_0 t)^{1/2} \tag{26}$$

is applied, where $\propto_0 = k(T_0)/\rho(T_0)c(T_0)$. The new form of the energy equation is:

$$(d/d\mu)\,(k\rho\; dT/d\mu) + 2\propto_0 c(T)\,\mu(dT/d\mu) = 0. \tag{27}$$

Initial conditions are:

$$T(0) = T_0 \tag{28}$$

$$T'(0) = G(T_0, \propto_s, k_s). \tag{29}$$

Finally, for convenience in the numerical computation, we convert to a finite interval for the independent variable, using:

$$\omega = 1 - \exp[-\,\delta\,\mu] \tag{30}$$

where δ is a numerical constant. The value of δ is chosen later so as to provide a suitably uniform change of temperature gradient on the interval $0 \leq \omega \leq 1$. This interval corresponds to $0 \leq \mu \leq \infty$ and to $0 \leq x \leq \infty$.

The energy equation now appears as:

$$(d/d\omega)\,[k\rho(1-\omega)(dT/d\omega)] - (2\propto_0/\delta^2)\,c(T)[\ln(1-\omega)]\,(dT/d\omega) = 0. \tag{31}$$

Initial condition (28) remains, and (29) takes the form:

$$T'(0) = G/\delta. \tag{32}$$

Variational Formulation

We are now in a position to put the problem in the canonical form of Eqs. (4)–(8). There will be an energy equation like (31) for each experiment, so that we may denote by $T^j(\omega)$, $j = 1, 2, \ldots, N$, the temperature of the jth experiment. Equation (31) is not of canonical form, however; and it is convenient to define new variables $y_1, y_2, \ldots, y_N$ representing the temperatures $T^1, T^2, \ldots, T^N$, respectively. We also choose $y_{N+1}, y_{N+2}, \ldots, y_{2N}$ to represent the set of temperature derivatives, $dT^1/d\omega$, $dT^2/d\omega$, $\ldots$, $dT^N/d\omega$, and $y_{2N+1}, y_{2N+2}, \ldots, y_{3N}$ to denote the products of thermal conductivity and gas density, $k(T^1)\rho(T^1), k(T^2)\rho(T^2), \ldots, k(T^N)\rho(T^N)$, respectively. Finally we define:

$$q_j = (d/d\omega)\,[k(T^j)\rho(T^j)]. \tag{33}$$

These definitions permit us to write the set of state equations, where

$j = 1, 2, \ldots, N$, as follows:

$$y_j' = y_{N+j} \tag{34}$$

$$y_j(0) = T_0{}^j \tag{35}$$

$$y'_{N+j} = [y_{N+j}/(1-\omega)y_{2N+j}]\,[(2\alpha_0/\delta^2)c(y_j)\ln(1-\omega) + y_{2N+j} - q_j(1-\omega)] \tag{36}$$

$$y_{N+j}(0) = G^j/\delta \tag{37}$$

$$y'_{2N+j} = q_j \tag{38}$$

$$y_{2N+j}(0) = k(T_0{}^j)\rho(T_0{}^j). \tag{39}$$

These are to be solved so as to minimize:

$$V = \sum_{j=1}^{N} [y_j(1) - T_\infty{}^j]^2. \tag{40}$$

To assist in the choice of the control variables, $q_1, q_2, \ldots, q_N$ which make V minimum, we introduce the adjoint sets β_j, β_{N+j} and β_{2N+j}, $j = 1, \ldots, N$. These, together with the state variables and control variables, comprise the variables of the Hamiltonian:

$$H(y_1, \ldots, y_{3N}; q_1, \ldots, q_N; \beta_1, \ldots, \beta_{3N})$$

$$= \sum_{j=1}^{N} (\beta_j y_j' + \beta_{N+j} y'_{N+j} + \beta_{2N+j} y'_{2N+j})$$

$$= \sum_{j=1}^{N} \{\beta_j y_{N+j} + \beta_{N+j}\,[y_{N+j}/(1-\omega)y_{2N+j}] \times [(2\alpha_0/\delta^2)c(y_j)\ln(1-\omega) + y_{2N+j} - q_j(1-\omega)] + \beta_{2N+j} q_j\}. \tag{41}$$

Differential equations for the adjoint variables, analogous to (7) with final conditions analogous to (8), are:

$$\beta_j' = \beta_{N+j}[y_{N+j}/(1-\omega)y_{2N+j}]\,[(2\alpha_0/\delta^2)\,(dc/dy_j)\ln(1-\omega)] \tag{42}$$

$$\beta_j(1) = 2[y_j(1) - T_\infty{}^j] \tag{43}$$

$$\beta'_{N+j} = \beta_j + [\beta_{N+j}/(1-\omega)y_{2N+j}]\,[2\alpha_0/\delta^2)c(y_j)\ln(1-\omega) + y_{2N+j} - q_j(1-\omega)] \tag{44}$$

$$\beta_{N+j}(1) = 0. \tag{45}$$

$$\beta'_{2N+j} = [\beta_{N+j}y_{N+j}/(1-\omega)y_{2N+j}]\{1 - (1/y_{2N+j}) \times [(2\alpha_0/\delta^2)c(y_j)\ln(1-\omega) + y_{2N+j} - q_j(1-\omega)] \tag{46}$$

$$\beta_{2N+j}(1) = 0. \tag{47}$$

The condition analogous to (9) is:

$$\partial H/\partial q_j = 0, \qquad j = 1, \ldots, N.$$

That is:

$$\beta_{N+j}y_{N+j} = \beta_{2N+j}y_{2N+j}, \qquad 0 \le \omega \le 1. \tag{48}$$

A Computational Procedure

The problem to be solved is defined by (34) to (39) and by (42) to (48). There are $7N$ equations to be solved for the $7N$ variables. $y_j(\omega)$, $y_{N+j}(\omega)$, $y_{2N+j}(\omega)$, $\beta_j(\omega)$, $\beta_{N+j}(\omega)$, $\beta_{2N+j}(\omega)$, $q_j(\omega)$, $j = 1, 2, \ldots, N$, on the interval $0 \le \omega \le 1$. Values of the y's are given at $\omega = 0$ and those of the β's are known at $\omega = 1$. The problem is of mixed, boundary—value type, therefore; and a trial-and-error solution is indicated.

One approach to the solution, which is followed here, is to guess the functions $q_j(\omega)$, solve for the y's and the β's and then determine whether (48) is satisfied. If (48) is not satisfied, the amount of mismatch in (48) can be used to determine an improved set $q_j(\omega)$. The procedure can then be repeated until (48) is satisfied.

This is an awkward procedure because N functions $q_j(\omega)$ must be guessed at the beginning. To avoid this awkwardness, we take advantage of the fact that thermal conductivities of dilute gases frequently vary with temperature approximately according to the form:

$$k(T) = aT^b \tag{49}$$

where a and b are pressure-independent quantities for which approximate values can be assigned. If in addition the gas can be considered ideal, following (12), we can write:

$$y_{2N+j}(\omega) = (P_j a/R)y_j^{b-1}. \tag{50}$$

The conditions (38) take the form:

$$y'_{2N+j}(\omega) = (b-1)(P_j a/R)y_j^{b-2}y_{N+j} \tag{51}$$

and the initial conditions become:

$$y_{2N+j}(0) = (P_j a/R)\,(T_0{}^j)^{b-1}. \tag{52}$$

It is not desirable to suppose that $k(T)$ must take the form (49); we would like the numerical solution to give $k(T)$ without need for its functional form being postulated *a priori*. It is convenient, however, to use (49) to (52) to get the solution started.

We define the set $q_j{}^0(\omega)$, $q_j{}^1(\omega)$, $q_j{}^2(\omega)$, etc., of successively improved estimates of $q_j(\omega)$. Computation steps may then be itemized as follows:

1. Select numbers a and b on the basis of whatever prior information is available.

2. Solve (34) to (37) together with (51) and (52) using an appropriate grid on $0 \leq \omega \leq 1$. We will call the resulting quantities:

$$y_j{}^0(\omega), \quad y_{N+j}^0(\omega), \quad y_{2N+j}^0(\omega).$$

3. Compute $q_j{}^0(\omega)$, from (51). That is:

$$q_j{}^0(\omega) = (b-1)\,(P_j a/R)\,(y_j{}^0)^{b-2} y_{N+j}^0. \tag{53}$$

4. Solve (42) to (47) backward from $\omega = 1$ on the same grid as that used in step 2. Use values of the y's from step 2. We denote the resulting adjoint functions:

$$\beta_j{}^0(\omega), \quad \beta_{N+j}^0(\omega), \quad \beta_{2N+j}^0(\omega).$$

5. Select the new set $q_j{}^1(\omega)$ from the recurrence formula (54), taking $r = 0$,

$$q_j^{r+1}(\omega) = q_j^r(\omega)\,[1 + A^{-1}\,|\,\beta_{N+j}^0(\omega) y_{N+j}^0(\omega) - \beta_{2N+j}^0(\omega) y_{2N+j}^0(\omega)\,|\,]. \tag{54}$$

Here A is an accelerating constant for which a value is to be chosen on the basis of preliminary numerical experimentation and for which the sign is chosen from the direction of inequality in (48).

6. Solve (34) to (39) for the new sets $y_j{}^1(\omega)$, $y_{N+j}^1(\omega)$, $y_{2N+j}^1(\omega)$. Use $q_j{}^1(\omega)$ in (38) and $y_{2N+j}^0(0)$ in (39).

7. Repeat steps 4, 5, and 6 until $q_j{}^r(\omega)$ satisfies a suitable convergence test.

Some Numerical Results

The variational computation is illustrated here for the case of a single experiment ($N = 1$) carried out with argon at about 7 atm, total pressure using data from Tandy [7] and Werner [8]. Further results will be reported in another place; additional computations for the single-experiment case are available [15].

Preliminary numerical integrations were carried out for $q = 0$. (This corresponds to a thermal conductivity proportional to temperature.) On the basis of these tests a grid with 100 increments on $0 \leq \omega \leq 1$ was selected for all subsequent integrations. A modified Euler's method was used for the three initial steps followed by Hamming's multiple-point method for the remainder of the grid [16].

A typical temperature profile after the first iteration is shown in Fig. 3. The region near $\omega = 1$ is shown magnified in Fig. 4. Truncation errors

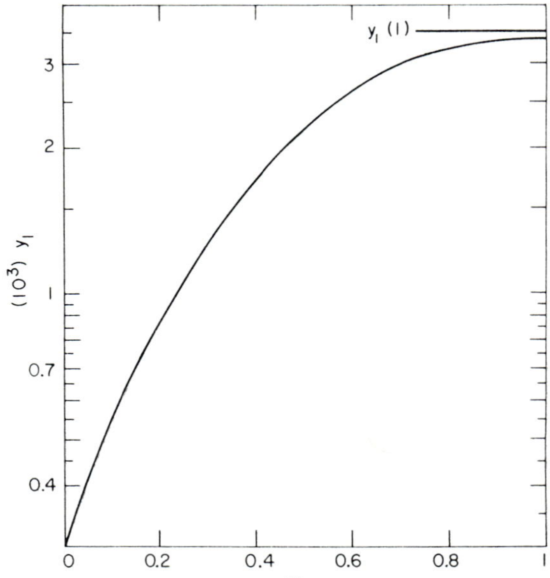

FIG. 3. Calculated temperature profile for heat conduction through argon at 7 atm.

cause the calculated profile to exhibit a maximum near $\omega = 1$. The position of the maximum, called ω_m, and the temperature difference at ω_m, $T_\infty - y_1(\omega_m)$, depend upon the value of δ used in Eq. (30). This is illus-

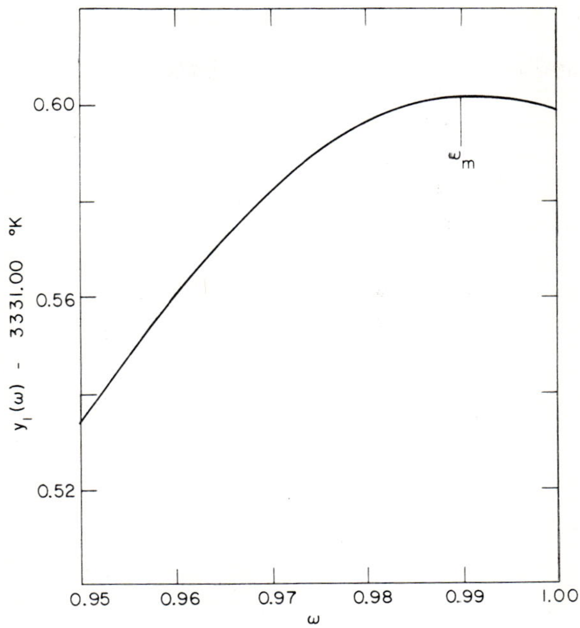

FIG. 4. Temperature profile far from the interface, showing the position of the calculated maximum temperature.

Table I

DEPENDENCE OF POINT OF MAXIMUM TEMPERATURE ON CHOICE OF δ (Eq. 30)

δ (Eq. 30)	Position of max. temperature (ω_m)	Temperature deviation at ω_m, $T_\infty - y_1(\omega_m)$, °K
60	0.85	198.7
70	0.91	191.9
80	0.92	185.6
100	0.94	174.7
110	0.95	170.0
150	0.95	42.6
200	0.98	87.5

strated in Table I. If δ is too small, the maximum occurs too early in the integration. The resulting profile is unrealistic physically, for one expects the temperature to rise monotonically to a maximum at or near $\omega = 1$. If δ is too large, the maximum may occur at a reasonable point; but the temperature deviation will be unrealistically large. There is thus a suitable compromise value of δ (150 in the example of Table I) for which ω_m occurs between $\omega = 0.95$ and $\omega = 1$ and for which the temperature deviation is about as small as one could hope for in the first iteration.

The forward integration for the temperature, y_1, the temperature gradient, y_2, and the product of density and thermal conductivity, y_3, were carried out on $0 \leq \omega \leq \omega_m$. Backward integrations for β_1, β_2, and β_3 were started at ω_m. The acceleration factor A (Eq. 54) was normally selected so that after three or four iterations the temperature deviation, $T_\infty - y_1(\omega_m)$, was less than 1° K.

Progress of the calculation of temperature profile through four iterations using $A = 2(10^6)$ is shown in Fig. 5. The point of maximum temperature rise, ω_m, was at $\omega_m = 0.99$, in the first iteration. It decreased to 0.97 for each of the last three iterations. The calculation was carried out for an initial guess:

$$q^0(\omega) = -6(10^{-7})(1 - \omega)^2. \tag{55}$$

Final temperatures at representative grid points together with experimental thermal conductivities as well as literature values of thermal conductivities of argon are shown in Table II. Deviations between values of thermal conductivity are due primarily to experimental errors.

Discussion

The problem considered has been idealized considerably with respect to physical phenomena. For limited conditions the thermal conductivity obtained is in good agreement with data from other kind of experiments, however, as has been demonstrated by others [4–6] and as can be seen from the results shown in Table II.

The additional physical details which might be included have to do with such things as the finite rate of response of the interface resistance thermometer, the variations of solid-phase physical properties with temperature, the possible pressure gradient in the gas, and viscous effects in

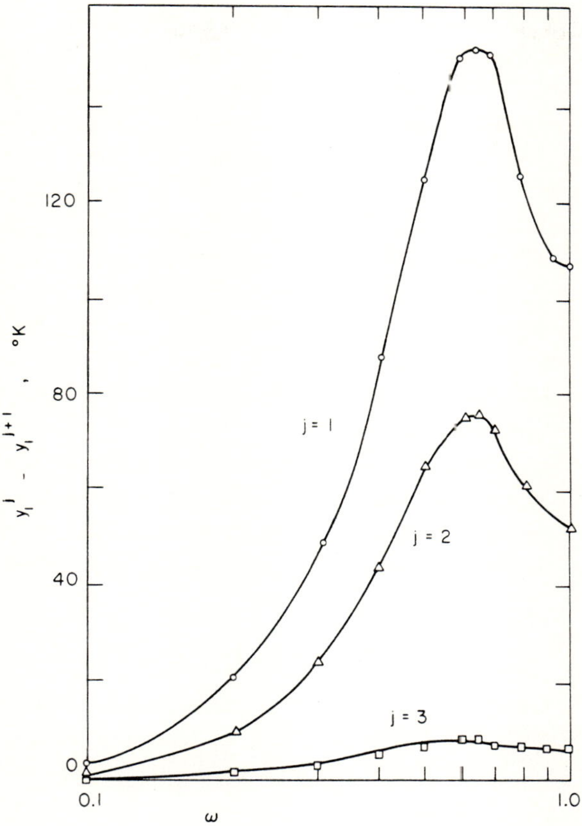

FIG. 5. Changes of the temperature profile through four iterations.

the gas leading to interaction of the reflected shock with the boundary layer in the gas phase.

One might also wish to carry out experiments in which chemical reactions and diffusional phenomena occur. For example, a study of diffusion coefficients has been described in which a component of the gas diffuses to the endwall of a shocktube and reacts there with a liquid film [17]. This is in some respects analogous to the thermal conduction experiment referred to here.

In all of these extensions additional describing equations like (1) must be added to the statement of the problem. Ordinarily these will not be

Table II

TEMPERATURES AND THERMAL CONDUCTIVITIES AFTER FOUR ITERATIONS[a]

ω	Temperature, $(y_1(\omega), °K)$	Thermal conductivity, (cal/cm sec °K)10^5	
		$k(y)_1$.	Literature value [6]
0.01	315.7	4.42	4.45
0.05	340.4	4.68	4.6
0.10	475.2	5.97	5.95
0.15	629.4	7.27	7.25
0.20	804.3	8.59	8.7
0.25	1000	9.95	10.2
0.30	1218	11.4	11.7
0.35	1456	12.8	13.2
0.40	1710	14.3	14.7
0.45	1976	15.8	16.1
0.50	2247	17.3	17.5
0.55	2515	18.7	18.8
0.60	2767	20.0	20.0
0.65	2994	21.2	21.1
0.70	3183	22.2	22.0
0.75	3327	22.9	22.7
0.80	3422	23.3	23.1
0.90	3494	23.6	23.5
0.97(ω_m)	3498	23.6	23.5

[a] Argon at 7 atm pressure.

susceptible to transformation to ordinary differential equation form, except by finite-difference methods. The scope of a problem is greatly enlarged when this is the case. Thus, if one elects to difference (1) over the range of the variable x, he obtains a first-order ordinary differential equation analogous to (4) for each increment of $x_0 \leq x \leq x_1$. If there are, say, M increments and one is considering N experiments, the total equations to be dealt with is of the order of $7NM$. This can be a formidable number. A relatively small-scale problem of this type has been successfully treated, however [3]. There seems to be no intrinsic difficulty in the scale of the problem and practical computation should be greatly facilitated as special digital computer equipment for solving sets of differential equations becomes more readily available.

Acknowledgment

This work was supported in part by the National Science Foundation through their grant GP-2549.

References

1. Bliss, G. A., "Lectures on the Calculus of Variations." Univ. Chicago Press, Chicago, Illinois, 1946.
2. Katz, Stanley, A General Minimum Principle for End-Point Control Problems, *J. Electron. Control* **16**, 189 (1964).
3. Denn, M. M., Gray, R. D., Jr., and Ferron, J. R., Optimization in a Class of Distributed-Parameter Systems, *Ind. Eng. Chem. Fundamentals* **5**, 59 (1966).
4. Smiley, E. F., "The Measurement of the Thermal Conductivity of Gases at High Temperatures with a Shocktube." Catholic Univ. America Press, Washington, D.C., 1957.
5. Camac, M., and Feinberg, R. M., Thermal Conductivity of Argon at High Temperatures, Res. Rep. 168, Avco Everett Res. Lab., Everett, Massachusetts, (March, 1963).
6. Lauver, M. R., Evaluation of Shocktube Heat-Transfer Experiments to Measure Thermal Conductivity of Argon from 700° to 8600° K, NASA TN D-2117 (February, 1964).
7. Tandy, D. C., The Measurement of the Thermal Conductivity of Argon and Carbon Dioxide with a Shocktube, MChE Thesis, University of Delaware, Newark, Delaware (1964).
8. Werner, P. R., The Thermal Conductivity of Carbon Dioxide and Nitrogen at High Temperatures as Determined by Shocktube Measurements, MChE Thesis, University of Delaware, Newark, Delaware (1965).
9. Pontryagin, L. S., Boltyanskii, V. A., Gamkrelidze, R. V., and Mischhenko, E. F., "Mathematical Theory of Optimal Processes." Wiley, New York, 1962.
10. Merriam, C. W., III, "Optimization Theory and the Design of Feedback Control Systems." McGraw-Hill, New York, 1964.
11. Mak, W. H., Sodium Line Reversal Temperature Measurements in Shocktube Flows, UTIA Tech. Note 66, Inst. Aerospace Studies, Univ. Toronto, Toronto, Ontario, Canada (May, 1963).
12. Bradley, J. N., "Shock Waves in Chemistry and Physics." Wiley, New York, 1962.
13. Gaydon, A. G., and Hurle, I. R., "The Shock Tube in High-Temperature Chemical Physics." Reinhold, New York, 1963.
14. Walenta, Z. A., Analogue Networks for High Heat-Transfer Rate Measurements UTIAS Tech. Note 84, Inst. Aerospace Studies, Univ. Toronto, Toronto, Ontario, Canada (November, 1964).
15. Fujimura, K., Measurement of Thermal Conductivity of Gases by Means of a Shocktube, MChE Thesis, Univ. Delaware, Newark, Delaware (1966).

16. Lapidus, Leon, "Digital Computation for Chemical Engineers", pp. 88, 105. McGraw-Hill, New York, 1962.
17. Dunham, P. G., and Ferron, J. R., Shocktube Studies of Gas-Liquid Reactions and of Gas-Phase Diffusion, AFML–TR–65–416, Air Force Materials Lab. Wright-Patterson A.F. Base, Ohio.

A Synergetic Approach to Problems of Nonlinear Dispersive Wave Propagation and Interaction†

NORMAN J. ZABUSKY

Bell Telephone Laboratories
Whippany, New Jersey

I. Introduction

I believe that we can mark the beginning of the mathematical study of nonlinear continua with Riemann's classic paper on nonlinear wave propagation, published in 1860. In the century which followed, no decisive progress was made and few general results were obtained for nonlinear systems, compared with the progress made in other branches of (linear!) mathematics and physics.

Nonlinear problems are naturally difficult and the methods of solution are highly restricted or "rigid," in that they are usually rendered useless if the problem posed is modified only slightly.

The sixties will probably be looked upon as the decade in which the pace of progress in the study of nonlinear continua accelerated; and perhaps this year will mark the mean time for this event. This Symposium on Methods of Solution for Nonlinear Partial Differential Equations, the publication of the new Journal of Nonlinear Mechanics, and the School of Nonlinear Mathematics and Physics which Professor M. D. Kruskal of Princeton University and I have organized for 1966, in addition to numerous conferences and symposia on specialized nonlinear and turbulent phenomena are some indications of the attention being given to the subject. This maturation is due, in part, to the ability to do physical experiments with intense coherent and incoherent sources and, in large

† This paper is dedicated to Professor L. Biermann on his sixtieth birthday. I would like to express my gratitude to Professor Biermann and his Institute of Astrophysics of the Max-Planck-Institute for Physics and Astrophysics for the support they gave to the International School of Nonlinear Mathematics and Physics held in Munich during the summer of 1966.

measure, to the development of sophisticated electronic digital computers.

The synergetic approach to nonlinear problems was first formulated by von Neumann and it was enunciated in the works of von Neumann and his collaborators in the mid-forties (in particular Goldstine and von Neumann [1]). The name was coined by Ulam in his excellent monograph "A Collection of Mathematical Problems" [2].

At a time when electronic computers were in their infancy, von Neumann appreciated the impact they would have on our understanding of two essentially nonlinear problems of fluid flow: the problem of shock waves and the problem of turbulence. Many of us are aware of the developments of the last twenty years, and in any case, we will be hearing much about them during these three days. However, we should not forget that von Neumann envisioned the computer as a tool for probing the hard front which bounds particular areas of nonlinear mathematics and physics.

II. The Synergetic Approach

The synergetic approach to nonlinear mathematical and physical problems can be defined as the simultaneous use of conventional analysis and computer numerical mathematics to obtain solutions to judiciously posed problems concerning the mathematical and physical content of a set of equations.

When faced with a complex and highly tangled mathematical representation, as almost all nonlinear problems are, one uses the computer as an heuristic probe, to force an opening into a region of difficulty. The numerical "answers" to this problem, if properly assimilated, can provide the prepared investigator with a new level of insight which may permit him to resolve one area of trouble by conventional analysis—usually some form of uniform asymptotic analysis, and rarely an exact solution. In attempting to complete this analysis or understand completely the details of the computer solution, the investigator will uncover a group of problems which will lead him further into the maze. Oftentimes he will have to back away from an impasse, review the problem, and probe another aspect of the equations. He must be constantly on the lookout for the new and small phenomena generated by the computer, for an understanding of these often will provide the pathways to an eventual analytical and global understanding of the properties of his equations.

I will illustrate the synergetic approach and the richness of mathematical and physical phenomena it has produced by presenting some work that Kruskal and I have done over the past few years. We have undertaken a study of wave propagation and interaction in a one-dimensional nonlinear —or as the physicists say, anharmonic—lattice, because of peculiar *recurrences* observed in the numerical solutions for such a problem.

Our progress was slow at first, partly because we did not have the sophisticated computer-output facilities that are now available. You will soon see some examples of computer-made movies which allow the investigator to make a quick yet detailed survey of his results. Deem, of Bell Laboratories at Whippany, has assisted us for the past year with the analysis, computer programming and film production. If you were to look over some of the summary papers [3–5] that Kruskal and I have given over the past few years at symposia like this one, you would get a good idea of the synergetic method. These talks and papers show the rich mathematical byproducts we have derived, and also show the various obstacles we have had to overcome and have yet to overcome.

III. The Nonlinear One-Dimensional Lattice

Debye [6] in 1914 suggested that the finiteness of the thermal conductivity of an anharmonic (or nonlinear) nonconducting lattice was due essentially to its nonlinearity. In other words, he suggested that the phonons or fundamental wave propagation modes would interact because of the nonlinearity, thereby inhibiting the propagation of energy. The net effect of many such nonlinear interactions or "phonon collisions" would manifest itself in a finite transport coefficient; that is, if there is a "mean" energy flux through a region, there will be an energy gradient across this region. The problem of deducing a finite thermal conductivity for an anharmonic lattice from its microscopic properties has challenged theoretical physicists for the last fifty years, and as Peierls [7] observed in his penetrating review paper, "It seems there is no problem in modern physics for which there are on record as many false starts, and as many theories which overlook some essential feature, as in the problem of the thermal conductivity of nonconducting crystals." It is undoubtedly this unsatisfactory state of affairs which motivated Fermi, Pasta, and Ulam [8], or FPU for short, to undertake a numerical study of the one-dimensional anharmonic lattice. They hypothesized with Debye that if one started from some smooth state, the nonlinear interactions would cause the energy to flow to all higher modes until eventually all the modes would have the same energy (in a time-average sense). The system would then be in "thermal" equilibrium. The time interval required to arrive at this equilibrium state—the relaxation time—would then provide a measure for the thermal conductivity.

This physical intuition or hypothesis, which led FPU to set up a computational program, represents in our case the beginning of the synergetic approach for this mathematical model. For convenience in presentation, we designate it as Hypothesis 1 or H1 and summarize it by:

H1: *Given a one-dimensional lattice with nearest neighboring masses coupled by nonlinear springs. If the energy of the system is initially in a smooth state, the energy will eventually be shared equally among all the degrees of freedom of the system.*

We anticipate some of our future work by observing that Hypothesis 1 gives no consideration to the size of the initial condition or equivalently to the size of the microscopic nonlinearity.

The mathematical model used by FPU to describe their one-dimensional lattice of length L is shown in Fig. 1. It consists of N-1 identical masses

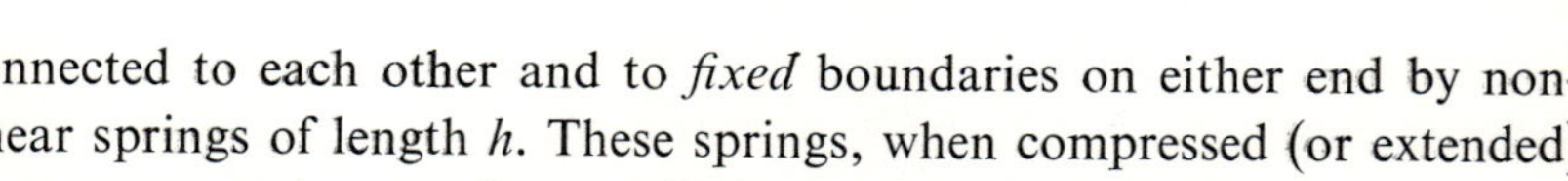

FIG. 1. A one-dimensional lattice of equimass particles with nearest neighbor nonlinear interactions. The Fermi–Pasta–Ulam problem treated a lattice of length $L = Nh$ and fixed boundary conditions.

connected to each other and to *fixed* boundaries on either end by nonlinear springs of length h. These springs, when compressed (or extended) by an amount Δ, exert forces which are given by the relation

$$F = \kappa(\Delta + \alpha\,\Delta^{p+1}), \tag{3.1}$$

where Δ represents the extension or compression of the spring, κ is the "linear" spring constant, α measures the strength of the nonlinearity and is taken >0, and p is an integer, usually 1 or 2 (the quadratic and cubic nonlinearity, respectively). The potential energy stored in such a spring is given by

$$E = \frac{1}{2}\kappa\,\Delta^2 + \frac{\kappa\alpha}{(p+2)}\,\Delta^{p+2}. \tag{3.2}$$

We will always work in the range where

$$1 > \alpha\,|\,\Delta^p\,|, \tag{3.3}$$

so that $F/\kappa\Delta$ and E are always > 0.

The equations governing the dynamics of this lattice are

$$\omega_0^{-2}\ddot{y}_i = (y_{i+1} - 2y_i + y_{i-1}) + \alpha[(y_{i+1} - y_i)^{p+1} - (y_i - y_{i-1})^{p+1}], \tag{3.4}$$
$$i = 1, 2, \ldots, N-1, \quad \text{and} \quad y_0 = y_N = 0,$$

where y_i is the displacement of the ith mass and $\omega_0^2 = \kappa/m$ is the resonant frequency of a simple linear spring-mass system. For $p = 1$, the equation takes the form

$$\ddot{y}_i = \omega_0^2(y_{i+1} - 2y_i + y_{i-1})\,[1 + \alpha(y_{i+1} - y_{i-1})], \tag{3.5}$$
$$i = 1, 2, \ldots, N-1, \quad \text{and} \quad y_0 = y_N = 0.$$

Fermi *et al.* chose for their initial condition

$$y_i\,|_0 = a\sin(i\pi/N), \quad \dot{y}_i\,|_0 = 0, \tag{3.6}$$

and considered the problem with fixed boundary conditions.

Much to their surprise, the system did not approach "equilibrium", that is, sharing of energy equally among all the N-1 modes of the system, but rather exhibited long-time "recurrences" and energy sharing only among the low modes of the system. This is shown in Fig. 2 where we observe the energy starting in mode 1, and, through the nonlinear interaction, being shared among several other modes, and finally returning almost completely to mode 1 after 158 linear periods or one recurrence period, t_R. Although FPU varied N from 16 to 64 and α from 1/4 to 1, they found no qualitative difference in the modal behavior; only the recurrence time varied, as we discussed elsewhere.† The quality of the numerical calcula-

† See Zabusky [3]. In Section 4a, I discussed the variation of t_R with N and α, and using data from five runs I was able to deduce the relation (Eq. 4.2)

$$t_R/t_L = 1.42(N^{1.38}/\alpha^{0.5}).$$

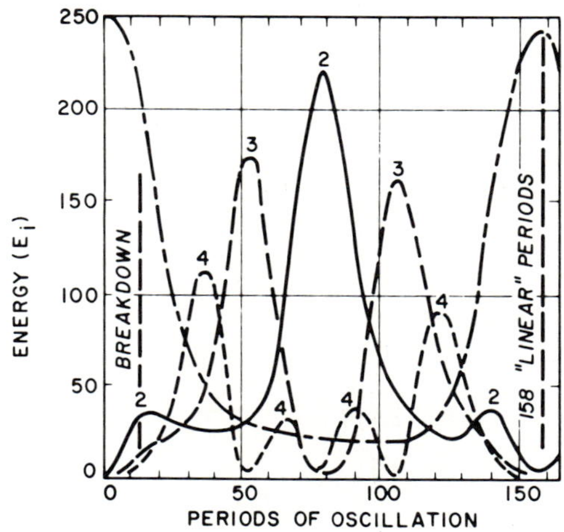

FIG. 2. The energy (arbitrary units) in the low modes of the nonlinear lattice plotted as a function of normalized time. Here, the nonlinearity is quadratic ($p = 1$); $N = 32$, $\alpha = 1/4$, and a first mode initial condition (3.6) with $a = 1$ was used. (From Fermi *et al.* [8], Fig. 1).

tions (that is, the effect of round-off errors) was gaged by observing the constancy of quantities which are analytically time-invariant, namely, the total momentum P and total energy H,

$$P = \sum_{i=1}^{N-1} m\,\dot{y}_i$$
$$H = \sum_{i=1}^{N-1} \frac{1}{2}\left[m\,\dot{y}_i^{\,2} + \kappa(y_{i+1} - y_i)^2 + \frac{\kappa\alpha}{3}(y_{i+1} - y_i)^3\right]. \tag{3.7}$$

In all cases we will present (FPU's work and our work) round-off error played a negligible role.

So, for the parameter configuration chosen (including amplitude), Hypothesis 1 is not true. One could be concerned with the lack of an equilibrium state and attempt to force it on the system [9], either through a gross modification of the spring and mass constants or by the appeal to one's physical intuition, that real three-dimensional crystals have many more degrees of freedom and, therefore, more available trajectories in phase space. Let us, however, follow the direction indicated by the computer solutions and seek an answer to the recurrence problem. Is it a "mathematical-physical" effect or is it purely a mathematical nonessential resulting from the manner in which we discretized time or applied our boundary conditions? Perhaps it is a combination of both?

There are at least two ways of proceeding toward an analytic understanding. The first is the physicist's conventional approach—or perhaps I should say the conventional physicist's approach—namely, that of decomposing the displacement y_i into $N-1$ interacting normal modes and then applying perturbation theory. To our minds† this has not been a fruitful approach. Instead, we have adopted a "continuum" approach, because we observed in Fig. 2 that the high modes of the system did not participate in the dynamics. Thus we are led to Hypothesis 2.

H2: *The recurrence phenomena and the detailed modal history can be explained by a continuum model.*

We assume that the displacements y_{i+1} and y_{i-1} can be written in terms of the displacement y_i and its derivatives by means of the Taylor series

$$y_{i\pm 1} = [y \pm hy_x + (h^2/2)y_{xx} \pm (h^3/6)y_{xxx} + (h^4/24)y_{xxxx} + O(h^5)]_i, \quad (3.8)$$

and substituting in (3.5) and omitting the subscript i we obtain

$$c^{-2}y_{tt} = (1 + \varepsilon y_x)y_{xx} + (h^2/12)y_{xxxx} + (\varepsilon h^2/6)y_{xxx}y_{xx} + O(h^4) + O(\varepsilon h^4), \quad (3.9)$$

where $\varepsilon = 2\alpha h$ and $c^2 = h^2\omega_0^2$ is the propagation speed of waves in the lowest-continuum linear limit, viz., $\varepsilon = h = 0$. In our case we normalize the length of the lattice so that $L = Nh = 1$. If $\alpha = O(1)$, then $\varepsilon = O(h) \gg h^2$, and the lowest continuum limit is the nonlinear hyperbolic partial differential equation

$$y_{tt} = c^2(1 + \varepsilon y_x)y_{xx}. \quad (3.10)$$

The initial and boundary conditions corresponding to the FPU problem are

$$y(x, 0) = a \sin \pi x, \qquad y_t(x, 0) = 0, \qquad y(0, t) = y(1, t) = 0. \quad (3.11)$$

The Hamiltonian for (3.10) is

$$H = \frac{1}{2}\int_0^1 dx\left[y_t^{\,2} + c^2 y_x^{\,2} + \frac{\varepsilon c^2}{3}\, y_x^{\,3}\right].$$

† See Zabusky [3]. A brief discussion of Ford's analysis is given in Section 4a. A brief discussion of Jackson's work is given in Kruskal [4].

The modulus of the ratio of the nonlinear energy density to the linear energy density is maximized where $| y_x |$ is a maximum, and for the initial conditions given in (3.11) is

$$\max | (\text{nonlinear energy density})/(\text{linear energy density}) | = \varepsilon a \pi/3. \quad (312)$$

The solutions of (3.10) "break down" or become multivalued after a time t_B

$$t_B/t_L = 2/\varepsilon a \pi^2 \quad (3.13)$$

which is inversely proportional to the ratio given in (3.12). Here $t_L = 2/c$ is the time required for one oscillation of the corresponding linear string ($\varepsilon = 0$). This result was established when the author found an exact solution to (3.10) with conditions (3.11) [10]. Thus, the lowest continuum nonlinear limit describes the evolution of the waveform of the discrete string at most up until this time. To obtain Fig. 2 FPU used $\alpha = 1/4$, $N = 1/h = 32$, and $a = 1$. Therefore, $\varepsilon = 1/64$ and breakdown is 12.95 linear periods. One can show [11] that t_B is slightly less than the time at which the second-mode energy has its first maximum. FPU did not realize that this continuum representation would not describe the discrete representation for all times. In fact, in their original memo they presented graphs of $y_i(t)$ vs. i† and not $(y_{i+1} - y_i)$ vs. i—which would have been much more suggestive of the breakdown phenomena.

At this point it is worth emphasizing that the benefits of the synergetic method are highly dependent on the type of graph drawn. As in a laboratory experiment, one must be on guard for the small effect. Very often one views the result of a computation according to some preconceived ideas—and for nonlinear problems this can be dangerous. In fact, we have found that for this and similar problems, the description according to interacting normal modes has usually been unrewarding—as far as making decisive progress.

At this stage, Kruskal and I applied some of the ideas contained in his work on an asymptotic theory for Hamiltonian systems having almost periodic solutions [12], and we developed a uniform asymptotic method for treating equations like (3.10) [11]—a mathematical byproduct along

† See Fermi *et al.* [8]. Figure 8 gives $y_i(t)$ vs. i.

the synergetic road. Essentially, we represented (3.10) by two coupled partial differential equations for two Riemann invariant quantities.

In this paper we showed that the lowest FPU modal energies agreed up until breakdown with those obtained from a Fourier decomposition of the waveform $y(x, t)$. The analysis proved, trivially, that we would also get breakdown if we took one Riemann invariant finite and the other zero and replaced the fixed boundary conditions by *periodic* or cyclic boundary conditions. The analysis also showed that, although the phenomena are nonlinear, signals propagating (or functions evolving) along the different characteristics did not interact with each other to lowest order. This suggested that the recurrence phenomena would be preserved for these analytically easier-to-treat initial-boundary conditions. A computer simulation for these conditions gave recurrence and we were on our way to a better analytic understanding.

The formation of a multivalued y_x (and therefore singular higher derivatives) after t_B, suggested at once that some of the higher-order terms omitted from (3.9) would play a role in determining the evolution of the system beyond breakdown.

In a review (Zabusky [3], Section 6) given in 1962, I suggested that the fourth derivative term alone could account for the recurrence time. This suggestion was based on a time-scaling argument from which a recurrence time was estimated that agreed qualitatively with some of the computations then available. Thus, we focused our attention on the partial differential equation

$$0 = c^{-2}y_{tt} - (1 + \varepsilon y_x)y_{xx} - (h^2/12)y_{xxxx}. \tag{3.14}$$

We emphasize that this additional term is the first term in an expansion that takes into consideration the discreteness or graininess of the medium. Said another way, if we linearize the equation we get a partial differential equation (like that for a beam under tension) for which the speed of propagation of a packet of waves is different than that of a single wave—that is, the medium is *dispersive*.

To study (3.14), we introduce quantities which were the Riemann invariants of (3.10), namely

$$\begin{aligned} r_{\pm} &= \tfrac{1}{2}\{\pm c^{-1}y_t + (2/3\varepsilon)[(1 + \varepsilon y_x)^{3/2} - 1]\} \\ &= \tfrac{1}{2}\{\pm c^{-1}y_t + y_x + (\varepsilon/4)y_x{}^2 + \cdots\} \end{aligned} \tag{3.15}$$

and obtain the coupled equations

$$c^{-1}(r_\pm)_t \mp [1 + \tfrac{1}{2}\varepsilon(r_+ + r_-)]\,(r_\pm)_x \mp (h^2/24)\,(r_+ + r_-)_{xxx}$$
$$= O(\varepsilon^2) + O(\varepsilon h^2). \qquad (3.16)$$

If we take $r_+ \big|_0 = 0$ and assume that $r_- \big|_0$ is a function having zero mean, it can be shown that r_+ is always $O(\varepsilon^2)$. Thus, we can omit it from (3.16) and we are left with a third-order equation for $r_- \equiv u$. If we transform into a frame of reference moving uniformly with velocity c, that is, $x \to (x - ct)$ and $t \to t$, we obtain the Korteweg–de Vries equation [13]

$$u_\tau + uu_x + \delta^2 u_{xxx} = 0, \qquad (3.17)$$

where

$$\tau = \varepsilon ct/2 \qquad \text{and} \qquad \delta^2 = h^2/12\varepsilon. \qquad (3.18)$$

We call δ^{-2} the "dispersion number" in analogy with the Reynolds number of dissipative fluid dynamics. That is, if u is normalized to unity, then large δ^{-2} corresponds to the nonlinear range and small δ^{-2} to dispersive range.

This brings us to Hypothesis 3.

H3: *The Korteweg–de Vries equation*

$$u_\tau + uu_x + \delta^2 u_{xxx} = 0$$

can describe propagation of finite amplitude waves on a lossless nonlinear lattice excited by a progressive wave initial condition.

Again, note that we have not restricted the finiteness of the initial excitation. We will soon be in a position to do so.

This equation was first obtained by Korteweg and de Vries as the quintessential equation for describing wave propagation on shallow water. Gardner and Morikawa [14] showed that it was also the limiting equation for describing wave propagation perpendicular to a uniform magnetic field in a cold lossless (collisionless) plasma. They have extended their theory to consider wave propagation in a *warm* plasma [15]. Morton has extended the theory to include wave propagation *oblique* to a uniform magnetic field [16]. The Korteweg–de Vries equation is applicable to both cases. In the former case δ^2 increases weakly with the temperature; in the latter case δ^2 is a function of the angle of propagation with respect to the magnetic field, and changes sign at certain angles.

IV. Solitons, the Korteweg–de Vries Equation, and Some Computational Results

The study of nonlinear higher-order differential equations in two and three dimensions will provide the most interesting and rewarding problems for computational mathematics of the future. However, as you will see, we still have much to learn about one-dimensional nonlinear wave phenomena.

For example, the Burgers equation

$$u_\tau + uu_x - \mu u_{xx} = 0, \tag{4.1}$$

has received considerable attention because it is a model for one-dimensional, pressureless and viscous or lossy fluid. By a clever transformation technique Hopf [17] and Cole [18], almost simultaneously, found an exact solution to (4.1). To date, no one has succeeded in doing the same for the Korteweg–de Vries equation. This state of affairs will probably exist for a very long time. It is undoubtedly related to the fact that the solutions of (3.17) with the nonlinear term omitted are asymptotically oscillatory in one direction of space and time, whereas the solutions of (4.1) with nonlinear terms omitted are asymptotically monotonic in space and time.

Well, I imply that we searched and failed to find an exact solution. We didn't have much luck either with straightforward asymptotic methods or with applications of "conventional" shock wave theory—only because we didn't know how to penetrate to the core. We then turned to the computer and simulated (3.17) using a numerical scheme which conserves momentum and "almost" conserves energy. We found breakdown, recurrence, and complete agreement with the modal energies of the full lattice equations!

This amounts, almost, to a *proof* of the equivalence of two mathematical representations. I will let future mathematicians worry about the degree of rigor embodied in this computational proof. To verify equivalence, one will undoubtedly have to introduce quantities which are very sensitive to the calculation. These figures of merit should be readily related to quantities for which it is easy to make analytical estimates. At present we use momentum and energy conservation. I should also emphasize that a numerical solution of (3.17) requires far less time than does a numerical solution of (3.16) (with $r_+ = 0$), or equivalently, the original lattice problem (3.5). This follows because in (3.17) the time scale τ includes ε, the nonlinear measure, whereas in the other problems it does not, and hence one must compute a "long" time before nonlinear effects manifest themselves.

This result was reported by my colleague [4] a little over two years ago, and as one can see from his paper we still did not understand the "why" of recurrence. Only when we looked at many graphs of the functions $u(x, t)$, $u_x(x, t)$, and $u_{xxx}(x, t)$ did we realize what was essential—namely, the *solitary-wave pulse* solutions of the time-independent problem. For reasons to be introduced shortly we have coined the word *soliton* to describe the solitary-wave pulse.

It is well known that nonlinear dispersive media viewed in an appropriate reference frame can support steady progressive waveforms as the solitary-wave pulse and the cnoidal wave [19]. The former, for which $u \to U_\infty$ as $|x| \to \infty$, will concern us now. We will consider a generalized from of the Korteweg–de Vries equation, namely

$$u_\tau \pm u^p u_x + \delta^2 u_{xxx} = 0, \tag{4.2}$$

where the $\pm$ sign corresponds to $\alpha = \pm 1$ in (3.4), and the exponent p in (4.2) and (3.4) also correspond. In our discussion we will mainly consider $\alpha = 1$ and $p = 1$.

If we replace u by $U(x - C\tau)$ where C is a constant, Eq. (4.2) becomes

$$-CU_x \pm U^p U_x + \delta^2 U_{xxx} = 0. \tag{4.3}$$

Equation (4.3) integrates directly

$$-CU \pm (p + 1)^{-1} U^{p+1} + \delta^2 U_{xx} = L. \tag{4.4}$$

Multiplying by U_x, we integrate again, rearrange and obtain the first-order nonlinear differential equation

$$\delta^2 U_x{}^2 - P_p(U) = 0, \tag{4.5}$$

where

$$P_p(U) = \mp 2[(p + 1)(p + 2)]^{-1} U^{p+2} + CU^2 + 2(LU + M), \tag{4.6}$$

and L and M are constants of integration related to C and U_∞. For convenience we consider a solitary wave on a zero brackgound, or $U(^\infty) = U_x(_\infty) = 0$ and therefore $L = M = 0$. Thus, for $p = 1$ and $\alpha = +1$

$$\delta^2 U_x{}^2 = -\tfrac{1}{3} U^2(U - 3C), \tag{4.7}$$

and one evaluates C by inspection as

$$C = A_C/3. \tag{4.8}$$

A_C is the amplitude of this "compressive" solitary wave and is always >0. This is shown graphically in Fig. 3. If one integrates (4.7) one obtains

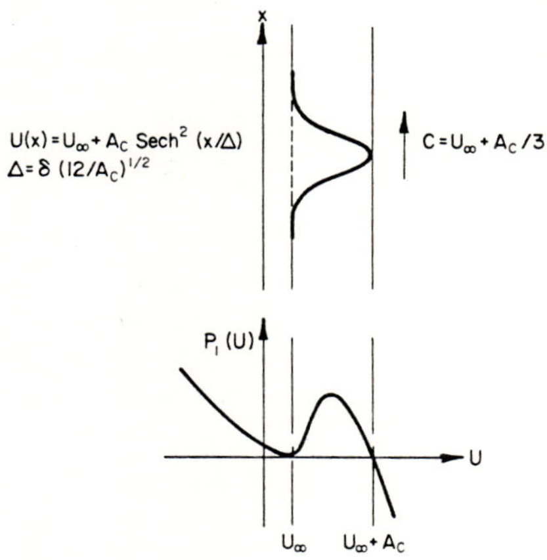

FIG. 3. The compressive soliton solution for the Korteweg–de Vries equation.

the result (with U_∞ restored here)

$$U = U_\infty + A_C \operatorname{sech}^2(x/\Delta), \tag{4.9}$$

where

$$C = U_\infty + A_C/3 \qquad \text{and} \qquad \Delta^2 = 12\delta^2/A_C. \tag{4.10}$$

Note that the pulse width Δ is inversely proportional to the square root of the amplitude and the amplitude is independent of the background, U_∞.

For third-order differential equations like (4.2) the character and number of solitary waves depend on whether $(p + 2)$ is odd or even. If $p + 2$ is *odd* we can have only *one* type of solitary wave (compressive as in Eq. (4.9) *or* rarefactive), while if $p + 2$ is *even* we can have both types of solitary waves. For example, if $U_\infty = U_x(\infty) = 0$ then Eq. (4.5) becomes

$$\delta^2 U_x^{\,2} = -U^2[2U^p/(p+1)(p+2) - C] \tag{4.11}$$

where we have chosen the minus sign ($\alpha = +1$). One possible solitary wave has the speed

$$C = 2A^p/(p+1)(p+2), \tag{4.12}$$

where A is the amplitude.

One normalizes (4.11) by introducing

$$v = U/[C(p+1)(p+2)/2]^{1/p} \qquad \text{and} \qquad \zeta = \sqrt{C}x/\delta. \tag{4.13}$$

Integrating and transforming back to dimensional quantities yields

$$U = A[\operatorname{sech}(x/\Delta)]^{2/p}, \tag{4.14}$$

where

$$\Delta = 2\delta/pC^{1/2} = \delta\left[\frac{2(p+1)(p+2)}{A^p p^2}\right]^{1/2}. \tag{4.15}$$

That is, the larger p is, the more slowly does the pulse amplitude decay toward the background. These pulses will be difficult to deal with numerically.

For $p = 2$, (4.14) and (4.15) give two possible solitary waves, the compressive with $A_C > 0$ and the "rarefactive" with $-A_R < 0$, as shown in Fig. 4a. In general, $A_C \neq A_R$ if $U_\infty \neq 0$.

For $p = 2$ there are two special cases where the *amplitude* of the solitary pulse is related to the background value U_∞. For $\alpha = +1$, we can have a "long-range" solitary pulse if $U_\infty \neq 0$, that is, where three of the four zeroes of P_2 coalesce as shown in Fig. 4b or

$$P_2 = -\tfrac{1}{6}(U - U_\infty)^3(U - A_C). \tag{4.16}$$

If one expands this and compares with P_2 as defined in (4.6) (taking $\alpha = +1$), one finds

$$A_C = -3U_\infty \qquad \text{and} \qquad C = U_\infty^{\,2}. \tag{4.17}$$

If we integrate the equation for U_x^2 we obtain

$$U = U_\infty - 4U_\infty[1 + \tfrac{2}{3}(xU_\infty/\delta)^2]^{-1}. \tag{4.18}$$

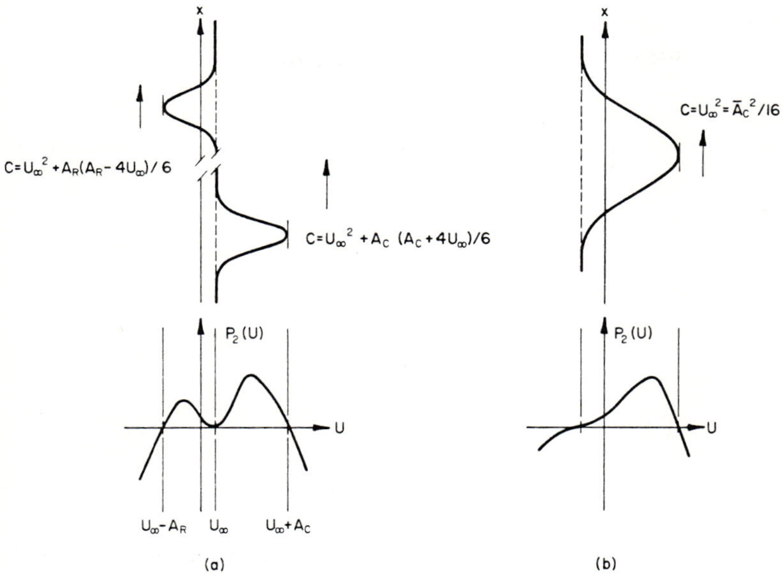

FIG. 4. Soliton solutions for the general Korteweg–de Vries equation with $p = 2$. (a) Compressive and rarefactive solitons. (b) Long-range compressive soliton.

Unlike the previous pulses this falls off algebraically as $|x| \to \infty$. We could also generalize by replacing u^p in (4.2) by the arbitrary function $A(u)$ and adding a fifth derivative

$$u_\tau + A(u)u_x + \delta^2 u_{xxx} + \beta\, \delta^4 u_{xxxxx} = 0. \tag{4.19}$$

In these generalizations, if we are to arrive at a better understanding of nonlinear dispersive wave propagation, we must contend with the variety of possible stationary states and the question of whether they are *stable* when they are subjected to infinitesimal and finite disturbances.

Now that we have a grounding in solitary-wave pulses or steady solutions localized about some point on the infinite line, I will show you some of the time-dependent computer solutions. To exhibit the synergetic approach as a *modus operandi*, I should discuss the solution starting from a smooth initial state, as for example a sinusoidal function of large wavelength. However, pedagogically it is easier to describe what happens when we solve (3.17) using a single soliton as an initial condition and then build up to the example which allowed us to validate Hypothesis 3.

A. Single Soliton

If we solve (3.17) with the initial condition (4.9) it is easy to verify whether or not the soliton state remains steady in a uniformly moving frame of reference. This serves as a critical test of the computational scheme.

We used a soliton of height $A = 100.0$ and placed it on a zero background ($U_\infty = 0$). We chose $\delta = 0.02$ which gave a width-at-half-maximum of 12 lattice sites. We used *periodic* boundary conditions† over 200 lattice sites. During the run of duration 0.0024 time units the soliton propagated through a normalized distance of 0.07965 (approximately 80 lattice sites or 6.64 widthts-at-half-maximum corresponding to a mean velocity of 33.18. This agrees well with the theoretical value of 100/3. The amplitude did not deviate by more than 0.19 from its original value of 100.0 and the energy decreased by 0.0065% from its original value. Had we made the soliton width larger by choosing a larger δ (thereby decreasing discretization errors) and the periodic interval larger (thereby introducing smaller effects from the overlap of soliton "tails"), we would have obtained more invariant results for amplitude and energy. The results satisfied us. In fact this calculation tells us that wide soliton states are "asymptotically" stationary solutions for systems of nonlinear coupled *ordinary* differential equations—the lattice.

B. Two Interacting Solitons

Because solitons are nonlinear dispersive wave entities their speed of propagation is a function (in this case a linear function) of their amplitude. Hence, if we initially place two solitons on the infinite line with the larger to the left of and separated from the smaller, then at some future time they will interact or "focus" at the same point in space. To elucidate the nature of the interaction of two solitons we solved (3.17) again with $\delta = 0.02$ for two cases. In the first case, $A_1 = 180.0$, $A_2 = 80.0$ ($A_1/A_2 = 2.25$), and the pulses were separated by 62 lattice sites (6.9 widths-at-half-maximum of No. 1) and placed on a background $U_\infty = -26\frac{2}{3}$ so that No. 2 remained fixed. In Fig. 5a the trajectory and amplitude histories of the two solitons are given and we see that the larger overtakes the

† To avoid difficulties associated with boundary conditions, all computations for the Korteweg–de Vries equation were made with *periodic* boundary conditions.

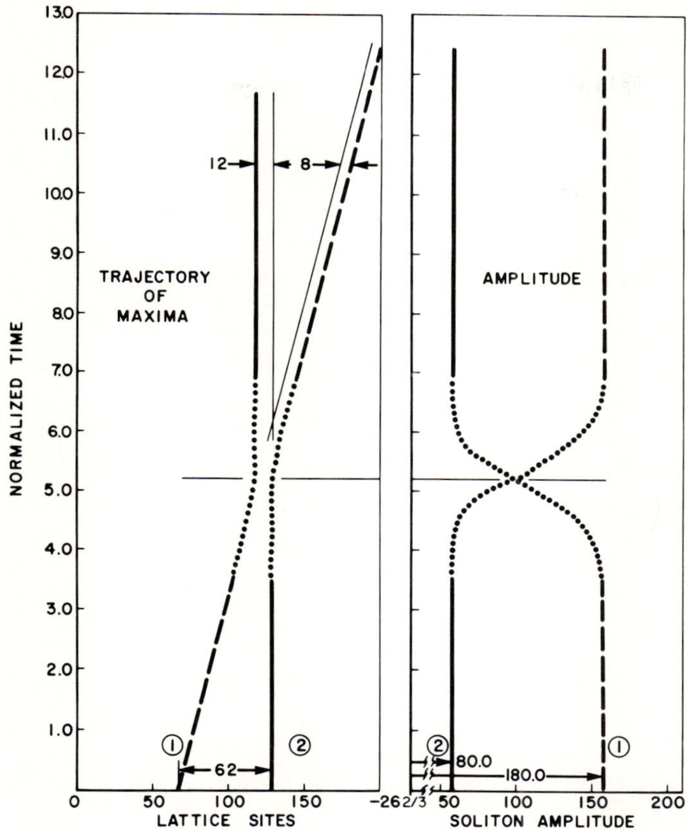

FIG. 5a.

FIG. 5. Interaction of two solitons for the Korteweg–de Vries equation, ($p = 1$), $\delta = 0.02$—Trajectory of maxima diagram and amplitude variation. (a) Two different solitons, $A_1/A_2 = 2.25$. (b) Two comparable solitons, $A_1/A_2 = 1.10$.

smaller, "accelerates" through it and "emerges" unaffected. After the interaction when solitons 1 and 2 are separated by 62 lattice sites their amplitudes differ from their original values by 0.065% and 0.487%, respectively. During the run the energy decreased monotonically by 0.751×10^{-4}% from its original value. In the second case we examined the behavior of two nearly equal solitons, $A_1 = 110$ and $A_2 = 100$. The pulses were separated by 55 lattice sites (6.1 widths-at-half-maximum of No. 1) and placed on a background $U_\infty = -33\frac{1}{3}$ so that No. 2 remains fixed. In Fig. 5b we see that they approach until their "tails" overlap slightly and then "exchange places"—that is, they do not have to overlap appreciably for the larger to exchange amplitudes and hence positions with the smaller. In comparing Fig. 5a with Fig. 5b note that: the time

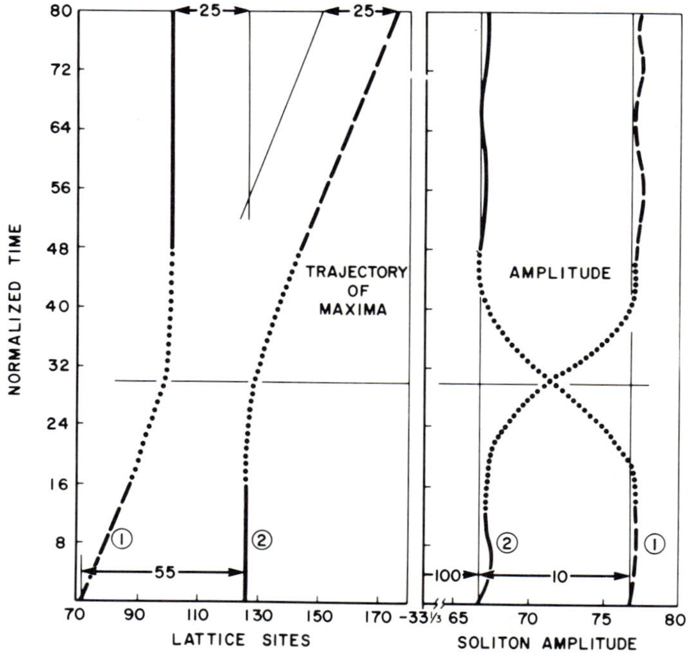

FIG. 5b.

scales differ by a factor of 8 : 1; the lattice site scales differ by a factor of 2 : 1; and the amplitude scales differ by a factor of 10 : 1. The amplified scale accounts for the apparently large initial oscillation that one sees in Fig. 5b. This oscillation is probably due to the fact that the almost equal-amplitude solitons were initially too close together.

It should be emphasized that these solitons emerged from interaction with one another, still preserving their initial identity to high numerical accuracy.

C. Sinusoidal Long–Wavelength Initial Condition

Now we can proceed to a more general case, one which was chronologically first, and ask what happens when we start with a sinusoidal curve

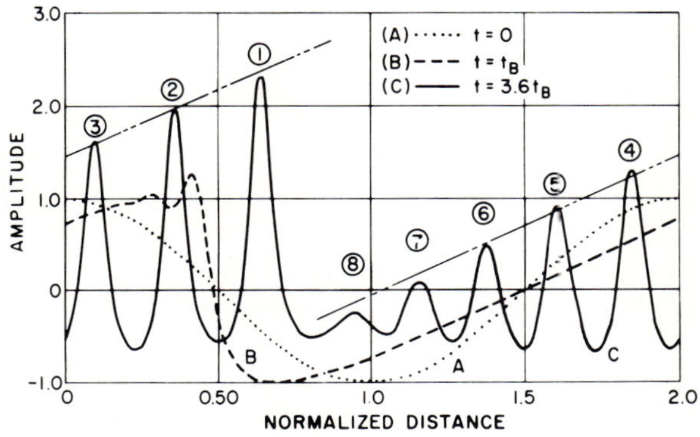

FIG. 6. Solutions of the Korteweg–de Vries equation, ($p = 1$), $\delta = 0.022$, at three different times.

whose amplitude-wavelength ratio of 1/2 is much smaller than $(\sqrt{12}\,\delta)^{-1}$. Figure 6 shows the solution of (3.17) with $\delta = 0.0222$, at three different times. Curve A is the initial condition—the cosine function

$$u\,\big|_0 = \cos \pi x, \tag{4.20}$$

curve B is the solution at t_B or the time at which solutions of (3.17) with $\delta^2 = 0$ would become multivalued. Note the beginnings of an oscillatory structure to the left of the front. This is due to the third derivative term. Curve C, at $t = 3.6\ t_B$, shows a fully developed oscillatory structure or, viewed another way, a collection of pulses.

Each of these pulses is a *soliton*, because when one is "reasonably" well separated from neighboring pulses, it moves with a speed and has a width that is given by (4.10) [20]. Furthermore, although many solitons can interact simultaneously, they emerge from the interaction still preserving their initial identity. This is graphically shown in Fig. 7, where the trajectory of the maxima of nine solitons is plotted as a function of time. The time axis begins at $0.1\ t_R = 3.04\ t_B$. Table I gives the amplitude and position of eight solitons at $t = 3.6\ t_B$ and also the velocity of each soliton as determined from Fig. 7.

You can see that each soliton moves along a straight line trajectory except when it interacts momentarily with other solitons. At this time they "accelerate" through one another and then emerge from the inter-

Table I

SOLITON PROPERTIES — KORTEWEG–DE VRIES EQUATION WITH $\delta = 0.0222$ AT $t = 3.6\, t_B = 0.118\, t_R$

Soliton No.	1	2	3	4
Amplitude (A_c)	2.85	2.62	2.43	2.16
Position	0.64	0.36	0.090	1.84
Velocity (C)	$+3.57 t_R^{-1}$	$1.73 t_R^{-1}$	0	$-1.56 t_R^{-1}$
Soliton No.	**5**	**6**	**7**	**8**
Amplitude (A_c)	1.83	1.45	1.00	0.495
Position	1.61	1.38	1.16	0.94
Velocity (C)	$-2.66 t_R^{-1}$	$-4.30 t_R^{-1}$	$-5.68 t_R^{-1}$	$-6.97 t_R^{-1}$

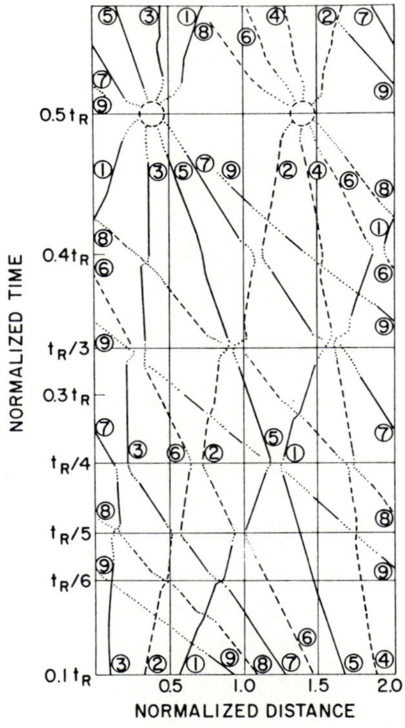

FIG. 7. Soliton trajectories on a space-time diagram beginning at $t = 0.1\, t_R = 3.04\, t_B$.

action unaffected. For example, when solitons 1, 4, 7 and 2, 5, 8 interact or focus, one is at $\frac{1}{3}t_R$ or $2t_R/3$ (not shown). At $\frac{1}{2}t_R$ all the odd solitons focus at $x = 0.385$ and the even at $x = 1.385$. At t_R (not shown) all nine solitons focus together and reconstruct a state almost identical to the original sinusoidal state. The negative-slope zero-crossing of the slightly distorted cosine function translated to $x = 1.33$ from its original value of $x = \frac{1}{2}$ (Eq. 4.20).

To dramatize the synergetic approach I will now show you a computer-made movie which will assuredly clarify and reinforce what I have just related.†

To obtain a proper mathematical description of recurrence it may help to think of the recurrence time as composed of time intervals describing three different phenomena. In the first interval the smooth solution breaks; in the second interval solitons form; and in the third interval the solitons interact and exchange positions. That is, after some time the largest soliton is to the left of the smallest and the intervening amplitude array is almost linear. Thus

$$t_R = 2t_B + 2t_F + t_{Ex}, \tag{4.21}$$

where the factor of 2 before t_B and t_F arises, because after exchange we go through the reverse process and "unform" the solitons and "unbreak" the steep waveform.

An approximation to the exchange time is obtained by making a calculation in the same spirit in which one determines a Poincaré recurrence time for noninteracting particles on a ring. Our problem is easier than the

† The motion picture, "Formation, Propagation, and Interaction of Solitons (Numerical Solutions of Differential Equations Describing Wave Motion in Nonlinear Dispersive Media)," was conceived by Zabusky, Kruskal and Deem and produced at the Bell Telephone Laboratories, Whippany, New Jersey during 1965. Part I, Numerical Solutions of the Korteweg–de Vries Equation $u_t + uu_x + \delta^2\, u_{xxx} = 0$, with Periodic Boundary Conditions, was produced in March, 1965. Part II, Interaction of a Compressive and Rarefactive Soliton — Solutions of the Eq. $u_t + u^2 u_x + \delta^2\, u_{xxx} = 0$, with Periodic Boundary Conditions, was produced in December, 1965. Part III, Numerical Solutions for an Anharmonic Lattice Initially Excited by an Intense Localized Pulse, was produced during November–December, 1965. This 16 mm, silent, black-and-white film runs 35 minutes, and is available on loan from the Bell Telephone Laboratories, Inc., Film Library, Murray Hill, New Jersey, 07971.

general case, for the solitons when fully formed are each separated by approximately d_r, and furthermore, their maximum values fall almost on a straight line so that there is a constant relative velocity C_r between neighboring solitons. For d_r we use the average value of the distance between solitons 1–7 as given in Table I or $d_r = 0.247$. For C_r we use the average value of the relative velocities between solitons 1–7 as given in Table I or $C_r = 1.54\ t_R^{-1}$. If we transform into a frame of reference in which the "center" soliton is fixed (No. 4 in Fig. 7), then those solitons to the right continue moving in this direction and the solitons to the left continue moving to the left. Thus, the exchange time is approximately the time taken for soliton 3 and soliton 5 to exchange places or

$$t_{Ex} \approx (2 - 2d_r)/C_r = 1.506\ t_R/1.54 = 0.978\ t_R. \tag{4.22}$$

For this run the exchange time is actually $0.8\ t_R$. During this time solitons 2 and 6 exchange places but move one more complete revolution than do 3 and 5; similarly, solitons 1 and 7 exchange places, moving two more complete revolutions than do solitons 3 and 5.

The analytical description of the solution of (3.17) in the formation time interval ($t_B < t < 3.6\ t_B$) is in a primitive state at present. Kruskal and I have introduced a macroscopic-microscopic averaging procedure to describe this time interval but have not succeeded in validating it.

Note that it was important to use periodic boundary conditions. This allowed the solitons to "circulate" until they were arrayed for focusing.

We now realize that had we chosen an initial condition for the *lattice* problem which was neither a pure standing wave nor a pure progressive wave, we would have obtained a different focussing time for the solitons propagating along the two characteristic directons. Therefore, the recurrence would not have been as good as the one we observed.

We have made several other runs with different initial conditions to elucidate soliton properties including: a "counterflowing" state or two constant states of equal magnitude and opposite sign joined by a "wide" smooth transition; and a sinusoidal-plus-random state, where the energy content of the random part was 9% of the total energy. These results have interesting properties and will be described in detail in a forthcoming work on our numerical methods and results.

V. Synergetics—Future Directions

In our presentation of the hypotheses and in our account of the numerical calculations we have alluded to three significant directions along the synergetic road. In two of these, namely the properties of dispersive shock waves and the properties of interacting solitons, enough information is available to provide support for a careful analytical investigation. In the following paragraphs the progress made in these areas will be outlined. In the third case, namely the effect of very large amplitude excitations on the subsequent motion of a lattice, we obviously do not have sufficient insight. Also, some recent heuristic work by Izrailev and Chirikov and by Chirikov [21] determined the conditions under which nonlinear dynamical systems will change from "regular" to "stochastic" behavior. Speaking qualitatively, they showed that a nonlinear system whose associated linear spectrum is *discrete* can exhibit "stochastic" behavior if the energy in adjacent spatial modes is sufficiently large to cause the frequency or phase variations to "overlap." They applied this consideration to the nonlinear lattice and showed that the energy threshold from "regular" to "stochastic" behavior was lower at the highest wavenumbers. This led Deem and myself to return to the lattice model (3.5) with periodic boundary conditions and we used an initial condition where the energy resides in the high-frequency or "optical" modes of the system. This is accomplished by constructing a localized disturbance where alternate identical masses were displaced along *two smooth* curves.

For small and moderate amplitudes of excitation, that is, where

$$\bar{\eta} = \max \{\alpha \mid y_{i+1} - y_i \mid \}_{t=0} < 0.10,$$

we found that two smooth pulses radiate from the localized high-frequency disturbance, and propagate at the acoustic speed, one in each "characteristic" direction. We have found that we can analytically describe the dynamics following this initial condition by using a continuum limiting procedure involving *two smooth curves* [22].

For large amplitudes of excitation, where

$$0.1 < \bar{\eta} < 0.5,$$

one finds that two smooth pulses are generated and propagate away as previously. Somewhat later two three-curve states are generated, and

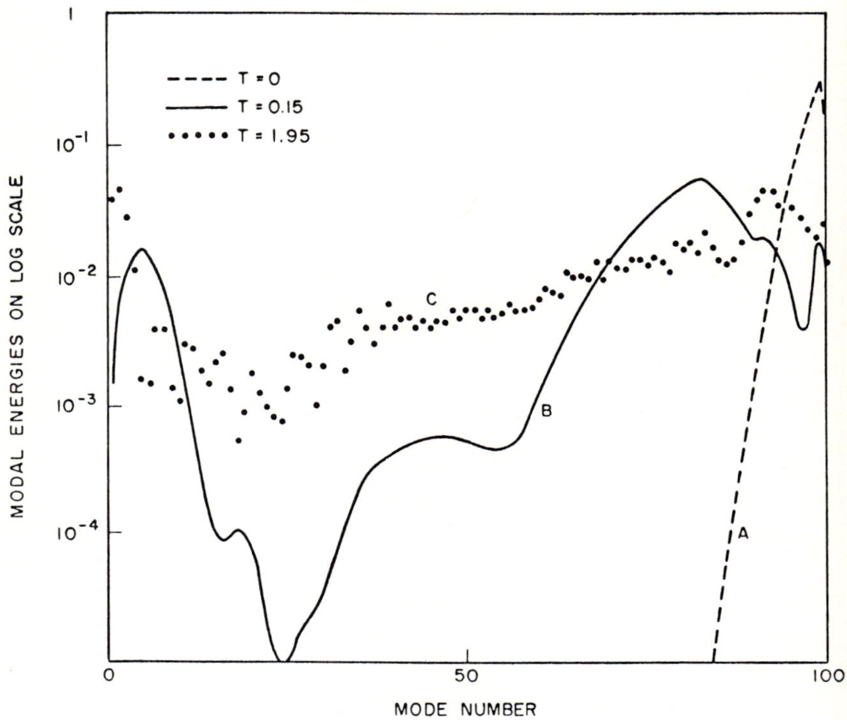

FIG. 8. The values of the modal energies for a lattice with $N = 100$ and a cubic nonlinearity, excited initially by a localized disturbance where alternate masses lie along two different smooth curves. Curve A, $(t/t_L) = 0$; Curve B, $(t/t_L) = 0.15$ and Curve C, $(t/t_L) = 1.95$. Modal energies normalized by total initial energy = 284.27. $\bar{\eta} = \max \{\alpha \mid y_{i+1} - y_i \mid\}_{t=0} = 0.326$.

each propagates along a characteristic direction at a velocity approximately equal to one-half the acoustic speed. That is, a pulse is formed where the motion of every third particle is correlated. For example, the modal energy spectrum for an initial condition where $\bar{\eta} = 0.326$ is given in Fig. 8. One sees: the energy initially in the high frequency modes $(t/t_L = 0)$; the energy populating the low frequency modes, indicating that a smooth state is generated $(t/t_L = 0.15)$; and the energy distributed almost equally among all modes $(t/t_L = 1.95)$.

A. Analytical Directions

Soon after we obtained the original numerical solutions of (3.17), we realized that it would be a fruitless exercise to attempt to apply conventional shock wave theory. We were dealing with a reversible system which supports a highly oscillatory waveform to the left of a uniformly translating front, and we soon concluded that new mathematical methods would have to be introduced.

Over two years ago Kruskal and I [23] (and simultaneously G. B. Whitham [24] introduced a mathematical procedure to describe the formation of the oscillations in the time interval after the breaking and before solitons are fully formed,

$$t_B < t < t_F .$$

In this uniform asymptotic procedure, which has features of the WKB method, we separate out the highly oscillatory behavior and treat instead quantities which describe it, e.g., mean value, wave-length, and amplitude. It is more convenient to apply this procedure to equations in conservation form equivalent to the Korteweg-de Vries equation.

1. Equations in Conservation Form

The Korteweg–de Vries equation is usually derived by an asymptotic procedure from a set of partial differential equations describing a dispersive medium in which momentum and energy are conserved. Therefore it can be written in two *equivalent* forms

$$0 = (u)_t + [\tfrac{1}{2}u^2 + \delta^2 u_{xx}]_x, \tag{5.1}$$

$$0 = (\tfrac{1}{2}u^2)_t + [\tfrac{1}{3}u^3 + \delta^2(uu_{xx} - \tfrac{1}{2}u_x^{\,2})]_x, \tag{5.2}$$

where the latter is obtained by multiplying the former by u and regrouping. Since Whitham needed a third conservation law for this method, he sought and found the relation

$$0 = (\tfrac{1}{3}u^3 - \delta^2 u_x^{\,2})_t + [\tfrac{1}{4}u^4 + \delta^2(u^2 u_{xx} - 2uu_x^{\,2}) + \delta^4(-2u_x u_{xxx} + u_{xx}^2)]_x \tag{5.3}$$

We will soon see that the third conserved quantity will tell us something of the structure of the solution.

As a matter of curiosity, Kruskal and I looked for and found that there exist a fourth and a fifth conservation equation, namely

$$0 = \left(\frac{u^4}{4} - 3\delta^2 uu_x{}^2 + \frac{9}{5}\delta^4 u_{xx}^2\right)_t + \left[\frac{u^5}{5} - \delta^2\left(\frac{9}{2}u^2u_x{}^2 - u^3u_{xx}\right)\right.$$
$$\left. + \delta^4\left(\frac{24}{5}uu_{xx}^2 - 6uu_xu_{xxx} + 3u_x{}^2u_{xx}\right) + \frac{9}{5}\delta^6(2u_{xx}u_{xxxx} - u_{xxx}^2)\right]_x, \tag{5.4}$$

$$0 = \left(\frac{1}{5}u^5 - 6\,\delta^2 u^2u_x{}^2 + \frac{36}{5}\delta^4 uu_{xx}^2 - \frac{108}{35}\delta^6 u_{xxx}^2\right)_t$$
$$+ \left[\frac{u^6}{6} - \delta^2(-u^4u_{xx} + 8u^3u_x{}^2) + \delta^4\left(12uu_x{}^2u_{xx} - 12u^2u_xu_{xxx}\right.\right.$$
$$\left. - 3u_x{}^4 + \frac{66}{5}u^2u_{xx}^2\right) - \frac{36}{5}\delta^6\left(\frac{10}{7}uu_{xxx}^2 + 2u_xu_{xx}u_{xxx}\right.$$
$$\left.\left. - 2uu_{xx}u_{xxxx} - \frac{1}{7}u_{xx}^3\right) + \frac{108}{35}\delta^8(u_{xxxx}^2 - 2u_{xxx}u_{xxxxx})\right]_x. \tag{5.5}$$

Recently, Miura of Princeton continued the calculations and found a 6th, 7th, 8th, 9th and 10th conservation equation.

If one associates the number of solitons with the number of conservation laws, it is reasonable to conjecture that one can find an infinite number of partial differential equations in conservation form *equivalent* to the Korteweg–de Vries equation. We will use these conservation equations to study properties of this equation.

We have also found [25] a formal procedure for calculating the conserved quantities, *if they exist*. Furthermore, for a generalized form of the Korteweg–de Vries equation, (4.2) with $p = 2$, we have found six equivalent equations in conservation form. For $p > 2$ we have been able to find *only three* equations in conservation form, and the conserved densities are

$$(u_x), \quad (\tfrac{1}{2}u^2), \quad \text{and} \quad \left(\frac{1}{p+2}u^{p+2} - \frac{p+1}{2}u_x{}^2\right).$$

2. A Uniform Asymptotic Procedure for Oscillations in Nonlinear Dispersive Media

To formally separate the high-frequency oscillations from the smooth or "slowly varying" behavior, we replace

$$u(x, \tau') \to U(\xi, \tau, \theta(\xi, \tau)), \tag{5.6}$$

where ξ, and τ are finitely varying quantities and δ is considered small. U as a function of θ describes the highly oscillatory properties of the solution with a unity period, or

$$U(\xi, \tau, \theta + 1) = U(\xi, \tau, \theta), \tag{5.7}$$

and we choose a zero reference point by the definition

$$U_\theta(\xi, \tau, \theta)\,|_{\theta=0} \equiv 0 \qquad \text{where} \qquad U_{\theta\theta}(\xi, \tau, \theta)\,|_{\theta=0} > 0. \tag{5.8}$$

The slowly varying quantities are obtained by "averaging" U, that is, integrating with respect to θ over a period of unity.
If we substitute (5.6) into the Korteweg–de Vries equation

$$u_{\tau'} + uu_x + \delta^2 u_{xxx} = 0,$$

and use the identities

$$\partial_{\tau'} = \partial_\tau - \delta^{-1}\Omega\partial_\theta \qquad \partial_x = \partial_\xi + \delta^{-1}K\partial_\theta, \tag{5.9}$$

we obtain the partial differential equation

$$\begin{aligned} 0 = {} & \delta^{-1}[-\Omega U_\theta + KUU_\theta + K^3 U_{\theta\theta\theta}] \\ & + [U_\tau + UU_\xi + 3KK_\xi U_{\theta\theta} + 3K^2 U_{\theta\theta\xi}] \\ & + \delta\,[K_{\xi\xi}U_\theta + 3K_\xi U_{\theta\xi} + 3KU_{\theta\xi\xi}] + \delta^2\,[U_{\xi\xi\xi}], \end{aligned} \tag{5.10}$$

where

$$\Omega = -\delta\theta_\tau, \qquad K = \delta\theta_\xi \tag{5.11}$$

are finitely varying quantities analogous to a frequency and wave number respectively, and obviously the consistency condition follows

$$\Omega_\xi + K_\tau = 0. \tag{5.12}$$

To lowest order in the perturbation expansion of the variables (Ω, K, U), we obtain an *ordinary* differential equation

$$0 = -\Omega U_\theta + KUU_\theta + K^3 U_{\theta\theta\theta}. \tag{5.13}$$

We have suppressed the superscript$^{(0)}$ on U here and in the following.

Equation (5.13) is similar to the stationary Korteweg–de Vries equation (4.3) with $p = 1$ and the upper sign, except that it has coefficients dependent on the parameters ξ, τ. We can integrate (5.13) as we did the Korteweg–de Vries equation previously and obtain

$$K^2 U_{\theta\theta} = -\tfrac{1}{2}U^2 + lU + m. \tag{5.14}$$

Multiplying by U_θ and integrating again

$$K^2 {U_\theta}^2 = -\tfrac{1}{3}U^3 + lU^2 + 2(mU + n), \tag{5.15}$$

where

$$l = l(\xi, \tau) = \Omega/K, \tag{5.16}$$

is a "phase" velocity and $m(\xi, \tau)$ and $n(\xi, \tau)$ are parameters added after integration with respect to θ.

The partial differential equations relating the slowly varying quantities are obtained by first replacing the $t = \tau'$ and x derivatives in (5.1), (5.2), and (5.3) by the derivatives given in (5.9) and then integrating with respect to θ over a periodic interval (0,1). To lowest order one obtains

$$(P_1)_\tau + (\tfrac{1}{2} P_2)_\xi = 0, \tag{5.17}$$

$$(\tfrac{1}{2} P_2)_\tau + (\tfrac{1}{3} P_3 - \tfrac{3}{2} Q_0)_\xi = 0, \tag{5.18}$$

$$(\tfrac{1}{3}P_3 - Q_0)_\tau + (\tfrac{1}{4}P_4 - 4Q_1 + 3R_0)_\xi = 0, \tag{5.19}$$

where

$$P_n(\xi, \tau) = \int_0^1 U^n(\xi, \tau, \theta)\, d\theta,$$

$$Q_n(\xi, \tau) = K^2 \int_0^1 U^n {U_\theta}^2\, d\theta, \qquad \text{and} \qquad R_n(\xi, \tau) = K^4 \int_0^1 U^n U_{\theta\theta}^2\, d\theta. \tag{5.20}$$

Now to this order in δ we have ten unknowns P_1, P_2, P_3, P_4, Q_0, Q_1, R_0, l, m, and n related by a set of three first-order partial differential equations in conservation form and one implicit relation

$$P_1 = \frac{\int_0^1 U\, d\theta}{\int_0^1 d\theta} = \frac{\int U\, dU/U_\theta}{\int_0^1 dU/U_\theta} = \frac{\int_0^1 U\, dU/Z}{\int_0^1 dU/Z} \tag{5.21}$$

where $U_\theta = Z(U, l, m, n)$ is obtained from (5.15).

To close the set we need six more relations among the variables and these are obtained; one from (5.13), three from (5.14), and two from (5.15). The set of equations can be simplified if we introduce the physically meaningful quantities: the mean-square height of the oscillations

$$q = P_2 - P^2, \qquad (P_1 = P); \tag{5.22}$$

and the skewness

$$j = -[(P_3/6) - q(l + \tfrac{1}{2}P) - P^3]. \tag{5.23}$$

For example, the partial differential equations become

$$\begin{aligned} 0 &= P_\tau + \tfrac{1}{2}(P^2 + q)_\xi \\ 0 &= \tfrac{1}{2}(P^2 + q)_t + (\tfrac{1}{3}P^3 + Pq + \lambda q + \tfrac{5}{2}j)_\xi \\ 0 &= (\tfrac{1}{3}P^3 + Pq + j)_t + (\tfrac{1}{4}P^4 + \tfrac{1}{4}q^2 + \tfrac{3}{2}P^2q + 2\lambda Pq + 6Pj - 3\lambda^2 q - 6\lambda j)_\xi \end{aligned} \tag{5.24}$$

where $\lambda = l + P$, and the implicit relation can be written in the form $\lambda = f(q, j)$. Whitham stated [24] that this set of equations is hyperbolic for arbitrary amplitude. [Kruskal and Miura have since obtained an elegant proof of the same result.]

3. Shock Waves

It is well known that if one is dealing with systems of nonlinear hyperbolic partial differential equations in *conservation form*, one can study an important class of stationary discontinuous solutions—the shocks—by assuming that in a uniformly moving frame of reference solutions jump from one value to another over a very small region [26]. Thus, one replaces

$$\partial_t f + \partial_x g = 0$$

by

$$-V[f] + [g] = 0, \tag{5.25}$$

where $[f]$ represents the jump in the quantity f across the line of discontinuity and V is the speed of the discontinuity.

With this large number of conservation laws we find ourselves in a paradoxical situation because we have not been able to show that the jump relations obtained by applying the prescription given in (5.25) are formally dependent. In fact, having so many conservation laws and knowing some of the properties of the numerical solutions, in particular the possibility of having recurrence, we are led to introduce the concept of a "reversible" shock. Unlike a conventional shock we will have three characteristics entering and three leaving the shock "front," and we must then use four conservation laws to obtain the shock velocity and the three discontinuous quantities, e.g., P, K^{-1}, and q. A brief survey of this work and some of the remaining paradoxes has been given by Kruskal [5].

4. Interaction of Solitons

Kruskal and I have sought to "understand" and use these conservation quantities by pursuing another direction. Our results to date indicate that the present approach may be more significant for nonlinear dispersive systems than the classical approach referred to in the previous paragraphs.

If we assume, as we did for the numerical computations, that $u(x, t)$ is periodic over a large spatial interval L, then by integrating the equations in conservation form we define the exact invariants, e.g.,

$$\begin{aligned} I_1 &= \int_0^L u\,dx \\ I_2 &= \int_0^L \frac{1}{2} u^2\,dx \\ I_3 &= \int_0^L \left(\frac{1}{3}u^3 - \delta^2 u_x{}^2\right) dx \\ I_4 &= \int_0^L \left(\frac{1}{4}u^4 - 3\delta^2 u u_x{}^2 + \frac{9}{5}\delta^4 u_{xx}^2\right) dx \end{aligned} \tag{5.26}$$

The periodicity assumption does not restrict the generality of our final conclusions. We will now show that the set of invariants can be used to obtain a set of independent nonlinear ordinary differential equations which describe properties of the partial differential equation as the solution evolves in time.

We recall that one can obtain the equations of motion for a conservative system by minimizing the action functional $A(y_t, y_x, y_{xxx}, \ldots)$, that is,

$$\hat{\delta} A = \hat{\delta} \int_0^T dt \int_{\text{volume}} dV\, \mathscr{L}(y_t, y_x, y_{xx}, \ldots) \equiv 0,$$

where $\mathscr{L}$ is the Lagrangian density of the system and $\hat{\delta}$ is the variational operator of the calculus of variations [27]. In the same spirit, we seek equations of "motion" by applying the variational operator and seeking the extremum of one of the invariants, I_n, subject to constraints on all the lower invariants. Thus, if we vary I_3, subject to constraints on I_1 and I_2, we obtain

$$0 = \hat{\delta}(I_3 + \mu' I_2 + \lambda' I_1) = \int_0^L dx[2\,\delta^2\, u_{xx} + u^2 + \mu' u + \lambda']\,\hat{\delta}u. \tag{5.27}$$

Thus, if $\hat{\delta}u$ is an arbitrary variation, then the quantity in brackets must vanish, or

$$0 = 2\,\delta^2\, u_{xx} + u^2 + \mu' u + \lambda'. \tag{5.28}$$

We recognize (5.28) as the first integral of the stationary Korteweg–de Vries equation with μ' related to the speed C (or the amplitude) and λ' to the background U_∞. For convenience we set $\lambda' = 0$ here and in what follows, because this corresponds to a zero background and it does not affect our results. We can summarize the procedure by saying that the result of extremizing I_3 leads to an ordinary differential equation which has one parameter (μ') and therefore describes one type of solitary wave or one type of periodic wave. Hence, if we extremize I_4 subject to constraints on I_3 and I_2, or

$$0 = \hat{\delta}(I_4 + \nu I_3 + \mu I_2), \tag{5.29}$$

then we will obtain an ordinary differential equation with two parameters and we may be able to describe a function representing two solitary waves in various degrees of interaction, that is, a function having at most two maxima or alternatively a periodic wave train with maxima "asymptotically" tangent to two amplitudes. If we perform the calculation indicated in (5.29) we obtain the fourth-order nonlinear ordinary differential equation

$$0 = (18/5)\,\delta^4\, u_{xxxx} + \delta^2(3u_x{}^2 + 6uu_{xx} + 2\nu u_{xx}) + u^3 + \nu u^2 + \mu u. \tag{5.30}$$

The two parameters ν and μ can be related to amplitudes of two "well-separated" solitons by replacing the fourth derivative in (5.30) by the result of differentiating the stationary Korteweg–de Vries equation, or

$$\delta^2\, u_{xxxx} = -u_x{}^2 + uu_{xx} - Cu_{xx}. \tag{5.31}$$

We then replace the second derivative, wherever it appears, by the first integral of the stationary Korteweg–de Vries equation, and upon combining we obtain

$$\delta^2 u_x{}^2 + u^3/3 - Cu^2 - (5u/3)\left[\mu + 2\nu C + \frac{18}{5}C^2\right] = 0. \tag{5.32}$$

If (5.32) is to represent the equation for well separated solitons on a zero background, then the coefficient of u (the bracket) must be set equal to zero. This defines a set of linear "amplitude-value" equations for the parameters μ and ν, in terms of the speeds C_1 and C_2

$$\nu = -(3/5)(A_1 + A_2) \qquad \text{and} \qquad \mu = (2/5)A_1A_2, \tag{5.33}$$

where we recall that $C_i = A_i/3$.

To verify our hypothesis on the existence of a two-amplitude solitary wave solution, Deem solved (5.30) numerically, with $A_1 = 150$, $A_2 = 75$, and $\delta = 0.2$. We started the calculation at $x_s = -0.45$ with initial conditions for U and its first three derivatives derived from the function

$$U(x) = A_1 \operatorname{sech}^2 [x/\Delta_1] + A_2 \operatorname{sech}^2 [(x - x_0)/\Delta_2], \tag{5.34}$$

where $\Delta_1 = 0.05657$, $\Delta_2 = 0.08000$, and $x_0 = 0.25$, and to our delight we obtained the result shown in Fig. 9. In essence, one sees superimposed two periodic wave trains composed of solitons of different amplitude. Table II gives the numerical values of the locations and values of the maxima and minima of this function.

Because of roundoff error (we were using floating point arithmetic and carrying eight significant figures) we did not get *only* two pulses as we had hoped by calculating with initial conditions derived from (5.34). However, the amplitude of the pulses agreed quite well with what we had selected *a priori.* The largest maximum value of wave train 1 was 149.9254 at $x = 2.860$ and the smallest maximum of wave train 2 was 77.7924 at $x = 0.145$. These maxima deviate from the *a priori* selected values of 150.0 and 75.0, respectively, because in this case the solitons are never sufficiently well separated and thus they are always interacting with one another. Note that because of the periodic nature of the solution we fortuitously obtained much more information on the behavior of interacting solitons than if our solution had exhibited only two maxima.

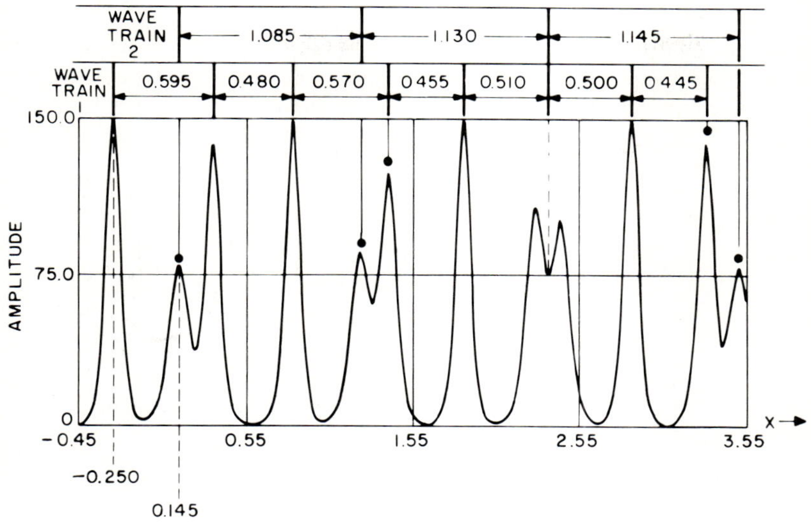

FIG. 9. Solutions of the nonlinear fourth-order ordinary differential equation obtained by extremizing the fourth invariant subject to constraints on the second and third invariant.

We expect this concept to generalize to higher order. That is, if we extremize I_n with constraints on the lower invariants, we will obtain a 2(n − 2)th order differential equation with $n - 2$ parameters. Having solutions to these ordinary differential equations of high order we may be able to determine how any *long* wavelength state will decompose into solitons. We will choose a differential equation of appropriate order, so that it has enough parameters to minimize the "error" between the smooth function one wishes to approximate and that which forms through interaction of a large number of solitons.

A fruitful area of future investigation for nonlinear dispersive systems should involve the study of the relationship between the existence of stable solitons and the existence of conservation laws. If one shows analytically that a large number of conservation laws exists, should one expect numerical solutions of the same equations to show a corresponding large number of solitons interacting and preserving their initial states? Conversely, if there are only three conservation laws should one expect solitons to be unstable to small perturbations?

Table II

LOCATIONS AND VALUES FOR MAXIMA AND MINIMA CORRESPONDING TO TWO DIFFERENT WAVE TRAINS OF INTERACTING SOLITONS (SOLUTION OF THE FOURTH ORDER ORDINARY DIFFERENTIAL EQUATION (5.30))

Maxima		Minima	
Value	Location	Value	Location
149.9037	−0.250	2.5241	−0.070
77.7924	0.145	38.4531	0.240
139.5605	0.345	0.3982	0.595
149.6433	0.825	2.0674	1.010
85.9145	1.230	61.2380	1.305
125.2700	1.395	0.7694	1.640
149.7470	1.850	1.3250	2.050
107.5319	2.285	74.7154	2.360
101.2966	2.435	1.5602	2.670
149.9254	2.860	0.6907	3.075
138.4137	3.305	40.8721	3.410
78.2943	3.505		

Acknowledgments

Many of the concepts and methods presented here were developed in collaboration with M. D. Kruskal. Recent original contributions of R. Miura and G. S. Deem have materially aided in the understanding of these phenomena.

References

1. GOLDSTINE, H. H., and VON NEUMANN, J., On the Principles of Large Scale Computing Machines, *in*: "Collected Works of John von Neumann" (A. Taub, ed.), Vol. 5, pp. 1–32. Macmillan, New York, 1963. The material in this paper was first given as a talk on May 15, 1946. In the opening sections the authors discuss the difficulty of nonlinear problems and describe how they propose to use the digital computer to break the deadlock. Also see: Recent Theories in Turbulence, *in*: "Collected Works of John von Neumann" (A. Taub, ed.), Vol. 6, pp. 437–472. This paper was issued as a report in 1949. On p. 469, we find a lucid formulation of the synergetic approach.
2. ULAM, S. M., "A Collection of Mathematical Problems." Wiley (Interscience), New York, 1960. See Chap. VII, Sect. 8: Physical Systems, Nonlinear Problems; and Chap. VIII, Sect. 10: Computing Machines as a Heuristic Aid—Synergesis.

3. ZABUSKY, N. J., Phenomena Associated with the Oscillations of a Nonlinear Model String (The Problem of Fermi, Pasta, and Ulam), *in*: "Proceedings of the Conference on Mathematical Models in the Physical Sciences" (Stefan Drobot, ed.), pp. 99–133. Prentice-Hall, Englewood Cliffs, New Jersey, 1963.
4. KRUSKAL, M. D., Asymptotology in Numerical Computation: Progress and Plans on the Fermi–Pasta–Ulam Problem. *in*; "Proceedings of the I.B.M. Scientific Computing Symposium—Large Scale Physical Problems in Physics." Held on December 9-11, 1963, at the I.B.M. Thomas Watson Research Lab.
5. KRUSKAL, M. D., Reversible Shock Waves, *Symp. Dynamics of Fluids and Plasmas, Univ. Maryland* (October 7–9, 1965).
6. DEBYE, P., "Vorträge über die Kinetische Theorie der Materie und der Elektrizität." Leipzig, Germany, 1914.
7. PEIERLS, R. E., Quantum Theory of Solids, *in*: "Theoretical Physics in the Twentieth Century" (M. Fierz and V. F. Weisskopf, eds.), p. 140. Wiley (Interscience), New York, 1961.
8. FERMI, E., PASTA, J. R., and ULAM, S. M., Studies of Nonlinear Problems, I, Los Alamos Rept. LA-1940 (May, 1955), (unpublished). Also, "Collected Works of E. Fermi." Vol. II, pp. 978–988. Univ. of Chicago Press, 1965.
9. NORTHCOTE, R. S., and POTTS, R. B., Energy Sharing and Equilibrium for Nonlinear Systems, *J. Math. Phys.* **5**, 383–398 (1964). The authors study a one-dimensional lattice, each of whose springs has a "hard core," that is, the potential energy stored in the spring is

$$\phi = \infty \qquad r \leq d$$
$$\phi = \phi_0 + \tfrac{1}{2}\kappa(x - a)^2 \qquad r > d,$$

where d is the radius of the hard core and the spring's natural length is $a-d$.
10. ZABUSKY, N. J., Exact Solution for the Vibrations of a Nonlinear Continuous Model String, *J. Math. Phys.* **3**, 1028–1039 (1962). P. D. Lax presented simpler methods for computing breakdown times for hyperbolic systems of equations of which (3.10) is a special case: Development of Singularities of Solutions of Nonlinear Hyperbolic Partial Differential Equations, *J. Math. Phys.* **5**, 611 (1964).
11. KRUSKAL, M. D., and ZABUSKY, N. J., Stroboscopic Perturbation Procedure for Treating a Class of Nonlinear Wave Equations, *J. Math. Phys.* **5**, 231–244 (1964). Section 5C discusses the behavior of the waveform and modal energy near breakdown.
12. KRUSKAL, M. D., Asymptotic Theory of Hamiltonian and Other Systems with All Solutions Nearly Periodic, *J. Math. Phys.* **3**, 806–828 (1962).
13. KORTEWEG, D. J., and DE VRIES, G., On the Change of Form of Long Waves Advancing in a Rectangular Channel, and on a New Type of Long Stationary Waves, *Phil. Mag.* **39**, 422–443 (1895).
14. GARDNER, C. S., and MORIKAWA, G. K., Similarity in the Asymptotic Behavior of Collision-Free Hydromagnetic Waves and Water Waves, Courant Inst. of Math. Sci. Rept. NYO 9082 (May, 1960).

15. Gardner, C. S., and Morikawa, G. K., The Effect of Temperature on the Width of a Small-Amplitude Solitary Wave in a Collision-Free Plasma, *Comm. Pure Appl. Math.* **18**, 35–49 (1965).
16. Morton, K. W., Finite Amplitude Compression Waves in a Collision-Free Plasma, *Phys. Fluids* **7**, 1800–1815 (1964).
17. Hopf, E., The Partial Differential Equation $u_t + uu_x = \mu u_{xx}$, *Comm. Pure Appl. Math.* **3**, 201 (1950).
18. Cole, J. D., On a Quasilinear Equation Occurring in Aerodynamics, *Quart. Appl. Math.* **9**, 225 (1951).
19. Lamb, H., "Hydrodynamics," 6th ed. Dover, New York, 1932. Chapter IX, Section 250 contains a discussion of various steady waveforms including those introduced by Stokes and the cnoidal waves of Korteweg and de Vries.
20. Zabusky, N. J., and Kruskal, M. D., Interaction of Solitons in a Collisionless Plasma and the Recurrence of Initial States, *Phys. Rev. Letters* **15**, 240–243 (1965).
21. Izrailev, F. M., and Chirikov, B. V., Statistical Properties of a Nonlinear String, *Soviet Phys.-Doklady* **11**, 30 (1966). Chirikov, V., When Does a Dynamical System Exhibit Stochasticity? Preprint. Paper presented in Sect. 6 at the *International Congress of Mathematicians, Moscow*, (August 16–26, 1966).
22. Deem, G. S., and Zabusky, N. J., Formation and Radiation of Solitons from Localized Intense Excitations of a Lossless Lattice, *Bull. Am. Phys. Soc.* **11**, 47 (1966). To be published as "Dynamics of Nonlinear Lattices, I. Localized Optical Excitations, Acoustic Radiation, and Strong Nonlinear Behavior."
23. Kruskal, M. D., Ref. 4, p. 57. Details of the method were presented by the author and Kruskal at A.P.S. Division of Plasma Physics Meeting, November 4, 1964 in New York. See: Differential Equations in Conservation Form and Wave Number Jump for the Collisionless MHD Shock, *Bull. Am. Phys. Soc.* **10**, 223 (1965).
24. Whitham, G. B., Nonlinear Dispersive Waves, *Proc. Roy. Soc.* **A 283**, 238–261 (1965).
25. Miura, R., Kruskal, M. D., Gardner, C. F., and Zabusky, N. J., Equations in Conservation Form for Nonlinear Dispersive Systems: The Generalized Korteweg–de Vries Equation (to be published).
26. See, for example, Lax, P. D., Hyperbolic Systems of Conservation Laws, II, *Comm. Pure Appl. Math.* **3**, 201–230 (1950) and Rozhdestvenskii, B. L., Discontinuous Solutions of Hyperbolic Systems of Quasilinear Equations, *Uspehi Mat. Nauk* **15**, No. 6, 59–117 (1960) (in Russian); translated as *Russian Math. Surveys* **15**, No. 6, 53–111 (1960).
27. For example, see Gelfand, I. M., and Fomin, S. V., "Calculus of Variations," Section 21. Prentice-Hall, Englewood Cliffs, New Jersey, 1963.

Uniformization of Asymptotic Expansions†

GUIDO SANDRI

Aeronautical Research Associates of Princeton, Inc.
Princeton, New Jersey

We develop a general technique for uniformizing asymptotic expansions. A basic formula for uniformizing counterterms is constructed by nesting an increasing number of extensions. Two successive extensions are shown to contain the counterterms for both secular terms and for singular perturbation terms. Simple examples are used to illustrate the method. A general constructive procedure is outlined to determine the counterterms from the nonuniformities arising in the direct perturbation expansion. A relationship is established between the secular perturbation counterterms and the singular perturbation counterterms. Finally, a brief review is given of the main results obtained so far and of the open problems.

I. Introduction

We discuss a general approach to uniformizing the asymptotic expansions which occur in the solution of the differential or integrodifferential equations that arise in continuum, quantum, and statistical mechanics.

A. Heuristic Discussion

The purpose of our general method is that of uniformizing expansions in a small parameter. We conjecture that our method is sufficiently general to uniformize any asymptotic expansion. Below we illustrate what we mean by a uniformly valid approximation. Assume that we have a function f, of the real variable t. The function is allowed to have an arbitrary shape, e.g., see Fig. 1.1. Singularities, finite discontinuities and oscillations are

† Sponsored by the Air Force Office of Scientific Research of the Office of Aerospace Research under Contract No. AF 49(638)1461.

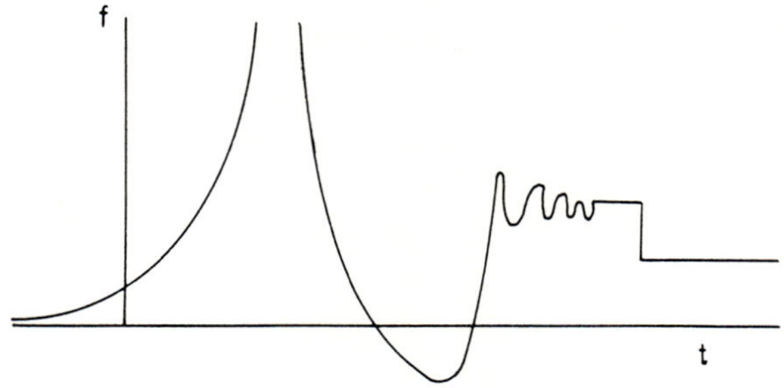

FIG. 1.1. Schematic representation of the function f.

allowed. We have a uniformly valid approximation (or, as mathematicians say [1], "representation") f^0, to order ε, (ε is a given number such that $\varepsilon \ll 1$) if and only if, for all t

$$f = f^0 + O(\varepsilon) \tag{1.1}$$

This means that f^0 must stay close to f (to order ε) for all the values of t which are of interest. In particular, f^0 must follow f when f has singularities, discontinuities, or oscillations. A schematic picture of a uniform and of a nonuniform approximation for f are shown in Fig. 1.2. We presuppose

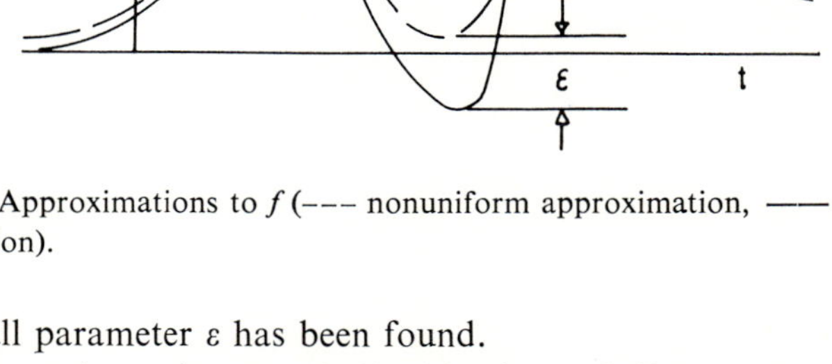

FIG. 1.2. Approximations to f (--- nonuniform approximation, —— uniform approximation).

that a small parameter ε has been found.

Our purpose is to give a method which is, we believe, very general and to show how known methods arise as special cases of our method. The purpose of showing how known methods arise as special cases is not only pedagogical. There is some merit in showing how a variety of phenomena (and some of these expansions *are* phenomena) arise from only a few principles. However, there is even a deeper interest in pursuing the discovery and formulation of the most general method. Namely, problems that are intractable by a given method are not less real than those that can be treated by the given method (in fact, often they are more real). We, thus, hope that new results will emerge from a general method, and it will be seen in Section III that new results have been discovered using our method. There is still much work to be done in obtaining precise conditions on the validity of any special form of the method, and we feel that our general framework should provide a fruitful basis for such a discussion.

We shall not be able to do justice to all the authors that have contributed to uniformizing asymptotic expansions. The literature on the subject is very rich and very diversified. Frequently, a parameter of the system under consideration is known to be small, but direct expansion in the parameter leads to serious misrepresentation of reality. It is this misrepresentation that is called a nonuniformity in the expansion. Three main approaches are frequently used to eliminate the nonuniformity. Some names are associated with each method for the purpose of orientation. The three approaches are closely related to the three types of problems discussed in Subsection B. The three approaches have been discussed often, and for purposes of orientation, we give below a few of the names associated with each of the approaches.

(1) Bogoliubov, Krylov, Mitropolski "Synchronization Methods" [2]; Sandri and Friedman, "Multiple Time Scales "[3, 4]. Typically applied to secular perturbation problems.

(2) Poincaré, Lighthill, Kuo, [5] "(P-L-K) Method for Expansion of Independent Variable." Typically applied to singular perturbation problems.

(3) Kaplun, Lagerstrom, Cole, Kruskal, van Dyke, [6] "Matched Asymptotic Expansions" (Inner and outer expansions). Typically applied to matching problems.

Methods related to type (1) above are the theory of "persistent effects" (van Hove *et al.* [7]) and the "sum over the most singular contributions" (Gell-Man and Brueckner [8], Prigogine [9]).

A basic tool in our context is asymptotic analysis (Erdelyi [1]) since for most real problems encountered one does not think about convergent series at all, but one discusses asymptotic expansions.

In this presentation we will use simple examples with ordinary differential equations to illustrate the ideas without plunging into lengthy calculations. We will, however, introduce the definitions and methods in an abstract framework in order to reveal the *mathematical mechanism* behind the rather intricate calculations that are carried out in practice.

A main theme is to demonstrate that when direct perturbation expansion fails, it is for good reason, namely, that the behavior of the unknown function is followed on an inappropriate scale. The precise nature of the failure is in fact the major clue for *constructing* (Section II) the proper variables ("clocks") with which the real phenomenon can be described naturally (i.e., uniformly).

Poincaré expanded the period of the lowest order term, in nearly recursive problems. Lighthill's expansion of the independent as well as of the dependent variables is useful in a more general context. Bogolubov and co-workers make all the constants, which appear in lowest order perturbation theory, slowly varying functions. This is equivalent to an expansion of the derivative. In our work we introduce a complete reparameterization of the lowest order term of the perturbation expansion.

B. Three Examples

The nonuniformities that arise in asymptotic expansions are, on the surface at least, of many different types. We find it useful to think of them in terms of three broad categories: (1) *Secular perturbation* problems for which one finds nonuniformity in the general solution of the equations in the direct perturbation expansion for large values of the independent variable. (2) *Singular perturbation* problems (our use of the term differs from van Dyke's ([6], p. 7)) in which the nonuniformity of the direct expansion occurs at a finite value of the independent variable. Singular perturbation with oscillatory character are particularly difficult and practically constitute a topic in itself. Problems of types (1) and (2) arise because of nonuniformities in the general solution of the direct expansion of the differential or integrodifferential equation under study. (3) *Matching* problems where one finds it impossible to satisfy the initial or boundary conditions with the approximate solutions given by perturbation theory.

The three types of problems are not unrelated, and the classification has only pragmatic use. In most real problems combinations of the three nonuniformities arise.

The three types of problems are illustrated by the following examples:

(1) *Secular perturbation.* Consider the equation

$$\frac{df}{dt} = -\varepsilon f, \qquad f(0) = 1 \tag{1.2}$$

Expanding $f = f^0 + \varepsilon f^1$ we find that

$$f^0 = 1, \quad f^1 = -t \tag{1.3}$$

clearly f^1 has secular behavior (see Fig. 1.3). This severely misrepresents

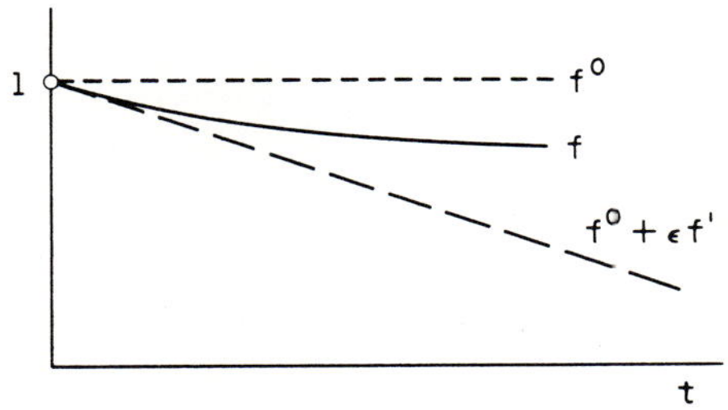

FIG. 1.3. The slow exponential decay.

the correct solution which is a simple exponential decay;

$$f = \exp[-\varepsilon t] \tag{1.4}$$

(2) *Singular perturbation.* Consider

$$(t + \varepsilon f)\frac{df}{dt} = -f, \qquad f(1) = 1 \tag{1.5}$$

Upon expanding f we obtain

$$f^0 = \frac{1}{t}, \qquad f^1 = \frac{1}{2t}\left(1 - \frac{1}{t^2}\right) \tag{1.6}$$

As typical with singular perturbation problems the first order correction is more singular than the lowest order result. See Fig. 1.4. The exact solution

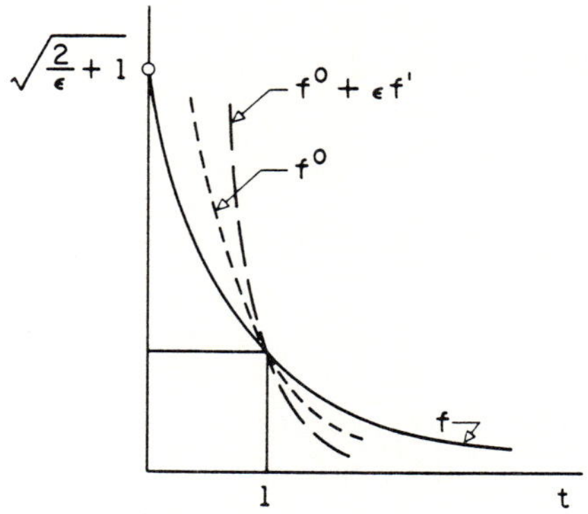

FIG. 1.4. Singular perturbation at the origin.

is

$$f = -\frac{t}{\varepsilon} + \sqrt{\frac{t^2}{\varepsilon^2} + \frac{2}{\varepsilon} + 1} \tag{1.7}$$

(3) *Matching problem*. Consider

$$\varepsilon \frac{d^2f}{dt^2} + A\frac{df}{dt} + Bf = 0, \qquad f(0) = 0, \qquad \dot{f}(0) = C \tag{1.8}$$

whose lowest order term prior to inclusion of the initial condition contains an arbitrary constant:

$$f^0 = \alpha \exp\left[-\frac{B}{A}t\right] \tag{1.9}$$

The initial conditions are met only at the cost of making the solution trivial. This is, of course, not the case with the exact solution of Eq. (1.8). which is

$$f(t) = c\frac{\exp[-\omega_- t] - \exp[-\omega_+ t]}{\omega_+ - \omega_-} \tag{1.10}$$

where

$$\omega_+ = \frac{A}{2\varepsilon} + \left(\frac{A^2}{4\varepsilon^2} - \frac{B}{\varepsilon}\right)^{1/2} \underset{\varepsilon\downarrow}{\sim} \frac{A}{\varepsilon} + \frac{B}{A} - \frac{B^2}{A^3}\varepsilon + O(\varepsilon^2) \tag{1.11}$$

$$\omega_- = \frac{A}{2\varepsilon} - \left(\frac{A^2}{4\varepsilon^2} - \frac{B}{\varepsilon}\right)^{1/2} \underset{\varepsilon\downarrow}{\sim} 0 + \frac{B}{A} + \frac{B^2}{A^3}\varepsilon + O(\varepsilon^2) \tag{1.12}$$

The situation is depicted in Fig. 1.5.

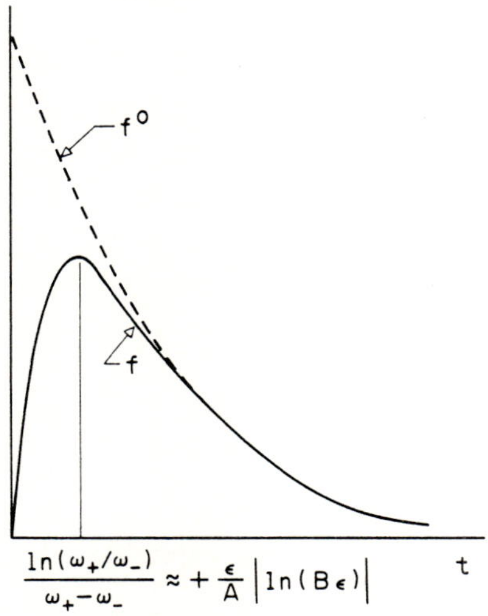

FIG. 1.5. The boundary layer model ("matching" problem).

We shall use the three examples, Eqs. (1.2), (1.5), and (1.8), to illustrate the details of our method and its specialization to the standard approaches to them indicated in Subsection A. The synchronization method is not suitable to problems of type (2) and (3) above; the multiple time scale method is not suited to type (2). The examples given are too simple to be a sensitive check of a method, however. All three can be solved with the method of extension and also with the matching of asymptotic expansions.

II. The Uniformization Method

In this section we give our general scheme for constructing the terms that must be added to the perturbation results in order to eliminate the nonuniformities that arise in perturbation expansions. We shall illustrate our general scheme with the simple examples given in Section I-B.

The mechanism underlying our scheme is that of introducing a set of independent variables (extension of the domain of the independent variable). The relation of the extended variables to the original independent variable is to be *determined* by the nature of the nonuniformities that occur in the direct perturbation expansion. The new variables are often called "clocks" for convenience, but it should be emphasized that the variables entering the problem are by no means restricted to a single real variable. Our procedure requires that the extended perturbation theory be carried out simultaneously with ordinary perturbation theory. In this manner the problem is so formulated that the nonuniformities can be eliminated by setting up counterterms that can absorb the singular contributions of the expansion in ε. The precise form of the counterterms is determined from the singularities themselves. We conjecture that, with the nesting principle given below, any singularity can be uniformized.

A. Formulation of the Method

Consider two topological vector spaces (T.V.S.) δ and $\boldsymbol{\delta}$ (read "delta extended") with

$$\dim \boldsymbol{\delta} \geq \dim \delta \tag{2.1}$$

Consider, also, a locally topological mapping f of δ into the T.V.S. ρ which contains the range of f. We say that **f** is an extension of f if and only if the sequence in Fig. 2.1 is commutative, where E is injective.

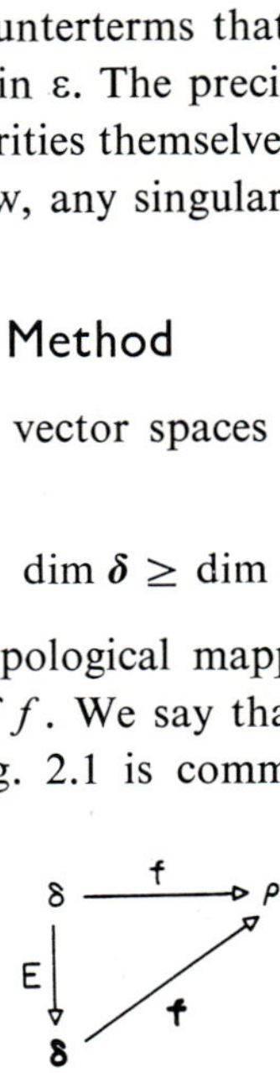

FIG. 2.1. Functional diagram for extension.

Mappings are single valued. They are denoted as follows:

surjective	onto
injective	one to one, into
bijective	surjective and injective, i.e., one to one and onto

Consider $t \in \delta$ and $\boldsymbol{\tau} \in \boldsymbol{\delta}$. We want to choose E in such a way that even though f does not have an asymptotic expansion of the separable form

$$\sum_n g_n(\varepsilon) f^{(n)}(t) \tag{2.2}$$

f does have a separable asymptotic expansion

$$\mathbf{f} \sim \sum_n G_n(\varepsilon)\, \mathbf{f}^{(n)}(\tau) \tag{2.3}$$

where $g_n(\varepsilon)$ and $G_n(\varepsilon)$ are asymptotic sequences. From Fig. 2.1 the composition of E with **f** gives f

$$\mathbf{f} \circ E = f \tag{2.4}$$

that is, there exists a representation of E

$$E: \qquad \tau = \tau(t) \tag{2.5}$$

such that the restriction of **f** to the image of δ under E coincides with f

$$\mathbf{f}(\tau(t)) = f(t) \tag{2.6}$$

We can *induce* the extensions of f by an appropriate extension of a homeomorphism H of δ into itself (Fig. 2.2.)

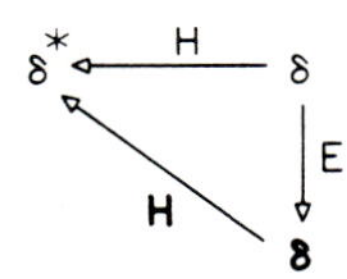

FIG. 2.2. Functional diagram for "inducing" extensions.

From Fig. 2.2 we find

$$f \circ H^{-1} \circ \mathbf{H} = \mathbf{f} \tag{2.7}$$

We see that there is a function $T(\tau)$ with $T \in H\delta = \delta^*$ which, from the commutativity of the sequence in Fig. 2.2 satisfies

$$T(\tau(t)) = t \tag{2.8}$$

and such that, from Eq. (2.7)

$$f(T(\tau)) = \mathbf{f}(\tau) \tag{2.9}$$

We call Eq. (2.7) (or Eq. 2.9) the "induction theorem". We say that while **f** is an extension of the dependent variable (f $\Rightarrow$ **f**), T is an extension of the independent variable t ($t \Rightarrow T$). We shall see that while the method (1) of Section I-A corresponds to an extension of the dependent variable, the method (2) of Section I-A (P-L-K method) corresponds to an extension of the independent variable. These two methods correspond to the construction of E and $\mathbf{H}$ as power series

$$E = \sum_n \varepsilon^n E^{(n)} \tag{2.10}$$

$$\mathbf{H} = \sum_n \varepsilon^n \mathbf{H}^{(n)} \tag{2.11}$$

Extensions can be nested as follows. Consider the individual cartesian components $\boldsymbol{\delta}_K$ of $\boldsymbol{\delta}$

$$\delta = \bigotimes_K \boldsymbol{\delta}_K \tag{2.12}$$

Inject $\boldsymbol{\delta}_K$ into a T.V.S. $\mathfrak{d}_K$ that satisfies

$$\dim \mathfrak{d}_K \geq \dim \boldsymbol{\delta}_K \tag{2.13}$$

with an appropriate mapping E_K'. We then have

$$\delta \overset{E}{\rightarrow} \boldsymbol{\delta} \overset{E'}{\rightarrow} \mathfrak{d} \tag{2.14}$$

where

$$E'\boldsymbol{\delta} = \bigotimes_K E_K' \, \boldsymbol{\delta}_K, \qquad \mathfrak{d} = \bigotimes_K \mathfrak{d}_K, \qquad \mathfrak{d}_K \supset E_K' \, \boldsymbol{\delta}_K \tag{2.15}$$

To exemplify the nesting principle we now construct a powerful formula (double extension formula) from which we shall deduce as special cases the methods (1) and (2) of Section A. We shall also see that the formula allows to treat a simple example of matching problem with a *single* expansion. We conjecture that, in fact, nested extensions always allow to uniformize matching-problem nonuniformities.

Consider the first order form of a differential equation

$$\frac{df}{dt} = F \tag{2.16}$$

where F depends on f derivatives of f with respect to parameters, ε and t. Change the independent variable from t to $s(t)$ (the stretching is to be determined). We can write Eq. (2.16) as $f(t) = f[s(t)]$

$$\frac{df}{ds}\frac{ds}{dt} = F \qquad \text{or} \qquad \frac{df}{ds} = F\frac{dt}{ds} \tag{2.17}$$

Perform the *first extension* on t

$$t \Rightarrow T(\tau_K), \qquad 1 \le K \le N \tag{2.18}$$

Equation (2.17) can be written on the image of the domain ($T = t$) as

$$\frac{df[s(T)]}{ds} = F\frac{dT}{ds} \tag{2.19}$$

Outside-of-the-image (of the domain), $T = T(\tau_K)$, extension of Eq. (2.19) gives

$$\frac{df}{ds} = F\sum_K \frac{\partial T}{\partial \tau_K}\frac{d\tau_K}{ds} \tag{2.20}$$

The second extension is chosen to be $s \to \{\sigma_K\}$. We have then the double extension formula (dot indicates derivative with respect to s):

$$\sum_K \dot{\sigma}_K \frac{\partial \mathbf{f}}{\partial \sigma_K} = \mathbf{F}\left\{\sum_{j,K} \frac{\partial \mathbf{T}}{\partial \tau_j}\frac{\partial \tau_j}{\partial \sigma_K}\dot{\sigma}_K\right\} \tag{2.21}$$

where $\mathbf{f}$, $\mathbf{F}$, τ_K and $\mathbf{T}$ are appropriately extended functions.

We shall see in the next section that the terms on the left side of Eq. (2.21) provide for the elimination of secular nonuniformities (Bogolubov, Sandri, Frieman terms) while the right-hand side provides, terms for elimination of singular nonuniformities (of the Poincaré, Lighthill, Kuo type).

The functional diagram corresponding to the double extension formula is shown in Fig. 2.3. We note that generalization to the continuous case is straightforward

$$\int d\lambda\, \dot{\sigma}(\lambda)\frac{\delta \mathbf{f}}{\delta \sigma(\lambda)} = F\left\{\iint d\mu\, d\nu\, \dot{\sigma}(\mu)\frac{\delta\tau(\nu)}{\delta\sigma(\mu)}\frac{\delta \mathbf{T}}{\delta\tau(\nu)}\right\} \tag{2.22}$$

All the quantities appearing in Eq. (2.2) are assumed to have a separable ε dependence. This permits, with appropriate choices of the asymptotic sequences, to equate term by term the two sides of Eq. (2.21).

The major requirement that we impose is now that the expansion Eq. (2.3) for $\mathbf{f}$ be uniform in ε, i.e.,

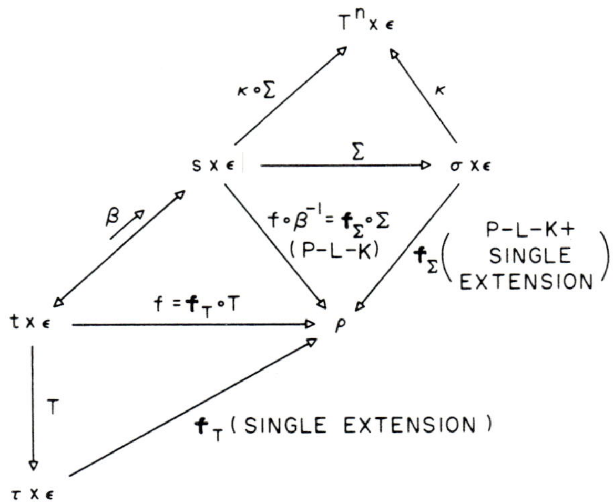

FIG. 2.3. Functional diagram for the double extension formula.

$$\frac{G_{n+1}\,\mathbf{f}^{(n+1)}}{G_n f^n} \underset{\varepsilon\downarrow}{\rightarrow} 0 \tag{2.23}$$

A general result [10] can be proven for Fourier transforms. If $\tilde{f}(v)$ and $\tilde{\mathbf{f}}(\mathbf{v})$are the Fourier transforms of $f(t)$ and any of its extensions $\mathbf{f}(\boldsymbol{\tau})$, then

$$f(v) = \int \Omega\,(v, \mathbf{v})\,\tilde{\mathbf{f}}(\mathbf{v})\,d\mathbf{v}$$

where

$$\Omega(v, \mathbf{v}) = \frac{1}{2\pi}\int dt\,\exp[-ivt + i\mathbf{v}\cdot\boldsymbol{\tau}(t)]$$

As a consequence of this theorem, for Fourier transformable functions, there is a linear transformation from $\mathbf{f}$ into f.

B. Special Cases

We now obtain, as special cases, several particular uniformizing formulae.

Case 1. Multiple time scale method. Set in the double extension formula Eq. (2.21)

$$s = T = t \tag{2.24}$$

Furthermore, make the linear choice represented by Eq. (2.10)

$$\tau_n = \varepsilon^n t \tag{2.25}$$

We then have the multiple time scale formula (we rename the σ_K's by τ_K)

$$\sum_K \varepsilon^K \frac{\partial \mathbf{f}}{\partial \tau_K} = \mathbf{F} \tag{2.26}$$

The uniformization of the exponential problem results in the exact solution being given by the first order result. We find, in fact, that

$$[\mathbf{f}^0] = \exp[-\tau_1] = \exp[-\varepsilon t]$$

The meaning of the extension is that the requirement of uniformity dictates a choice of time scale corresponding to a sufficiently slow clock to allow for a noticeable decay in lowest order. Direct perturbation expansion equation (1.3) gave no change at all for f^0. The Bogolubov formula follows from Eq. (2.26) for functions which are stationary in lowest order. The exact connection between Bogolubov's and our formula is given by ($\alpha = f^0$, $A = \dot{\alpha}$ in Bogolubov's notation)

$$\left[\frac{\partial \mathbf{f}}{\partial \tau_K}\right] = A^K \frac{\partial f}{\partial \alpha} \tag{2.27}$$

where the square bracket denotes restriction to the image of t.

Case 2. Singular perturbation. The P-L-K formula follows from setting.

$$\sigma_0 = s, \qquad \sigma_K = 0 \qquad (K > 0) \tag{2.28}$$

Our example Eq. (1.5) is again readily treated, and again the exact answer is obtained in first order with the result.

$$f = \frac{1}{s}, \qquad t = s + \varepsilon \frac{s}{2}\left(1 - \frac{1}{s^2}\right) \tag{2.29}$$

By eliminating s between the two equations in Eq. (2.29), one readily recovers the exact result, Eq. (1.7). The meaning of the extension in this case is that uniformity requires a choice of scale that pushes the singularity outside the physical domain

$$t(s) \xrightarrow[s \to 0]{} -\infty$$

and stretches the independent variable to follow its natural behavior.

Case 3. Matching problem. Equation (1.8) can be thought of as a simple model of boundary layer problem. Choosing $s = T = t$, the double extension formula becomes (renaming the σ_K's)

$$\sum_K \dot{\tau}_K \frac{\partial \mathbf{f}}{\partial \tau_K} = \mathbf{F} \tag{2.30}$$

Choose now

$$\tau_{-1} = \frac{t}{\varepsilon}, \qquad \tau_K = \varepsilon^K t \quad K \geq 0 \tag{2.31}$$

We readily obtain in lowest order

$$\mathbf{f}^0 = A(\tau_0) \exp[-\omega_+ \tau_{-1}] + B(\tau_0) \tag{2.32}$$

The first order calculation readily yields

$$A, B \,\alpha\, \exp\left[-\frac{B}{A} \tau_0\right] \tag{2.33}$$

The boundary conditions can clearly be satisfied by Eqs. (2.31) and (2.32). The zeroth order equation corresponding to Eq. (2.31) can be taken as analogous to the boundary layer equation, while the first order equations are the analog of the inviscid (Euler) equations which are valid for large distances from the body. With our method a single expansion suffices to yield both types of behavior.

Case 4. We note an interesting connection between the P-L-K formula and the multiple time scale formula. The P-L-K formula reads

$$\frac{df}{ds} = F\left(1 + \varepsilon \frac{dT_1}{ds} + \varepsilon^2 \frac{dT_2}{ds} + \cdots\right) \tag{2.34}$$

hence

$$\begin{aligned} F &= \frac{df}{ds}\left(1 + \varepsilon \frac{dT_1}{ds} + \varepsilon^2 \frac{dT_2}{ds} + \cdots\right)^{-1} \\ &= \frac{df}{ds}\left(1 - \varepsilon \frac{dT_1}{ds} - \varepsilon^2 \left[\frac{dT_2}{ds} - \left(\frac{dT_1}{ds}\right)^2\right] + \cdots\right) \end{aligned} \tag{2.35}$$

The multiple time scale formula on the other hand can be written as $\dot{\tau}_n = \varepsilon^{nH^{(n)}}$

$$F = \left[\frac{\partial \mathbf{f}}{\partial \tau_0} H^0\right] + \varepsilon\left[\frac{\partial \mathbf{f}}{\partial \tau_1} H^1\right] + \varepsilon^2\left[\frac{\partial \mathbf{f}}{\partial \tau_2} H^2\right] \tag{2.36}$$

Hence, equating powers of ε (provided that the restrictions do not introduce additional powers)

$$\frac{df^0}{ds} = \left[\frac{\partial \mathbf{f}^0}{\partial \tau_0} H^0\right] \tag{2.37}$$

$$\frac{df^1}{ds} - \frac{df^0}{ds}\frac{dT_1}{ds} = \left[\frac{\partial \mathbf{f}^1}{\partial \tau_0} H^0\right] + \left[\frac{\partial \mathbf{f}^0}{\partial \tau_1} H^1\right] \tag{2.38}$$

and so on. We, thus, see that secular and singular perturbation problems are related.

Case 5. We note that the van der Pol equation

$$\ddot{f} + \omega^2 f = \eta \dot{f} - \gamma f^2 \dot{f} \tag{2.39}$$

is readily treated and gives for large t

$$f \sim 2\,(\eta/\gamma)^{1/2} \cos(t + \phi) \tag{2.40}$$

C. Constructive Procedure

It is, or course, highly desirable to mechanize our extension procedure as much as possible. In particular, we would like to have iron cast rules that allow for the calculations to proceed without having to inject any special information at all. This means that we would like to be able to express the σ_K and τ_j in terms of the coefficients in the original differential equation. We illustrate the construction of the variables τ_K for the exponential example Eq. (1.2). Consider, for simplicity, two clocks, a fast one τ_0 and a slow one τ_1, which we have to determine as functions of t

$$\dot{\tau}_0 = g(t, \varepsilon), \qquad \dot{\tau}_1 = h(t, \varepsilon) = G(\varepsilon)H(t) \tag{2.41}$$

For short times, since the decay is slow, perturbation theory is adequate. We can, therefore, set

$$\tau_0 = t \tag{2.42}$$

The slow behavior is determined as follows. Extension of Eq. (1.2) gives

$$\frac{\partial \mathbf{f}}{\partial \tau_0} + G(\varepsilon)\dot{H}\frac{\partial \mathbf{f}}{\partial \tau_1} = -\varepsilon\, \mathbf{f}(\tau_0, \tau_1) \tag{2.43}$$

To eliminate the *first order* secularity we must choose

$$G(\varepsilon) = \varepsilon \tag{2.44}$$

Expanding **f** as

$$\mathbf{f} = \mathbf{f}^0 + \varepsilon\delta\mathbf{f} \tag{2.45}$$

we find

$$\frac{\partial \mathbf{f}^0}{\partial \tau_0} = 0, \qquad \frac{\partial \delta\mathbf{f}}{\partial \tau_0} + H(\tau_0)\frac{\partial \mathbf{f}^0}{\partial \tau_1} = -\,\mathbf{f}^0 \tag{2.46}$$

Performing the τ_0 integration in Eq. (2.46) and requiring

$$\varepsilon\, \delta\mathbf{f}/\mathbf{f}^0 \underset{\varepsilon\downarrow}{\to} 0 \tag{2.47}$$

for all τ_0, we find [11]

$$0 = 1 + \frac{\partial \log \mathbf{f}^0}{\partial \tau_1} H(\tau_0) \tag{2.48}$$

whence, by differentiation (since $\mathbf{f}^0$ is τ_0 independent)

$$H(\tau_0) = \text{constant}$$

thus, the "guess" that suggests $\tau_1 = \varepsilon t$ is, in fact, needed for a uniform expansion.

Cross differentiation of Eq. (2.46) shows that Eq. (2.48) is necessary as well as sufficient for uniformity [10] as a consequence of the requirement of compatibility of the two equations in Eq. (2.46).

III. Results and Open Problems

A. Fluid Dynamics

The major technique used to investigate the Navier–Stokes equations has been the matching of inner and outer expansions (Kaplun, Lagerstrom and Cole, and van Dyke). In Section II-B, Case 3 we showed above that a "model" of the Navier-Stokes equations can be treated uniformly by means of a single expansion with the method of extension. It remains to apply the method of extension to the full Navier-Stokes equations for realistic problems.

B. Quantum Theory

Interesting results have been obtained by applying the method of extension to the following problems:

(1) The Rayleigh–Schrödinger perturbation method for the energy eigenvalue ($E = \Sigma_k \varepsilon^k E^k$) can be interpreted in terms of a sequence of slower and slower time scales that correspond to successively more stationary (persistent) effects. The precise connection is

$$\left[i\hbar \frac{\partial \psi}{\partial \tau_K} \right] = E^K \tag{3.1}$$

(2) Resonance absorption—while direct perturbation theory yields

$$\psi_{\text{pert.th.}} \propto \frac{1}{E_2 - E_1 - \omega} \tag{3.2}$$

The correct resonant behavior requires a slow variable with the result [12]

$$\psi \propto \frac{1}{(1 + \varepsilon^2 \Delta^2)^{1/2}}, \qquad \varepsilon\Delta = E_2 - E_1 - \omega \tag{3.3}$$

(3) Decay of unstable states [13]—the Weisskopf–Wigner prescription can be made mathematically precise. For the Weiskopf–Wigner model (which is equivalent to the lowest sector of the Lee model) one obtains the decay on the slow scale τ_1 as

$$\frac{\partial \psi}{\partial \tau_1} = -\lambda\psi \tag{3.4}$$

with

$$\lambda = \text{Re} \int U^*(\mathbf{x}) \frac{1}{\nabla^2} U(\mathbf{x})\, d\mathbf{x} \tag{3.5}$$

where $U(\mathbf{x})$ is the form factor.

An interesting open problem is that of deriving the cross section formula in scattering theory without the unsatisfactory $\delta(0) \sim 1/(VT)$ argument usually given [14].

Another major open problem is that of obtaining an asymptotic expansion for a quantized field theory. The lowest order result (for scalar ϕ^3 theory) is a free field. The first order filed $\phi^{(1)} = \Delta_R * \phi^{(0)2}$ violates the axioms of field theory and it is much too singular to form the basis for higher approximations. In second order one finds the divergent coincidence of two δ singularities in the self-mass. The situation we suggest is analogous to that of the single equation, Eq. (1.5). A sufficiently singular result in lowest order makes it unsuitable for direct perturbation theory even if the singularity is per se acceptable.

C. Nonequilibrium Statistical Mechanics

The multiple time scale method has yielded two major results that go beyond Bogoliubov's theory. The Abelian result (see Sandri [3], Section 7) for the transient that preceeds the onset of kineticity has been obtained; namely, the one-particle distribution function decays on the fast time scale as

$$\frac{\partial \mathbf{f}^{(1)}}{\partial \tau_0} \underset{\tau_0 \uparrow}{\sim} \frac{1}{\tau_0{}^4} \tag{3.6}$$

A more difficult Tauberian result (see Sandri [3], Section 8) (determine the asymptotic behavior of the integrand given the asymptotic behavior of the integral) has also been obtained. If the kinetic regime is to take over, then the initial two-body correlation g must satisfy the Tauberian requirement

$$\int^{\infty} L e^{-v \cdot \nabla t} g \, dt < \infty \tag{3.7}$$

where L is a linear operator and v the relative velocity of the particle pair. The result ("absence of parallel motions") is

$$v^2 g \underset{v \downarrow}{\sim} v^{\eta} \tag{3.8}$$

for some $\eta > 0$. Both the Abelian and the Tauberian results have been established in the weak coupling limit. It remains to establish them for short range and plasma potentials. These results complete, in principle, the derivation from purely mechanical laws of the irreversible equations of kinetic theory to lowest order. A major open problem is that of uniformizing the asymptotic expansion of kinetic theory beyond the Boltzmann collision term. This problem corresponds to establishing kinetic theory of three-body effects.

Acknowledgments

We are much indebted to W. Hayes, M. Kruskal, and R. Sullivan for discussions and to A. Kritz for a critical reading of the manuscript.

References

1. ERDELYI, A., "Asymptotic Expansions." Dover, New York, New York, 1956.
2. KRYLOV, N., and BOGOLIUBOV, N., "Introduction to Nonlinear Mechanics." Princeton Univ. Press, Princeton, New Jersey, 1947. N. Bogoliubov and Y. Mitropolsky, "Asymptotic Methods in the Theory of Nonlinear Oscillations." Gordon and Breach, New York, New York, 1961).
3. SANDRI, G., The New Foundations of Statistical Dynamics, *Ann. Phys.* **24**, 330 (1963).
4. FRIEDMAN, E., *J. Math. Phys.* **4**, 410 (1963).
5. LIGHTHILL, M., PHIL. *Mag.* **40**, 1179 (1949); H. Tsien, Advanc. Appl. Mech. **4**, Chapter 6 (1956); K. Friedrichs, *Bull. Am. Math. Soc.* **61** (1955). This paper contains an excellent qualitative review of singular perturbation problems.
6. VAN DYKE, M., "Perturbation Methods in Fluid Mechanics." Academic Press, New York, 1964.
7. VAN HOVE L., HUGENHOLTZ, N., and HOWLAND, L., "Quantum Theory of Many Particle Systems." Benjamin, New York, 1961.
8. GELL-MANN, M., and BRUECKNER, K., *Phys. Rev.* **106**, 364 (1957).
9. PRIGOGINE, I., "Nonequilibrium Statistical Mechanics." Wiley (Interscience), New York, 1962.
10. SANDRI, G., *Nuovo Cimento* **36**, 67 (1965).
11. SANDRI, G., and SULLIVAN, R., *Nuovo Cimento* **37**, 1799 (1965).
12. CASE, K., unpublished.
13. BOLDT, E., and SANDRI, G., *Phys. Rev.* **135**, B1808 (1964).
14. We are indebted to S. DRELL for pointing out this interesting application of uniformizing techniques.

High Order Accurate Difference Methods in Hydrodynamics†

SAMUEL Z. BURSTEIN

Courant Institute of Mathematical Sciences
New York University
New York, New York

1. Introduction

We consider a class of problems in which the fluid is to exhibit no perceptible viscous or heat conducting effects. Much effort has been spent, in recent years, on the construction of finite difference approximations to the equations of hydrodynamics. We shall mention some current techniques being used in Lagrange calculations and then describe some aspects of Eulerian type calculations, i.e., the method of Lax and Wendroff. The methods discussed apply to problems in three independent variables.

2. Trends in Lagrange Calculations

One of the most successful methods used for the integration of the equations of motion of a compressible inviscid fluid was due to Robert Richtmyer and John von Neumann. The differential equations were written in Lagrangian form (see Richtmyer [1]) and centered differences were used to approximate the differentials. When the algorithm was applied to flow problems which contained shock waves, the numerical solution diverged from the true solution. The problem was remedied by adding an artificial viscous pressure to the pressure term in the momentum and energy equations. The resultant numerical solutions were smooth and converged to the exact solution. For flows which undergo large distortions, Lagrange calculations lead to inaccuracies. The distortions arise from the fact that unlike Eulerian calculations, where the mesh remains stationary, the Lagrangian mesh moves during the course of the calculation. More-

† Work supported by U.S. Atomic Energy Commission under Contract No. AT (30–1)–1480.

over, the motion of the mesh lines, if not monitored carefully, produce highly distorted zones which eventually lead to very inaccurate numerical results. Indeed, the calculation may become highly unstable as mesh points or lines become crowded locally. It has been evident for some time that for problems which exhibit large distortions, the numerical solution can only be successfully followed through the technique of realignment of the mesh after several time cycles or even after every computation cycle. Several methods have been proposed to achieve realignment of the Lagrange mesh. A remapping of the Lagrange calculation onto an Eulerian mesh is one such method.

This technique has been used by Noh of the Lawrence Radiation Laboratory in one and two dimensional problems. Figure 1 shows, in

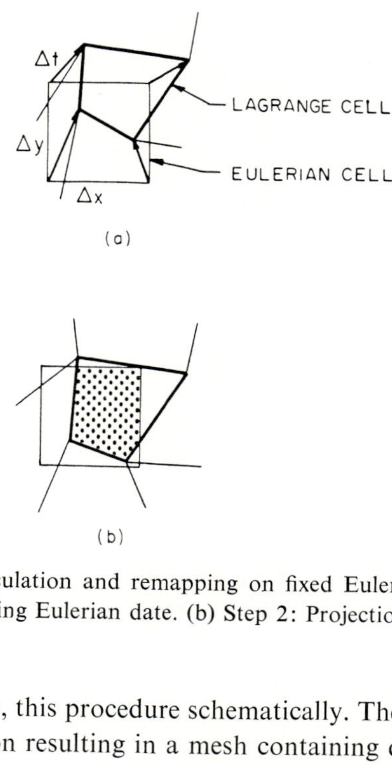

FIG. 1. Lagrange calculation and remapping on fixed Eulerian grid. (a) Step 1: Lagrange calculation using Eulerian date. (b) Step 2: Projection and recomputation of Eulerian data.

two space dimensions, this procedure schematically. The first step involves a Lagrange calculation resulting in a mesh containing quadrilaterals. The second step, requires the mapping of this Lagrange mesh onto the original Eulerian grid. It is the most difficult part of the method as it involves the automatic computation of the intersection of all Lagrange quadrilaterals with the Eulerian cell. The shaded area in Fig. 1 is the intersection of the primary Lagrange cell with the Eulerian cell. These areas are the weighting functions needed in order to determine the fluid properties on the Eulerian grid.

Straightening of the distorted Lagrange mesh has been proposed by Browne [3]. The technique is such that the Lagrange mesh rezoning would be independent of the finite difference approximation to the differential equation. The numerical technique results in a straightening of the distorted Lagrange mesh with a corresponding redistribution of the conserved variables. In [4], a method is given in which the class of problems considered can be most simply described by both Eulerian and Lagrangian subregions. The method of coupling the calculation at the interface of these subregions is an important aspect of the calculation. The results of several such calculations are presented in the paper by Noh [4] (the code, called CEL, stands for Coupled Eulerian Lagrangian).

3. Eulerian Calculations in Three Independent Variables

In this section we will describe a relatively new technique for the differencing of the equations of hydrodynamics, i.e., the method of Lax and Wendroff. The reader should also investigate the results of [5–13].

The differential equation is assumed to be written in divergence free form

$$D\phi = 0 \tag{1}$$

where D is the partial differential operator in three dimensional space time and ϕ is a vector valued function. If Eq. (1) describes the divergence free nature of the equations of fluid dyammics then ϕ has the vector components (w, f, g) so that Eq. (1) becomes

$$\frac{\partial w}{\partial t} + \frac{\partial f}{\partial x} + \frac{\partial g}{\partial y} = 0 \tag{2}$$

The vector w contains the physical components per unit volume which are preserved in nature, i.e.,

$$w = \begin{pmatrix} \rho \\ m \\ n \\ E \end{pmatrix} \quad \begin{matrix} \text{mass} \\ \text{momentum in } x \text{ direction} \\ \text{momentum in } y \text{ direction} \\ \text{total energy} \end{matrix}$$

and the vectors f and g denote the flux of these quantities in the x and y direction, respectively. In general, f and g depend on w in a nonlinear way. By using the chain rule, Eq. (2) may be rewritten in terms of the matrices $A = \partial f/\partial w$ and $B = \partial g/\partial w$:

$$\frac{\partial w}{\partial t} + A\frac{\partial w}{\partial x} + B\frac{\partial w}{\partial y} = 0 \tag{3}$$

We wish to obtain approximate solutions to Eq. (2) by spatial difference operators of the form

$$v(t + k) = Pv(t) \tag{4}$$

P is the spatial difference operator and v is a vector function defined on a lattice in three space which covers the volume of integration and is an approximation to the solution w. P is called a one step operator and can be written in terms of coefficient matrices a_i as

$$P = \sum_i a_i\, T^{ih_j} \tag{5}$$

The translation operator is denoted by T^{ih_j}; ih_j corresponds to the amount of translation in the jth direction. For the method of Lax and Wendroff $h_j = (\Delta x, \Delta y)$ and i runs through values $-1 \leq i \leq 1$. Hence the difference operator P uses the information contained at (x_i, y_j, t) as well as the eight nearest neighbors.

In the execution of Eq. (4), the matrices defined in Eq. (3) need repeated evaluation for proper centering of the difference equation at the mesh point $(x_i, y_j, t + \Delta t/2)$, i.e.,

$$v(t + k) = v(t) - \Delta t\left\{\left(\widehat{\frac{\partial f}{\partial x}} + \widehat{\frac{\partial g}{\partial y}}\right) - \frac{\Delta t}{2}\left(\overbrace{\frac{\partial}{\partial x}\left[A\left(\frac{\partial f}{\partial x} + \frac{\partial g}{\partial y}\right)\right] - \frac{\partial}{\partial y}\left[B\left(\frac{\partial f}{\partial x} + \frac{\partial g}{\partial y}\right)\right]}\right)\right\} \tag{6}$$

where $\widehat{\partial/\partial x}$, $\widehat{\partial/\partial y}$ denote a centered difference approximation to the derivatives in the x and y directions. The terms in the bracket correspond to centered difference approximations of the first and second time derivatives of the vector function v.

4. Two Step Lax–Wendroff Schemes

A two step Lax–Wendroff scheme is one in which temporary values are generated by a first order scheme. In the second step, these values are used to generate a solution which is second order accurate. There are several variations that one can use, and they all eliminate the need for introducing the matrices of Eq. (3). The method can be written in the form of

$$\textit{Step } 1: \tilde{v}(t + nk) = P_n v(t), \qquad n = 1/2 \quad \text{or} \quad 1 \tag{7a}$$

$$\textit{Step } 2: v(t + k) = Qv(t) + R_n \tilde{v}(t + nk), \qquad n = 1/2 \quad \text{or} \quad 1 \tag{7b}$$

Although we have indicated these schemes in a general way, it is clear that the solution depends upon the successive application of two distinct operators.

Usually the spatial difference operator P has only first order accuracy in determining the intermediate vector $\tilde{v}$. As indicated by the value of n, $\tilde{v}$ can be computed at either a half time step or full time step. The second step, requires the use of difference operators Q, R which result in Eq. (7b) being second order accurate (the truncation error is 0 (Δt^3)). In [1], Richtmyer presents a two step Lax–Wendroff scheme for which:

Step 1:

$$\begin{aligned} \tilde{v}(x_i, y_j, t + \Delta t) = {} & \tfrac{1}{4}(v_{i+1,j} + v_{i-1,j} + v_{i,j+1} + v_{i,j-1}) \\ & - \frac{\Delta t}{2\,\Delta x}[f(v_{i+1,j}) - f(v_{i-1,j})] \\ & - \frac{\Delta t}{2\,\Delta y}[g(v_{i,j+1}) - g(v_{i,j-1})] \end{aligned} \tag{8a}$$

Step 2:

$$\begin{aligned} v(x_i, y_j, t + 2\,\Delta t) = {} & v(x_i, y_j, t) - \frac{\Delta t}{\Delta x}[f(\tilde{v}_{i+1,j}) \\ & - f(v_{i-1,j})] - \frac{\Delta t}{\Delta y}[g(\tilde{v}_{i,j+1}) - g(\tilde{v}_{i,j-1})] \end{aligned} \tag{8b}$$

The right-hand side of Eq. (8a) is evaluated at time t. It is clear that step 2 is second order accurate since the space and time differences are both centered at $(x_i, y_j, t + \Delta t)$. Richtmyer investigated the amplification matrix of the combined system Eqs. (8a) and (8b) obtained by linearizing and substituting for the quantity v_{ij}^l the expression $v^l \exp(\sqrt{-1}\,\kappa_x i\,\Delta x + \sqrt{-1}\,\kappa_y j \Delta y)$. This corresponds to the Fourier transformation L_f applied to the system (7a) and (7b) to give

$$L_f \mathscr{P} v = H(\xi, \eta) L_f v \tag{9}$$

where $\xi = \kappa_x\,\Delta x$ and $\eta = \kappa_y\,\Delta y$.

Here $\mathscr{P}$ represents $Q + RP$. The wave numbers are $\kappa_x = 2\pi/L_x$ and $\kappa_y = 2\pi/L_y$ where L_x, L_y are the wave lengths in the x and y directions; the amplification matrix is $H(\xi, \eta)$. Richtmyer showed that the eigenvalues of H, corresponding to (8a) and (8b) are less than or equal to 1 if

$$(\,|\,\bar{u}^>\,|\, + c)\frac{\Delta t}{\Delta} \leq \frac{1}{\sqrt{2}}, \qquad \Delta = \Delta x = \Delta y \tag{10}$$

where $|\,\bar{u}^>\,|$ is the magnitude of the particle velocity and c is the adiabatic speed of sound. One point is evident, and that is the eigenvalue for $H(\pi, \pi)$ is exactly 1, obtained from inspection ($H = I + G\;(A \sin \xi + B \sin \eta)$ where G is a second order trigonometric polynomial in its argument). This case corresponds to the shortest wavelengths, represented in the solution $L_x = 2\,\Delta x$ and $L_y = 2\,\Delta y$; they will not be damped. As a result, in the solution of flows which contain discontinuities or complicated boundary conditions, this component of the solution may possibly grow in time and overwhelm the calculation (see Houghton *et al.* [9]).

We now present a variant of the system represented by (8a) and (8b). Here $n = 1$:

Step 1:

$$\begin{aligned}\tilde{v}(x_{i+1/2}, y_{j+1/2}, t + \Delta t) = {} & \tfrac{1}{4}(v_{i,j} + v_{i+1,j} + v_{i,j+1} + v_{i+1,j+1}) \\ & - \frac{\Delta t}{2\,\Delta x}\{f(v_{i+1,j}) - f(v_{i,j}) \\ & + f(v_{i+1,j+1}) - f(v_{i,j+1})\} \\ & - \frac{\Delta t}{2\,\Delta y}\{g(v_{i+1,j+1}) - g(v_{i+1,j}) \\ & + g(v_{i,j+1}) - g(v_{i,j})\}\end{aligned} \tag{11a}$$

Step 2:

$$v(x_i, y_j, t + \Delta t) = v(x_i, y_j, t) - \frac{\Delta t}{4\,\Delta x}\{f(v_{i+1,j}) - f(v_{i-1,j})$$
$$+ f(\tilde{v}_{i+1/2,j+1/2}) - f(\tilde{v}_{i-1/2,j+1/2})$$
$$+ f(\tilde{v}_{i+1/2,j-1/2}) - f(\tilde{v}_{i-1/2,j-1/2})\}$$
$$- \frac{\Delta t}{4\,\Delta y}\{g(v_{i,j+1}) - g(v_{i,j-1}) + g(\tilde{v}_{i+1/2,j+1/2})$$
$$- g(\tilde{v}_{i+1/2,j-1/2}) + g(\tilde{v}_{i-1/2,j+1/2}) - g(\tilde{v}_{i-1/2,j-1/2})\} \tag{11b}$$

The amplification matrix corresponding to system (11a) and (11b) with $\Delta t/\Delta x = \Delta t/\Delta y = \lambda$, can be given by

$$H(\xi, \eta) = I + i\lambda A \sin \xi\left(\frac{3 + \cos \eta}{4}\right) + B \sin \eta\left(\frac{3 + \cos \xi}{4}\right)$$
$$+ \frac{\lambda^2}{2}\{A[(1 - \cos \xi)(1 + \cos \eta)]^{1/2}$$
$$+ B[(1 - \cos \eta)(1 + \cos \xi)]^{1/2}\}^2 \tag{12}$$

The eigenvalues of this complex matrix have been computed numerically and are shown in Fig. 2. It should be noted that as in system (8a) and (8b) the absolute value of the maximum eigenvalue $|\mu(H(\pi, \pi))| = 1$. However the amplification matrix for the Lax–Wendroff one-step method (obtained by linearizing Eq. (6) and taking the Fourier transform) as given by

$$H(\xi, \eta) = I + i\lambda\{A \sin \xi + B \sin \eta\} + \lambda^2\{A^2(1 - \cos \xi)$$
$$+ \frac{(AB + BA)}{2} \sin \xi \sin \eta + B^2(1 - \cos \eta)\} \tag{13}$$

dampens all wave lengths except, of course, $\xi = \eta = 0$. Figure 3 shows the absolute value of the eigenvalues of (13). In both Figs. 2 and 3, the value of the mesh ratio is the largest allowable, i.e., if CFL corresponds to the Courant-Friedrichs-Lewy condition then

$$\lambda \quad \text{in Fig. 2} = 0.7550 \cdot \text{CFL} \tag{14a}$$
$$\lambda \quad \text{in Fig. 3} = 0.5406 \cdot \text{CFL} \tag{14b}$$

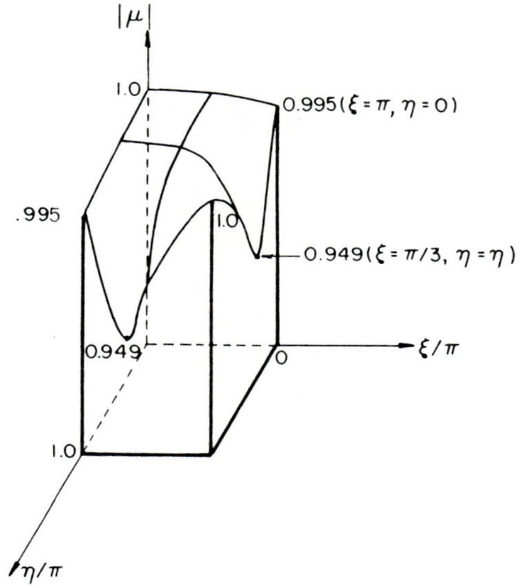

FIG. 2. Absolute eigenvalues $|\mu(H(\xi, \eta))|$ of Eq. (12), λ is the maximum allowable for $|\mu| \leq 1$.

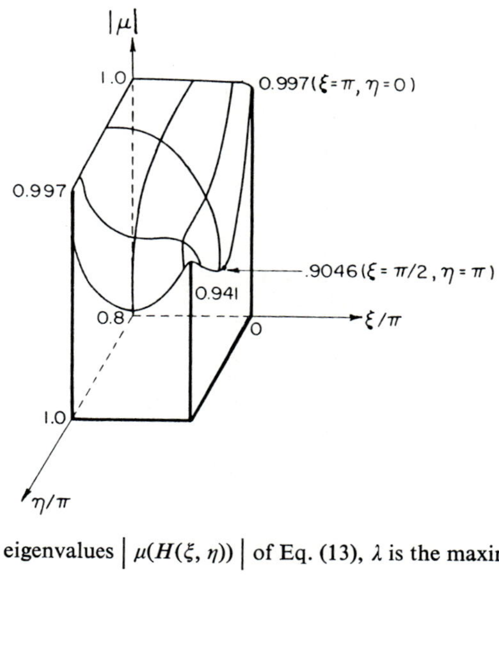

FIG. 3. Absolute eigenvalues $|\mu(H(\xi, \eta))|$ of Eq. (13), λ is the maximum allowable for $|\mu| \leq 1$.

Notice that conditon (14a) is very close to the value given by (10) for the system (8a) and (8b) and that (14b) is larger than the condition given in [6], i.e., $\lambda \leq 1/\sqrt{8}$ CFL. The reason for the difference in allowable step size is that Lax and Wendroff derived the result for (13) in the general case of first order symmetric hyperbolic systems of partial differential equations.

Before we finish this section, we present the two step method for $n = 1/2$ for the system (7a) and (7b). The scheme is similar to (8a) and (8b), in that the diffusing low order accuracy operator is first applied followed by the leap frog difference in the second step:

Step 1:

Compute $\tilde{v}(x_{i+1/2}, y_{j+1/2}, t + \Delta t/2)$ by Eq. (11a) but replace Δt with $\Delta t/2$. (15a)

Step 2:

$$
\begin{aligned}
v(x_i, y_j, t + \Delta t) = v(x_i, y_j, t) &- \frac{\Delta t}{2\,\Delta x}\{f(\tilde{v}_{i+1/2,j+1/2}) \\
&- f(\tilde{v}_{i-1/2,j+1/2}) + f(\tilde{v}_{i+1/2,j-1/2}) \\
&- f(\tilde{v}_{i-1/2,j-1/2}\} - \frac{\Delta t}{2\,\Delta y}\{g(\tilde{v}_{i+1/2,j+1/2}) \\
&- g(\tilde{v}_{i+1/2,j-1/2}) + g(\tilde{v}_{i-1/2,j+1/2}) - g(\tilde{v}_{i-1/2,j-1/2})\}
\end{aligned}
\tag{15b}
$$

By taking the Fourier transform of system (15), we find the amplification matrix is quite similar to that given by (12)

$$
\begin{aligned}
H(\xi, \eta) = I + i\lambda&\left\{A \sin \xi\left(\frac{1 + \cos \eta}{2}\right) + B \sin \eta\left(\frac{1 + \cos \xi}{2}\right)\right\} \\
&+ \frac{\lambda^2}{2}\{A[(1 - \cos \xi)(1 + \cos \eta)]^{1/2} \\
&+ B[(1 - \cos \eta)(1 + \cos \xi)]^{1/2}\}^2
\end{aligned}
\tag{16}
$$

Equation (16) also has the property that $H(\pi, \pi) = 1$ and it is clear that (16) although of second degree is not the same form as the trigonometric polynomial that Richtmyer obtained.

5. Instabilities of the Nonlinear Type

Although the several difference schemes described in the previous sections were shown formally (in the linear sense) to be stable for some suitable value of mesh ratio, recent numerical experiments have shown that under certain conditions instabilities may result when these algorithms are used. In [8] and [9] instabilities were encountered in calculations with three independent variables. In both of these calculations by Burstein and Houghton–Kasahara–Washington, the shortest wavelength component of the solution $L = 2\,\Delta x$ grew and swamped the smooth portion of the solution. In the latter calculation, the two step method given by Eqs. (8a) and (8b) was used. In [9] it is clearly shown that after approximately 1.9×10^4 time cycles, the $2\,\Delta x$ component of the solution grows, while the $4\,\Delta x$ component does not. In [8], the solutions of interest were not smooth, as in the weather model of [9], but contained discontinuities as part of the solution. This calculation was performed with the one step Lax–Wendroff equations and in the analysis carried out, a possible mechanism for instability was given. It was based on the observation that when some of the eigenvalues of the matrices A and B become zero (sonic lines, stagnation points) the damping of the linearized operator becomes neutral.

In a region where the solution varies most rapidly, so that the $2\,\Delta x$ component of solution is the most important, the absence of positive damping could lead to instability.

Richtmyer in [1] and Richtmyer and Morton in [11] have investigated the effect of boundary conditions on the stability of some difference equations. They also looked at the stability of certain difference schemes, including the Lax–Wendroff scheme, when used in simple initial-value problems.

Recently, Lax and Nirenberg have succeeded in proving that the norm of the scalar difference operators, which has Lipschitz continuous coefficients, satisfies

$$\|P\| \leq 1 + M\,\Delta t \tag{15}$$

where M is dependent on the Lipschitz constant provided that the amplification factor associated with P satisfies $|G(\xi)| \leq 1$ for all ξ. Lax describes a possible explanation for instability in [10] which is based on the growth of the Lipschitz constant of v. As the solution v becomes wrinkled, as a result of a slight growth in amplitude of the short wavelength component of the solution, the right-hand side of (15) will exceed one by a large amount and instability will result.

Unfortunately, the explanations which have been presented only give plausible mechanisms for nonlinear instability, rather than the actual cause.

6. Navier–Stokes Equations

The two step methods that have been presented lend themselves quite easily to the solution of the complete Navier–Stokes equations which can be written in the form

$$\frac{\partial w}{\partial t} + \frac{\partial f(w)}{\partial x} + \frac{\partial g(w)}{\partial y} + S(w, w_x, w_y, w_{xx}, \ldots) = 0 \tag{16}$$

The vector S contains the heat conduction and viscous dissipation terms which give rise to familiar phenomena in fluid dynamics, i.e., boundary layers, shock structure, etc. Equation (16) is no longer a purely hyperbolic equation. It is possible to replace S by $\mathrm{Re}^{-1}S$ where Re corresponds to the Reynolds number. When the Reynolds number is infinite, we have the case of inviscid flow. For very small Reynolds number the flow is highly viscous and (16) tends to be parabolic in nature. Correspondingly, we would expect the stability condition to be hyperbolic $\Delta t \sim \Delta x$ for Re large and parabolic $\Delta t \sim \Delta x^2$ for small values of Re. In [12] Gary found empirically that in one dimensional flow, the time step be chosen as the minimum of one half the parabolic or hyperbolic stability condition. It is possible to write Eq. (16) in a matrix form analogous to (3), but it is a complicated procedure. Instead, the two step methods are particularly suited for the solution of Eq. (16). Some success has been achieved by using a two step difference approximation to Eq. (16) [13]. Thommen centered all terms at $t + \frac{1}{2}\Delta t$ except S (which was centered at t). This author together with Rubin have recently carried out a series of numerical experiments (to be published) on the one dimensional Navier–Stokes equations. We have found that the numerical solution depends critically on how one centers the dissipative terms.

7. Conclusions

It is clear that much work remains in the field of successful numerical approximation to the equations of hydrodynamics for the solution of the mixed initial boundary value problems.

References

1. Richtmyer, R. D., A. Survey of Difference Methods for Nonsteady Fluid Dynamics, NCAR Tech. Note 63-2 (August, 1962).
2. Noh, W. F., Private communication.
3. Browne, Ph. L., Rezone—a Proposal for Accomplishing Rezoning in Two-Dimensional Lagrangian Hydrodynamics Problems, Los Alamos Scientific Lab. Rep. LA-3455-MS (September, 1965).
4. Noh, W. F., CEL: A Time-Dependent, Two-Space-Dimensional Coupled Eulerian–Lagrange Code, *Methods Computational Phys.* **3**, Chapter 9, (1964).
5. Lax, P. D., and Wendroff, B., Systems of Conservation Laws, *Comm. Pure Appl. Math.* **13**, 217–237 (1960).
6. Lax, P. D., and Wendroff, B., Difference Schemes for Hyperbolic Equations with High Order Accuracy, *Comm. Pure Appl. Math.* **7**, 381–398 (1964).
7. Burstein, S. Z., Numerical Methods in Multidimensional Shocked Flow, A.I.A.A. *J.* **2**, 211–217 (1964).
8. Burstein, S. Z., Finite Difference Calculations for Hydrodynamic Flows Containing Discontinuities, AEC Computing and Appl. Math. Center, NYO-1480-33 (September, 1965). Also to appear in *J. Computational Phys.*
9. Houghton, D., Kasahara, A., and Washington, W., Long Term Integration of the Barotropic Equations by the Lax–Wendroff Method, *Monthly Weather Rev.* **94**, No. 3 141–150, (1966).
10. Lax, P. D., Stability of Linear and Nonlinear Difference Schemes *in*: "Numerical Solution of Partial Differential Equations" (J. H. Bramble, ed.), Chapter 14. Academic Press, New York, 1966.
11. Richtmyer, R. D. and Morton, K. W., Stability Studies for Difference Equations: I. Nonlinear Stability, II. Coupled Sound and Heat Flow, NYO-1480-5 (August, 1964).
12. Gary, J., On Certain Finite Difference Schemes for the Equations of Hydrodynamics, NYO-1988 (March, 1962).
13. Thommen, H., U., Numerical Integration of the Navier–Stokes Equations, GDC-ERR-AN759 (August, 1965).

Nonlinear Problems in the Dynamics of Thin Shells†

WILLIAM A. NASH

Department of Engineering Science and Mechanics
University of Florida
Gainesville, Florida

For the past few decades it has been rather generally acknowledged that the problem of buckling of thin elastic shells must be investigated through use of nonlinear finite-deflection theory. This type of investigation stands in contrast to that used for the problem of elastic buckling of bars, rings, and plates, which may be successfully analyzed by classical infinitesimal-deflection theory. In the case of these latter bodies the results of infinitesimal-deflection theory are in satisfactory agreement with experimental evidence. In the case of shells, however, this agreement no longer exists. For example, for static axial compression of a circular cylindrical shell, experimental results range from 20% to 60% of the values predicted by linear small-deflection theory.

The first attempt at presenting a nonlinear theory was made by Donnell [1] in 1934. His theory was based upon finite deformations together with a consideration of the initial imperfections present in the axially compressed circular cylindrical shell. The computations involved were formidable for that day, with the result that the author introduced several drastic mathematical assumptions which rendered the final results somewhat less than satisfactory. Nevertheless, the beginnings had been made on a new approach that showed excellent promise of being in better agreement with experimental evidence. In 1941 this work was continued by von Karman and Tsien [2] for the same case of an axially compressed cylindrical shell. More recently, very satisfactory nonlinear analyses of axially compressed

† These results were obtained in the course of research sponsored partially by the U.S. Army Research Office, Durham, under Grant No. DA-ARO(D)-31-124-G372 with the University of Florida, and partially by the Air Force Flight Dynamics Laboratory, Wright-Patterson Air Force Base, under Contract No. AF33(657)-10220 with the General Electric Company.

cylindrical shells have been offered by Kempner [3] and Almroth [4]. An excellent summary and critique, as well as an investigation of the effects of internal pressurization on buckling of a statically loaded axially compressed cylindrical shell, has been given by Thielemann [5].

The problem of dynamic buckling of thin elastic shells has only come into prominence in the past few years. One of the first treatment is due to Vol'mir [6], who investigated the buckling of a shallow circular cylindrical panel subject to rapidly applied axial compression along its generators. The panel edges were taken to be simply supported. Finite-deflection equations, incorporating a consideration of initial imperfections present in the specimen, were formulated and an approximate solution to these nonlinear equations was obtained by the Galerkin procedure. The final results were presented in the form of dimensionless axial load versus time. It is interesting to observe that Vol'mir's solution indicates that the dynamically applied load increases to a value 65% greater than the upper critical load found from linear theory for the case of static compression of such a panel. After this peak load is reached, there occurs an abrupt drop in axial load, and for a very short interval of time the load actually becomes negative, corresponding to an extension of the panel. As time increases still further, the load fluctuates, indicating nonlinear oscillations of the panel.

Another study closely related to [6] has been presented by Agamirov and Vol'mir [7]. Here, the authors apply finite-deflection analysis, together with a consideration of initial imperfections, to the problem of dynamic buckling of a cylindrical shell subject to either hydrostatic pressure or axial compression. The analytic approach is identical to that employed by Vol'mir in [6]. However, the authors greatly simplified the mathematics by taking the values of certain parameters to be the same as formed for static analysis. This, of course, introduces an error of unknown magnitude. No numerical results were presented for the axial compression case. For hydrostatic loading the authors found a critical pressure approximately twice the upper critical load formed from linear theory for the case of static compression.

An experimental investigation of buckling of cylindrical shells by dynamically applied hydrostatic pressure has been reported by Vol'mir and Mineev [8]. The shells were duralumin, with a radius-thickness ratio of 112. For the case of lateral pressure increasing at a rate of 6500 atmospheres per second, the dynamic buckling load was 3.98 times the upper critical value found from static analysis. Also, the number of waves, around the circumference increased as the loading rate increased. It is to be observed that the dynamically applied hydrostatic pressure was applied to the specimen by suddenly opening a valve between a high-pressure fluid reservoir and a lower-pressure fluid surrounding the cylinder. Inspection of the experimental apparatus reveals that the dynamic pressure along the length of the shell cannot be constant, and, hence, an error of some unknown magnitude is present in the tests.

In 1960, Coppa [9] presented a phenomenological theory for the buckling of a circular cylindrical shell subject to an axially symmetric impact at one end. According to this theory, when a cylindrical shell is struck at one end by a rigid, infinite mass with a velocity V, a stress wave travels from the struck end at the dilatational wave velocity c. When the wave front has traveled a distance L_{cr} from the struck end, instability will occur in this region, provided that the velocity V is sufficiently great. The instability configuration may be of two types, either an axisymmetric ring-type buckle or a triangular configuration depending upon the ratio L_{cr}/D where D is the cylinder diameter. Once instability has begun, the unstable region is unable to transmit the axial force which initiated instability and, hence, the cylindrical region beyond the critical region remains stable. He further demonstrated that the postbuckling behavior for triangular buckling can be described by a continuous set of stable inextensional configurations having progressively shorter axial lengths, successively lower axial resistances, and the same number of circumferential waves as the critical configuration.

A refinement of the analysis presented in [7] has been offered by Kadashevich and Pertsev [10]. In this work the authors again consider the effect of a uniformly distributed transverse dynamic load on a cylindrical shell. However, in contrast to the earlier work, not only is the inertia of the formation of buckling waves taken into account but also the inertia of axisymmetric compression of the shell. This permits the authors to obtain the equations of motion of the shell for the case of more rapid application of the load than is admissible in the system of equations given in [7]. Lagrange's equations are employed to obtain the differential equations of motion, and these are solved by digital computer techniques. Another more recent study by Roth and Klosner [11] refines the analysis of [7] to include more free parameters.

Let us consider the case of a circular cylindrical shell struck at one end and completely restrained at the other. The mode of application of the force is such that the rate of end shortening is constant; i.e., $e = Vt$, where e represents end shortening and t denotes time. The coordinate system for such a shell is illustrated in Fig. 1, and consists of a coordinate x

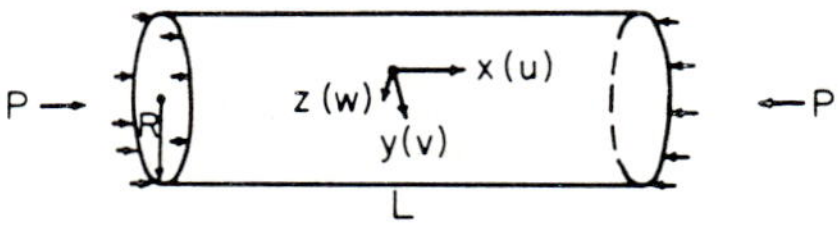

FIG. 1. Shell coordinates.

along a generator, y in the circumferential direction, and z in the inward radial direction. The components of displacement of a point in the middle surface of the shell in these directions are designated by u, v, and w respectively. Further, L denotes the length of the shell, R the radius to the middle surface, h the wall thickness, P the total axial compressive force, p the mean axial stress, and q the intensity of internal pressure.

We shall employ the following approximate nonlinear strain-displacement relations to describe the membrane strains:

$$\begin{aligned}
\varepsilon_x &= \frac{\partial u}{\partial x} + \frac{1}{2}\left(\frac{\partial w_1}{\partial x}\right)^2 - \frac{1}{2}\left(\frac{\partial w_0}{\partial x}\right)^2 \\
\varepsilon_y &= \frac{\partial v}{\partial y} + \frac{1}{2}\left(\frac{\partial w_1}{\partial y}\right) - \frac{1}{2}\left(\frac{\partial w_0}{\partial y}\right)^2 + \frac{w_1}{R} - \frac{w_0}{R} \qquad (1) \\
\varepsilon_{xy} &= \frac{\partial u}{\partial y} + \frac{\partial v}{\partial x} + \frac{\partial w_1}{\partial x}\frac{\partial w_1}{\partial y} - \frac{\partial w_0}{\partial x}\frac{\partial w_0}{\partial y}
\end{aligned}$$

In these equations, ε_x and ε_y denote normal strains of the middle surface and ε_{xy} the shearing strain of that surface. The initial imperfections of the shell are represented by w_0, the total deflection by w_1, and, consequently, the net deflection by $w = w_1 - w_0$. By appropriate differentiations and linear combinations of Eqs. (1) we obtain

$$\frac{\partial^2 \varepsilon_x}{\partial y^2} + \frac{\partial^2 \varepsilon_y}{\partial x^2} - \frac{\partial^2 \varepsilon_{xy}}{\partial x\, \partial y} = \left[\left(\frac{\partial^2 w_1}{\partial x\, \partial y}\right)^2 - \frac{\partial^2 w_1}{\partial x^2}\frac{\partial^2 w_1}{\partial y^2}\right] - \left[\left(\frac{\partial^2 w_0}{\partial x\, \partial y}\right)^2 - \frac{\partial^2 w_0}{\partial x^2}\frac{\partial^2 w_0}{\partial y^2}\right] - \frac{1}{R}\left[\frac{\partial^2 w_1}{\partial x^2} - \frac{\partial^2 w_0}{\partial x^2}\right] \tag{2}$$

Let us now introduce the Airy function $\phi\,(x, y)$ of the membrane stresses. This function is defined by the relations

$$\sigma_x = \frac{\partial^2 \phi}{\partial y^2}, \qquad \sigma_y = \frac{\partial^2 \phi}{\partial x^2}, \qquad \tau_{xy} = -\frac{\partial^2 \phi}{\partial x\, \partial y} \tag{3}$$

where σ_x and σ_y denote normal stresses of the middle surface and τ_{xy} represents the shearing stress of that surface. If the relations (3) together with the biaxial stress-strain relations are introduced into Eq. (2), the compatibility equation is obtained

$$\frac{1}{E}\nabla^4 \phi = \left[\left(\frac{\partial^2 w_1}{\partial x\, \partial y}\right)^2 - \frac{\partial^2 w_1}{\partial x^2}\frac{\partial^2 w_1}{\partial y^2}\right] - \left[\left(\frac{\partial^2 w_0}{\partial x\, \partial y}\right)^2 - \frac{\partial^2 w_0}{\partial x^2}\frac{\partial^2 w_0}{\partial y^2}\right] - \frac{1}{R}\left[\frac{\partial^2 w_1}{\partial x^2} - \frac{\partial^2 w_0}{\partial x^2}\right] \tag{4}$$

where

$$\nabla^4 = \frac{\partial^4}{\partial x^4} + 2\frac{\partial^4}{\partial x^2\, \partial y^2} + \frac{\partial^4}{\partial y^4}$$

and E represents Young's modulus.

We shall now consider an infinitesimal element of the shell, loaded by bending moments M_x and M_y and a twisting moment M_{xy}. Also, Q_x and Q_y represent the resultant transverse shearing forces. Summing moments about the x and y axes and neglecting higher order terms, we obtain

$$\frac{\partial M_x}{\partial y} + \frac{\partial M_{xy}}{\partial y} - Q_x = 0, \qquad \frac{\partial M_y}{\partial y} + \frac{\partial M_{xy}}{\partial x} - Q_y = 0 \tag{5}$$

Summing forces in the radial direction we obtain

$$\frac{\partial Q_x}{\partial x} + \frac{\partial Q_y}{\partial y} + \sigma_x h \frac{\partial^2 w_1}{\partial x^2} + \sigma_y h\left(\frac{1}{R} + \frac{\partial^2 w_1}{\partial y^2}\right)$$

$$+ 2\tau_{xy} h \frac{\partial^2 w_1}{\partial x\, \partial y} - \frac{\gamma}{g} h \frac{\partial^2 w}{\partial t^2} - q = 0 \tag{6}$$

Here γ denotes the weight density of the shell material and g represents the gravitational acceleration.

We shall employ the usual moment-curvature relations of shell theory:

$$M_x = -D\left(\frac{\partial^2 w}{\partial x^2} + \nu \frac{\partial^2 w}{\partial y^2}\right)$$

$$M_y = -D\left(\frac{\partial^2 w}{\partial y^2} + \nu \frac{\partial^2 w}{\partial x^2}\right) \tag{7}$$

$$M_{xy} = -\frac{Gh^3}{6} \frac{\partial^2 w}{\partial x\, \partial y}$$

where G represents the shear modulus of the material, ν is Poisson's ratio, and D is the flexural rigidity defined by

$$D = \frac{Eh^3}{12(1 - \nu^2)}$$

Here w, rather than w_1, appears in (7) because the moments depend only upon changes of curvature. If the relations (7) are now substituted in (5), we find Q_x and Q_y as functions of w. These results, together with the relations (3), may now be substituted in (6) to yield the equilibrium equation

$$\nabla^4(w_1 - w_0) = \frac{h}{D}\left[\frac{\partial^2 \phi}{\partial x^2} \frac{\partial^2 w_1}{\partial y^2} + \frac{\partial^2 \phi}{\partial y^2} \frac{\partial^2 w_1}{\partial x^2}\right.$$

$$\left. + 2 \frac{\partial^2 \phi}{\partial x\, \partial y} \frac{\partial^2 w_1}{\partial x\, \partial y} + \frac{1}{R} \frac{\partial^2 w_1}{\partial x^2} - \frac{\gamma}{g} \frac{\partial^2 w}{\partial t}\right] - \frac{q}{D} \tag{8}$$

We now select for w_0 and w_1 the functions

$$\begin{aligned} w_0 &= f_0 \sin \alpha x \sin \beta y + \psi_0 \sin^2 \alpha x \sin^2 \beta y \\ w_1 &= f_1 \sin \alpha x \sin \beta y + \psi_1 \sin^2 \alpha x \sin^2 \beta y \end{aligned} \tag{9}$$

where $\alpha = m\pi/L$ and $\beta = n/R$. Here m is the number of half waves in the axial direction of the shell and n the number of half waves in the circumferential direction. In Eqs. (9) f_0 and ψ_0 are initial deflection parameters of the shell, which will either be known or will be assumed. Further, f_1 and ψ_1 are time-dependent deflection parameters of the deformed shell. The configuration (9) implies diamond-shaped buckles. The relations (9) correspond to an elastically supported shell, one whose ends are neither clamped nor simply supported. However, as in the case of static buckling of shells, it is unlikely that boundary conditions are of extreme significance provided the length of the shell is greater than approximately 1.5 diameters.

If the relations (9) are substituted in the compatibility equation (4), we obtain the following solution of (4) for the Airy stress function

$$\begin{aligned}\phi = {} & K_1 \sin \alpha x \sin \beta y + K_2 \sin 3\alpha x \sin \beta y \\ & + K_3 \sin \alpha x \sin 3\beta y + K_4 \cos 2\alpha x + K_5 \cos 2\beta y \\ & + K_6 \cos 2\alpha x \cos 2\beta y + K_7 \cos 4\alpha x \\ & + K_8 \cos 4\beta y + K_9 \cos 4\alpha x \cos 2\beta y \\ & + K_{10} \cos 2\alpha x \cos 4\beta y - \frac{py^2}{2} + \frac{qR}{2h} x^2 \end{aligned} \tag{10}$$

where the K_1, K_2, ..., K_{10} represent various combinations of f_0, f_1, ψ_0, ψ_1, α, β, E, and R.

We have now found a stress function satisfying compatibility. Equilibrium must now be satisfied. If we substitute the above values of w_1, w_0, and ϕ in the equilibrium equation, it will not in general be satisfied. However, we may satisfy it approximately by the Galerkin technique. For this purpose, let

$$\begin{aligned} H = D\nabla^4(w_1 - w_0) - h\bigg[& \frac{\partial^2\phi}{\partial x^2}\frac{\partial^2 w_1}{\partial y^2} + \frac{\partial^2\phi}{\partial y^2}\frac{\partial^2 w_1}{\partial x^2} \\ & - 2\frac{\partial^2\phi}{\partial x\,\partial y}\frac{\partial^2 w_1}{\partial x\,\partial y} + \frac{1}{R}\frac{\partial^2\phi}{\partial x^2} - \frac{\gamma}{g}\frac{\partial^2 w}{\partial t^2}\bigg] + q \end{aligned} \tag{11}$$

That is, H is the residual of the equilibrium equation after approximate values of w_1, w_0, and ϕ have been substituted. To minimize H we employ the Galerkin method, viz.,

$$\int_0^{2\pi R}\int_0^L H \frac{\partial w}{\partial f_1}\, dx\, dy = 0$$
$$\int_0^{2\pi R}\int_0^L H \frac{\partial w}{\partial \psi_1}\, dx\, dy = 0 \tag{12}$$

These equations become

$$\int_0^{2\pi R}\int_0^L H \sin \alpha x \sin \beta y\, dx\, dy = 0$$
$$\int_0^{2\pi R}\int_0^L H \sin^2 \alpha x \sin^2 \beta y\, dx\, dy = 0 \tag{13}$$

For the above values of w_1, w_0, and ϕ, H may now be determined from Eqs. (11). This value of H, together with the above Galerkin equations, yields the coupled nonlinear differential equations in f_1 and ψ_1:

$$\frac{\gamma}{g}\frac{d^2 f_1}{dt^2} = V_0 f_1 t + V_1 f_1 + V_2 f_1^3 + V_3 \psi_1 + V_4 f_1 \psi_1 + V_5 f_1 \psi_1^2 + V_6$$
$$\frac{g\gamma}{4g}\frac{d^2 \psi_1}{dt^2} = W_0 \psi_1 t + W_1 \psi_1 + W_2 {\psi_1}^2 + W_3 {\psi_1}^3 \tag{14}$$
$$+ W_4 f_1 + W_5 {f_1}^2 + W_6 {f_1}^2 \psi_1 + W_7$$

In this last pair of coupled nonlinear differential equations, the condition of periodicity of circumferential displacement has been introduced, as well as the specified manner of end loading, i.e., constant rate of end shortening. The quantities v_0, v_1, ... and W_0, W_1, ... represent various combinations of D, h, R, E, α, β, f_0, ψ_0, q, and V.

Because of the complexity of the quantities involved in Eqs. (14), it is desirable to proceed from this stage onward only by numerical techniques. Since the Eqs. (14) are coupled nonlinear equations, no closed-form solution is known to exist. However, with given initial conditions, these two equations may be readily integrated numerically by the Runge–Kutta method. We shall investigate the initial conditions

$$
\begin{aligned}
f_1 &= f_0 & \frac{df_1}{dt} &= 0 \\
\psi_1 &= \psi_0 & \frac{d\psi_1}{dt} &= 0
\end{aligned}
\tag{15}
$$

The numerical solutions of Eqs. (14) have been carried out on an IBM 1620 for shells having the following constants:

$E = 10.6 \times 10^6 \,\#/in^2$ $\quad$ $h = 0.004\ in.$ and $0.008\ in.$

$\nu = 0.33$ $\quad$ $V = 11.5$, 23, and $46\ ft/sec$

$L = 22.8\ in.$ $\quad$ $f_0 = \psi_0 = h/2$

$R = 2.85\ in.$

Further, α was taken to be equal to β, which implies a wave aspect ratio of unity. Figures 2 and 3 illustrate the results of these computations. For

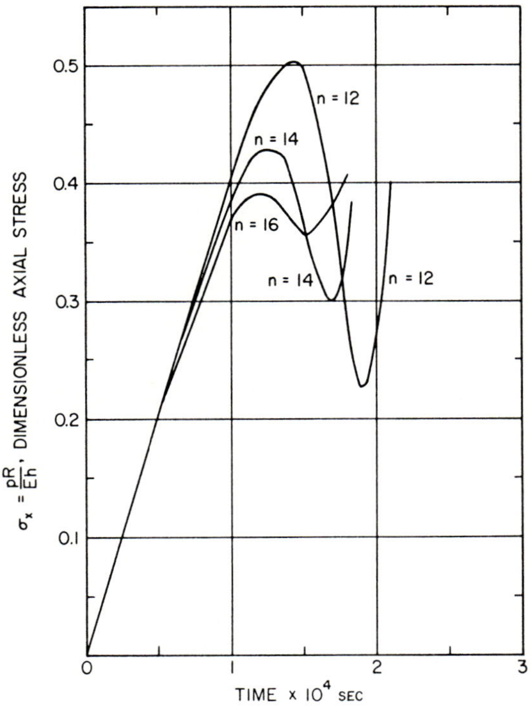

FIG. 2. Axial stress versus time ($V_0 = 23$ ft/sec, $h = 0.008''$, aluminum).

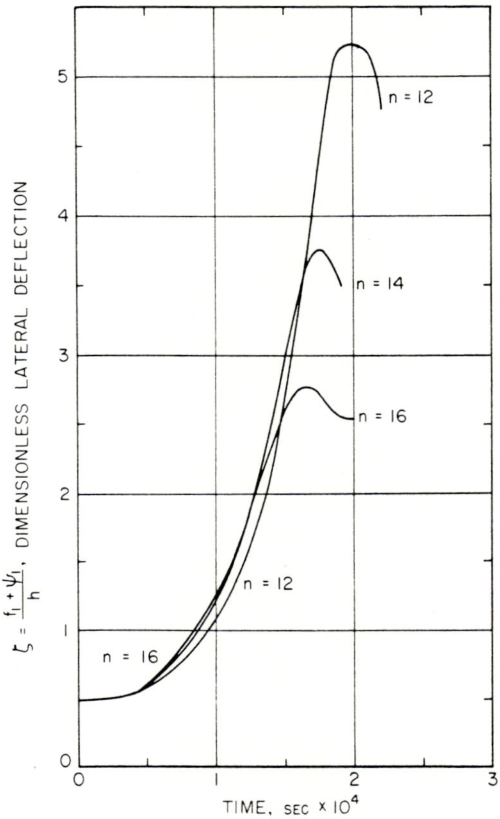

FIG. 3. Lateral deflection of the shell wall versus time ($V_0 = 23$ ft/sec, $h = 0.008''$, aluminum).

each set of parameters, both a dimensionless axial stress

$$\sigma = P^R/Eh$$

and a dimensionless lateral deflection

$$\zeta = (f_1 + \psi_1)/h$$

were plotted as functions of time. Since the rate of end shortening has been taken to be constant, this implies that these same two quantities are just as readily presented as functions of unit end shortening. It is to be observed from these figures that it is necessary to make a selection of which curve is the appropriate one to consider; i.e., for a given rate of end shortening and a given thickness shell, there is plotted a family of curves for various values of n, the number of half waves in the circumferential direction. The question arises as to which curve is the one that realistically describes the load-time relation of the shell. The criterion adopted was that the ζ-t curve first to deviate from the abscissa axis and first to reach its peak value is the one that should be considered.

On the basis of this criterion, inspection of the curves reveals certain extremely interesting results:

(a) With increasing rates of end shortening, the first maximum load, termed the buckling load, corresponding to the first maximum dimensionless axial stress, is attained sooner.

(b) With increasing rates of end shortening, the buckling load as defined in (a) is increased.

(c) With increasing rates of end shortening, the number of half waves that form around the circumference increases.

Let us next consider the case of a truncated conical shell struck axially at one end and completely restrained at the other. The coordinate system employed to analyze such a shell may be regarded as a generalization of that shown in Fig. 1; i.e., the position of a point on the middle surface is represented by the distance s measured from the vertex along a generator, together with the angle φ between the axial plane through the point and a coordinate reference plane. The semivertex angle of the cone is designated by γ. The components of displacement due to applied load in the direction of a generator, in the circumferential direction, and inward normal to the shell surface are designated by u, v, and w, respectively. The normal component of displacement existing because of initial imperfections is designated as w_0, the total normal component after loading by w_1, so that $w_1 = w_0 + w$. The strain-displacement relations are taken to be

$$\begin{aligned}
\varepsilon_s &= \frac{\partial u}{\partial s} + \frac{1}{2}\left[\left(\frac{\partial w_1}{\partial s}\right)^2 - \left(\frac{\partial w_0}{\partial s}\right)^2\right] \\
\varepsilon_\varphi &= \frac{1}{s}\left\{u - w\cot\gamma + \frac{\partial v}{\partial \psi} + \frac{1}{2s}\left[\left(\frac{\partial w_1}{\partial \psi}\right)^2 - \left(\frac{\partial w_0}{\partial \psi}\right)^2\right]\right\} \qquad (16) \\
\gamma_{s\varphi} &= s\frac{\partial}{\partial s}\left(\frac{v}{s}\right) + \frac{1}{s}\left(\frac{\partial u}{\partial s} + \frac{\partial w_1}{\partial s}\frac{\partial w_1}{\partial \psi} - \frac{\partial w_0}{\partial s}\frac{\partial w_0}{\partial \psi}\right)
\end{aligned}$$

where $\psi = \varphi \sin \gamma$.

The curvature relations are

$$\kappa_s = \frac{\partial^2 w}{\partial s^2}, \qquad \kappa_\varphi = \frac{1}{s}\left(\frac{\partial w}{\partial s} + \frac{1}{s}\frac{\partial w}{\partial \psi}\right), \qquad \kappa_{s\varphi} = \frac{\partial}{\partial s}\left(\frac{1}{s}\frac{\partial w}{\partial \psi}\right) \tag{17}$$

and the forces and moments per unit length of the middle surface are

$$N_s = \frac{Eh}{1 - \nu^2}(\varepsilon_s + \nu\varepsilon_\varphi) \qquad M_s = D(\kappa_s + \nu\kappa_\varphi)$$

$$N_\varphi = \frac{Eh}{1 - \nu^2}(\varepsilon_\varphi + \nu\varepsilon_s) \qquad M_\varphi = D(\kappa_\varphi + \nu\kappa_s)$$

$$S = \frac{Eh}{2(1 + \nu)}\gamma_{s\varphi} \qquad H = D(1 - \nu)\kappa_{s\varphi} \tag{18}$$

We may obtain the governing differential equations from Hamilton's principle for conservative systems in the form

$$\delta \int_{t_1}^{t_2} L\, dt = 0 \tag{19}$$

where $L = T - V$, T representing kinetic energy of the shell and V the potential energy. The former may be written in the form

$$T = \frac{1}{2}\rho h \int_0^{2\pi\sin\gamma} \int_{s_1}^{s_2} \left(\frac{\partial w_1}{\partial t}\right)^2 s\, ds\, d\psi \tag{20}$$

and the latter is composed of the strain energy U_i, stored in the body due to deformations and given by

$$U_i = \frac{1}{2}\int_0^{2\pi\sin\gamma} \int_{s_1}^{s_2} (N_s E_s + N_\varphi \varepsilon_\varphi + S\gamma_{s\varphi} + M_s\kappa_s + M_\varphi\kappa_\varphi + 2H\kappa_{s\varphi})\, s\, ds\, d\psi \tag{21}$$

together with the potential energy U_e of the external loads given by

$$U_e = -\int_0^{2\pi\sin\gamma} \left[s\left(N_s u + N_{s\varphi} v - Q_s w + M_s \frac{\partial w}{\partial s} + M_{s\varphi}\frac{1}{s}\frac{\partial w}{\partial \psi}\right)\right]_{s_1}^{s_2} d\psi \tag{22}$$

where h equals shell thickness, E is Young's modulus, ν is Poisson's ratio, ρ denotes mass density, and t represents time. Also, s_1 and s_2 are distances from the vertex to the smaller and larger ends of the cone. If the variations are performed and a stress function Φ introduced according to the definitions

$$N_s = \frac{1}{s}\left(\frac{\partial \Phi}{\partial s} + \frac{1}{s}\frac{\partial^2 \Phi}{\partial \psi^2}\right), \qquad N_\varphi = \frac{\partial^2 \Phi}{\partial s^2}, \qquad S = -\frac{\partial}{\partial s}\left(\frac{1}{s}\frac{\partial \Phi}{\partial \psi}\right) \tag{23}$$

we are led to the equation of motion

$$D\,\Delta\,\Delta\, w - \frac{1}{s} N_\varphi\left(\cot\gamma + \frac{\partial w_1}{\partial s} + \frac{1}{s}\frac{\partial^2 w_1}{\partial \psi^2}\right) - N_s \frac{\partial^2 w_1}{\partial s^2} - 2S\frac{\partial}{\partial s}\left(\frac{1}{s}\frac{\partial w_1}{\partial \psi}\right) + \rho h \frac{\partial^2 w_1}{\partial t^2} = 0 \tag{24}$$

and the compatibility equation

$$\begin{aligned}\Delta\,\Delta\,\Phi = -Eh\Bigg\{&\left[\frac{1}{s}(\cot\gamma)\frac{\partial^2 w_1}{\partial s^2} - \frac{1}{s^2}\left(\frac{\partial^2 w_1}{\partial s\,\partial\psi}\right)^2 + \frac{1}{s^2}\frac{\partial^2 w_1}{\partial s^2}\frac{\partial^2 w_1}{\partial \psi^2}\right.\\ &\left.+ \frac{2}{s^3}\frac{\partial w_1}{\partial \psi}\frac{\partial^2 w_1}{\partial s\,\partial\psi} - \frac{1}{s^4}\left(\frac{\partial w_1}{\partial \psi}\right)^2 + \frac{1}{s}\frac{\partial w_1}{\partial s}\frac{\partial^2 w_1}{\partial s^2}\right]\\ &- \left[\frac{1}{s}(\cot\gamma)\frac{\partial^2 w_0}{\partial s^2} - \frac{1}{s^2}\left(\frac{\partial^2 w_0}{\partial s\,\partial\psi}\right)^2 + \frac{1}{s^2}\frac{\partial^2 w_0}{\partial s^2}\frac{\partial^2 w_0}{\partial \psi^2} + \frac{2}{s^3}\frac{\partial w_0}{\partial \psi}\frac{\partial^2 w_0}{\partial s\,\partial\psi}\right.\\ &\left.- \frac{1}{s^4}\left(\frac{\partial w_0}{\partial \psi}\right)^2 + \frac{1}{s}\frac{\partial w_0}{\partial s}\frac{\partial^2 w_0}{\partial s^2}\right]\Bigg\}\end{aligned} \tag{25}$$

where

$$\Delta = \frac{\partial^2}{\partial s^2} + \frac{1}{s}\frac{\partial}{\partial s} + \frac{1}{s^2}\frac{\partial^2}{\partial \psi^2} \tag{26}$$

and D is the flexural rigidity of the shell. The problem of the time-dependent response of the conical shell to a time-varying axial load thus reduces to the simultaneous solution of Eqs. (24) and (25).

If the operator $\Delta\,\Delta$ is multiplied by s^4, we obtain an operator of the Euler type in the variable s. This can then be transformed into a linear operator with constant coefficients by the substitution

$$s = s_1 e^z \tag{27}$$

If, further, we introduce the transformations

$$\Phi = s_1 e^{2z} F, \qquad w_1 = s_1 e^z \tilde{w}_1 \cot \gamma, \qquad w_0 = s_1 e^z \tilde{w}_0 \cot \gamma \tag{28}$$

we are led to the transformed equations of motion and compatibility:

$$\begin{aligned} \tilde{w}^{IV} - 2\tilde{w}'' + \tilde{w} + 2\ddot{\tilde{w}}'' + 2\ddot{\tilde{w}} + \ddot{\ddot{\tilde{w}}} \\ = \frac{s_1{}^2}{D} e^{2z}\Big[N_\varphi(1 + \tilde{w}_1{}' + \tilde{w}_1 + \ddot{\tilde{w}}_1) + N_s(\tilde{w}_1{}'' + \tilde{w}_1{}') \\ + 2S\dot{\tilde{w}}_1{}' - s_1{}^2 e^{2z} \rho h \frac{\partial^2 \tilde{w}_1}{\partial t^2}\Big] \end{aligned} \tag{29}$$

$$\begin{aligned} F^{IV} + 4F''' + 4F'' + 2\ddot{F}'' + 4\ddot{F}' + 4\ddot{F} + 4\ddot{\ddot{F}} \\ = Eh \cot^2 \gamma \Big\{\Big[\dot{\tilde{w}}_1'^2 - (1 + \ddot{\tilde{w}}_1 + \tilde{w}_1{}' + \tilde{w}_1)(\tilde{w}_1'' + \tilde{w}_1{}')\Big] \\ - \Big[\dot{\tilde{w}}_0'^2 - (1 + \ddot{\tilde{w}}_0 + \tilde{w}_0{}' + \tilde{w}_0)(\tilde{w}_0'' + \tilde{w}_0)\Big]\Big\} \end{aligned} \tag{30}$$

where primes denote differentiation with respect to z, and dots with respect to ψ.

If the condition of periodicity of circumferential displacement is enforced, and an initial imperfection pattern of the form

$$w_0 = f_{01} \sin \alpha z \sin \beta\psi + f_{02} \sin^2 \alpha z \sin^2 \beta\psi \tag{31}$$

and a buckled configuration of the form

$$w_1 = f_1(t) \sin \alpha z \sin \beta\psi + f_2(t) \sin^2 \alpha z \sin^2 \beta\psi \tag{32}$$

assumed, then Eq. (30) may be solved for F and Eq. (29) solved by Galerkin's method. This has been done for an initial imperfection $f_{02} = 0.01 f_{01}$, $s_1/s_2 = 1.171$, $\gamma = 0.05$ radians, minimum radius of curvature at small end/shell thickness $= 1.75 \times 10^2$ by digital computer techniques, and the results are plotted in Fig. 4. There, the time-dependent axial load $P(t)$ has been rendered dimensionless by introducing a new variable

$$\tilde{q} = \sqrt{3(1 - \nu^2)}\left(\frac{P}{2\pi E h^2 \cos^2 \gamma}\right) \tag{33}$$

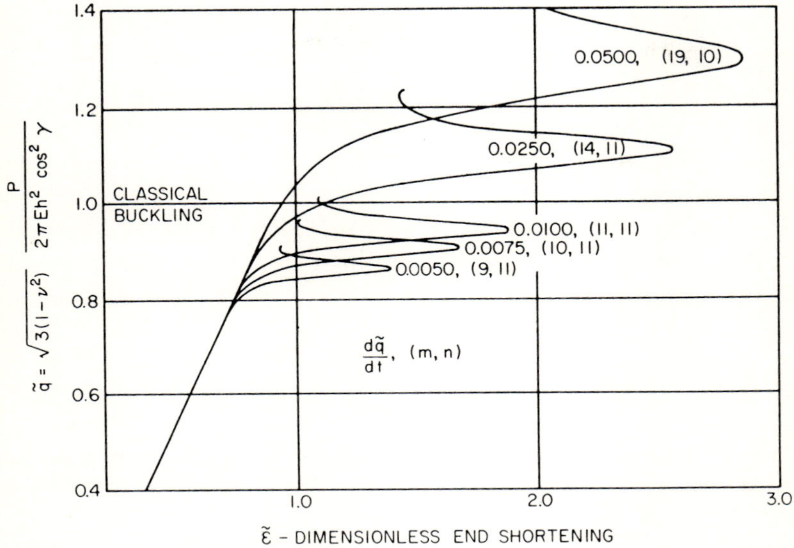

FIG. 4. Effect of loading rate.

The abscissa of that figure is $\tilde{\varepsilon}$ defined by the relation

$$\tilde{\varepsilon} = \frac{\text{mean end shortening}}{\dfrac{2\cot^2\gamma}{K}} \tag{34}$$

where

$$K = \sqrt{12(1-\nu^2)}\left(\frac{R_{\varphi 1}}{h}\right)\cot^2\gamma \tag{35}$$

and $R_{\varphi 1}$ = minimum radius of curvature at the small end of the cone. Further, for the various curves shown in that figure, the first number at the right of the curve indicates the dimensionless loading rate dq/dt, and the second and third numbers respectively correspond to the number of longitudinal and circumferential half waves forming upon buckling. Thus, Fig. 4 indicates axial load as a function of end shortening for various loading rates. As is evident from these plots, there is a rapid increase in end shortening followed by nonlinear oscillations. The area where the rapid increase in end shortening occurs is the region in which the shell loses stability. It is also evident that the load in the region of instability increases significantly with increasing load-rates, as does the number of longitudinal waves. Additional numerical results, not shown, have indicated that increasing the angle of conicity increases the load parameter in the region of instability; and, thus, the case of a cylindrical shell represents the lower bound for this critical parameter. Also, the load parameter decreases with increasing values of $R_{\varphi 1}/h$.

It would be of considerable interest to develop additional numerical data for other loading conditions. Further, there is an urgent need for comprehensive experimental data pertinent to dynamic buckling of conical shells.

References

1. Donnell, L. H., A new theory for the buckling of thin cylinders under axial compression and bending, *Trans. Am. Soc. Mech. Engrs.* **56**, No. 11, 795–806 (1934).
2. Von Karman, T., and Tsien, H. S., The buckling of thin cylindrical shells under axial compression, *J. Aeronaut. Sci.* **8**, No. 8, 303–312 (1941).
3. Kempner, J., Postbuckling behavior of axially compressed circular cylindrical shells, *J. Aeronaut. Sci.* **21**, No. 5, 329–335, 342 (1954).
4. Almroth, B. O., On the bucking of circular cylinders subjected to axial compression, LMSC Rept. 6-90-61-115, Missiles and Space Company, Lockheed Aircraft Corporation, Sunnyvale, California (July, 1962).
5. Thielemann, W. F., "New developments in the nonlinear theories of the buckling of thin cylindrical shells." Aeronautics and Astronautics, Proceedings of the Durand Centennial Conference, Pergamon Press, Oxford, 1960.
6. Vol'mir, A. S., On the stability of dynamically loaded cylindrical shells, *Dokl. Akad. Nauk SSSR* **123**, No. 5, 806-808 (1958). [Transl. in: *Soviet Phys. Doklady* **3**, No. 6, 1287–1289 (1958)].
7. Agamirov, V. L., and Vol'mir, A. S., Behavior of cylindrical shells under dynamic loading by hydrostatic pressure or by axial compression, *Izv. Akad. Nauk SSSR Otd. Tekhn. Nauk Mekhan. Mashinostr.* **3**, 78–83 (1959). [Transl. in: *J. Am. Rocket Soc.* **31**, No. 1, 98–101 (1961).
8. Vol'mir, A. S., and Mineev, V. E., An experimental investigation of the buckling of a shell under dynamic load, *Dokl. Akad. Nauk SSSR* **125**, No. 5, 1002–1003 (1959). [Transl. in: *Soviet Phys. Doklady* **4**, No. 2, 464–465 (1959).
9. Coppa, A. P., On the mechanism of buckling of a circular cylindrical shell under longitudinal impact, TIS Rept. R60SD494, General Electric Company (Proc. Tenth Intern. Congr. Appl. Mech., PP274-276, Mechanics (USSR)-Periodical Selection of Translations of Foreign Articles, No. 6), 1961.
10. Kadashevich, Yu. I., and Pertsev, A. K., Loss of stability of a cylindrical shell under dynamic loads, *Izv. Skad. Nauk SSSR Otd. Tekhn. Nauk Mekhan.*

Mashinostr. **3**, 30–33 (1960). Transl. in: *J. Am. Rocket Soc.* **32**, No. 1, 140–143 (1962).

11. Roth, R. S., and Klosner, J. M., Nonlinear response of cylindrical shells with initial imperfections subjected to dynamic Axial Loads, Rept. RAD-TM-63-43, Avco Corporation, July 31, 1963, Lawrence, Massachusetts.

Index